Phase Transitions
in Surface Films

NATO ADVANCED STUDY INSTITUTES SERIES

A series of edited volumes comprising multifaceted studies of contemporary scientific issues by some of the best scientific minds in the world, assembled in cooperation with NATO Scientific Affairs Division.

Series B: Physics

RECENT VOLUMES IN THIS SERIES

This series is published by an international board of publishers in conjunction with NATO Scientific Affairs Division

A Life Sciences	Plenum Publishing Corporation
B Physics	London and New York
C Mathematical and Physical Sciences	D. Reidel Publishing Company Dordrecht and Boston
D Behavioral and Social Sciences	Sijthoff International Publishing Company Leiden
E Applied Sciences	Noordhoff International Publishing Leiden

Phase Transitions in Surface Films

Edited by

J. G. Dash

University of Washington
Seattle, Washington

and

J. Ruvalds

University of Virginia
Charlottesville, Virginia

PLENUM PRESS • **NEW YORK AND LONDON**
Published in cooperation with NATO Scientific Affairs Division

Library of Congress Cataloging in Publication Data

Nato Advanced Study Institute on Phase Transitions in Surface Films, Erice, Italy, 1979.
   Phase transitions in surface films.

   Nato advanced study institutes series: Series B, Physics; v. 51)
   Includes index.
   1. Thin films–Congresses. 2. Surfaces (Physics)–Congresses. 3. Phase transformations
(Statistical physics)–Congresses. I. Dash, J. G., 1923-   II. Ruvalds, J. III. Title.
IV. Series.
QC176.82.N37 1970   541.3'435   79-28484
ISBN 0-306-40348-X

Proceedings of the NATO Advanced Study Institute on Phase Transitions in Surface
Films, held at the Ettore Majorana Centre for Scientific Culture, Erice, Sicily,
Italy, June 11–25, 1979.

© 1980 Plenum Press, New York
A Division of Plenum Publishing Corporation
227 West 17th Street, New York, N.Y. 10011

Preface

The Advanced Study Institute on Phase Transitions in Surface
Films was held at the Ettore Majorana Centre for Scientific Culture in Erice, Sicily, during June 11 to June 25, 1979. This
Institute was the second course of the International School of
Low Temperature Physics which was established at the Centre in
1977, with the guidance and inspiration of T. Regge and A.
Zichichi. The 1979 course selected a topic on one of the most
rapidly advancing fields of condensed matter physics in the
late 70's. The program of topics and speakers was developed with
the advice of the Organizing Committee, composed of J. Friedel,
N. D. Mermin, R. E. Peierls, T. Regge and J. Wheatley.

These two weeks were memorable for the range and depth of
the lectures and seminars, and the sustained high interest of
the students and faculty through a damanding schedule of over 5
hours a day of formal sessions. A large fraction of the leading
researchers in the field were there. It would have been
impossible to gather such a large group five years earlier, for
the field itself is hardly 10 years old.

Although the foundations of the thermodynamics of surface
films were laid down by Gibbs about 100 years ago, and experiments
on oil/water monolayers were carried out by Pockels and Rayleigh
at about the same time, the beginnings of the modern field were
much more recent.

In 1967 A. Thomy and X. Duval published the first of a
series of important papers on the vapor pressures of krypton
and other gases on exfoliated graphite. These studies revealed
a variety of distinct regimes now known to resemble two dimensional gases, liquids, and solid phases. The discovery in 1970
of a commercially available and particularly convenient form
of exfoliated graphite--Grafoil--has led to the application of a
wide variety of experimental techniques for monolayer studies.

The experimental advances have led to a great surge of

theoretical activity, stimulated in part by the challenge of
novel phenomena, and partly by the possibility of realizing states
of matter which had previously existed only in theoretical models.
There are several film regimes in which theory and experiment
are in remarkably close agreement, showing that in these cases
the models are faithful to the real systems. In some regimes
experiment and theory are not so close, causing both theorists
and experimentalists to revise and refine current techniques
and design new and more exacting tests. The activity has
attracted many newcomers to the field in the past few years.
This growth has been so rapid that it is unlikely for such a
large fraction of its practitioners to be present at a small
summer institute a few years hence. Therefore this collection
should be especially representative of the state of the field,
viewed by many of those who have helped to create it.

     The lectures, as contrasted with the seminars, were designed
to be of a tutorial or review nature, although the rapid growth
of the field makes it sometimes difficult to distinguish between
review and the presentation of new results. The chapters in
this collection are the written versions of the lectures.
They each deal with selected areas of monolayer physics, and
together they cover much of the field. Most of the topics that
are not covered in this collection were treated in seminars
and short papers during the two weeks at Erice. The complete
Programme of the institute is given in order to draw attention to
those topics that might otherwise be unnoticed by newcomers to
the field. One exception is the seminar on magnetic films given
by Melvin Pomerantz, published here because it was so novel and
intriguing that the editors decided to include it as a way of
bringing it greater attention. One of the short papers is also
included here at the suggestion of several participants since
it deals with a common concern: the characterization of substrate heterogeneity.

     Aside from the scientific programme of the Institute, it
was memorable for its location and facilities. The Ettore
Majorana Centre for Scientific Culture, with the gracious leader-
ship of Director A. Zichichi, hosts schools and institutes in
many disciplines each year, a large number of them returning
periodically every year or few years. It is easy to understand
why. The beauty of Erice, the warmth of its people, the comfort
of the living arrangements and efficiency of the Centre's staff
were important factors contributing to the success of the School.
The expertise and warm hospitality of A. Gabriele and S. Pinola
were particularly appreciated.

     We are grateful to several organizations for financial
assistance. The Scientific Affairs Division of NATO awarded a

grant that was crucial in bringing all of the lecturers and many of the seminar speakers and students from Europe and North America. The Italian government, through its Ministries of Education and of Scientific Research, the Italian Research Council, and the Sicilian Regional Government, gave support for several students' travel and local expenses, and provided for the Centre's facilities and staff. The U. S. National Science Foundation and the European Physical Society each provided two student fellow-ships providing for local expenses.

     Finally, but far from least important, we thank Beverly Dexter, of the University of Washington, and Jean Kea of the University of Virginia, for their assistance through the two years of preparation for the Institute and this volume.

J. G. Dash, J. Ruvalds
Editors

   The Centre for Scientific Culture provides a marvelous link
to the rich history and traditions of Erice, which was founded by
the legendary son of Venus and assimilated the influence of the
Phoenicians, Carthaginians, Romans, and various medieval cultures.
The courtyard of the Convent of San Rocco, depicted in the above
drawing by Helene Connolly, provided an inspirational setting
for the School which is named after the brilliant Sicilian
physicist Ettore Majorana. Following the distinguished tradition
of the international Schools of Physics founded originally by
A. Zichichi in 1963, the present Course was characterized by
vigorous discussions and highly stimulating contributions from
the participants.

J. G. Dash and J. Ruvalds   T. Regge   A. Zichichi
Directors of the Course   Director of the Director of the
                             School   Centre

Contents

AN INTRODUCTION TO THE THERMODYNAMICS OF SURFACES

Robert B. Griffiths

Department of Physics
Carnegie-Mellon University
Pittsburgh, Pa. USA   15213

Contents:

I.   Introduction

      Although I have worked on problems in ordinary "bulk" thermo-
dynamics for many years, it is only recently that I have begun to
study surfaces. There is, of course, an advantage to having a novice
give you an introduction to the subject: I know where a lot of the
conceptual difficulties and traps are located, since I have fallen
into them myself on numerous occasions. There is also a disadvantage
which you should keep in mind: my own understanding of the subject
is still quite limited and I may lead you into error. Professor Dash
has wisely given me only three lectures, and this puts a definite
upper limit on the amount of misinformation which I can transmit to
you.

      Ideally, one should approach the study of surface thermodynamics
only after acquiring a thorough knowledge of bulk thermodynamics,
since the former is, in a sense which will become clear later, heavily
dependent on the latter. As there is not sufficient time to give you
a review of bulk thermodynamics at the outset, I shall have to be
content with occasionally reminding you of things which (I hope!) you
already know.[1]

      My approach to the subject will follow the tradition introduced
by Gibbs[2], with a few modifications which I learned from Cahn[3].
However, for pedagogical purposes I prefer to begin with a somewhat
artificial model, a fluid against an ideal wall. This will yield a
number of "correct formulas" applicable to the more general case, so
it will not be a waste of time. Next I shall take up fluid-fluid
interfaces in some detail. The subjects of curved interfaces, fluid-
solid interfaces, and surface phase transition will be treated much
more briefly.

      Due to the limits of time, these lectures are devoted almost
entirely to formalism: deriving some fundamental equations and
explaining their physical significance. Of course, the only way to
really understand a formalism is to apply it to a number of examples.
A few problems are placed at the end for this purpose, but the really
serious student should acquire one of the many books devoted to the
subject of the physics or chemistry of surfaces[4].

II.   Fluid Against an Ideal Wall: Surface Excess Quantities

      Imagine a fluid of particles confined inside a cubical box of
edge ℓ by means of a potential $\psi(\vec{r})$ which is infinite outside the box.
Inside the box ψ depends only on the distance to the nearest wall,
and goes rapidly to zero when this distance exceeds some microscopic
length. Using statistical mechanics we could compute the density ρ
(moles per unit volume) of fluid as a function of $\vec{r}$. Let us suppose
that near one of the walls $\rho(z)$ has the form sketched in Figure 1,
where z is the perpendicular distance from the wall. We expect
$\rho(z)$ to approach its bulk value ρ^α
fairly rapidly with increasing z,
so that the two differ by a
significant amount only for z less
than some microscopic distance λ
which is much smaller than ℓ.
Clearly the properties of the fluid
will differ from those of the bulk
near the wall, and it is the task
of surface thermodynamics to provide a (rather coarse) description
of this difference.

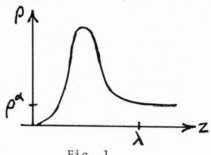

Fig. 1

      Imagine that the region inside the box, of volume $V=\ell^3$, is split
up into two regions, Fig. 2. One of them consists of points closer
than a distance z_0 to one of the
walls, and has a volume V^S (the
superscript s stands for surface),
while the remaining region, the
interior of the box, has volume
V^α, so that

$$V = V^\alpha + V^S. \qquad (2.1)$$

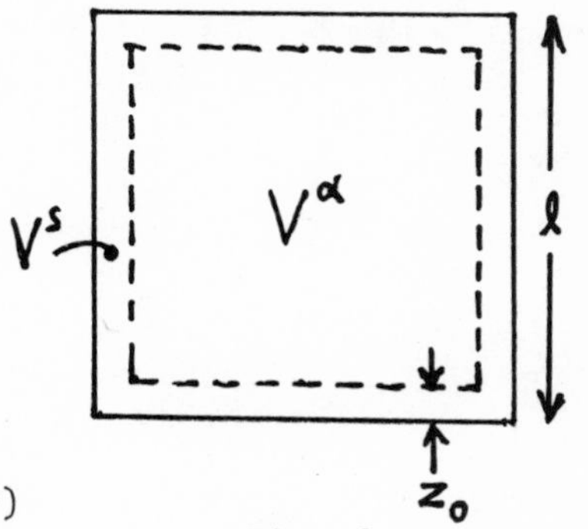

It will be convenient to regard
the total quantity N (moles) of
fluid in the box as consisting of
a "bulk" and a "surface" (or "wall")
contribution

Fig. 2

$$N = N^\alpha + N^S, \qquad (2.2)$$

where N^α is <u>by definition</u> equal to:

$$N^\alpha = \rho^\alpha V^\alpha. \qquad (2.3)$$

We shall assume that z_0 is a microscopic distance, much less than
ℓ.

      One should regard (2.2) as defining N^S as the difference between

N, which is a physical quantity, and N^α, which is the amount of
fluid which would occupy the volume V^α if its density were everywhere
equal to the bulk value ρ^α. Now if z_o is larger than λ, the actual
density of fluid in the inner region, Fig. 2, will differ from ρ^α by
a negligible amount, and thus N^S is the actual amount of fluid in the
outer region of volume V^S. However, this will no longer be the case
when z_o is small. In particular, when $z_o=0$ and thus $V^S=0$, N^S will
not (in general) vanish (see Problem 1).

The quantity N^S (and likewise V^S) is typical of the "surface
excess" quantities which one encounters in the Gibbs approach to
surface thermodynamics. It is "surface extensive" since it is proportional to the area $\mathbf{a}$ of the surface, $6\ell^2$, provided we ignore (as
we shall) corrections of order λ/ℓ or z_o/ℓ. It is defined as the
difference (whence the term "excess") between the actual value, N,
of an "extensive" thermodynamic variable and a value, N^α, assigned
to a bulk phase whose volume depends, to some degree, on an arbitrary convention (the choice of z_o).

The existence of such arbitrary conventions, and the associated
chameleon-like behavior of surface excess quantities when a convention
is changed, can prove to be a major conceptual hurdle for the beginning student, especially as there is nothing quite like it in bulk
thermodynamics. Hence it seems worthwhile to explore this topic in
a bit more detail.

Suppose that z_o is altered so that V^S changes by an amount ΔV^S.
Since V is fixed, V^α, (2.1), changes by an amount $-\Delta V^S$, and N^α, (2.3),
by an amount $-\rho^\alpha \Delta V^S$. But as N is fixed, the change in N^S, (2.2), is
given by

$$\Delta N^S = \rho^\alpha \Delta V^S. \tag{2.4}$$

Thus if we know the change in V^S when the convention is altered and
if we know the properties of the bulk fluid phase, we can compute
the change in N^S. The same is true for any other surface excess
quantity. Note, by the way, that the relative simplicity of (2.4)
would not be present if we were to define N^S as the actual amount of
fluid within the region V^S. In that case, ΔN^S would depend on the
details of $\rho(z)$ near the wall and not simply on a property of the
bulk fluid. The advantage of using (2.2) to define N^S is that it
does not depend on the actual density profile, and in this sense it
is a true "thermodynamic" variable, independent of details of the
microscopic structure.

Of course, the quantities of physical significance in surface
thermodynamics are those quantities which do not depend on an
arbitrary convention. It may be useful at this point to draw an
analogy with electrodynamics. The physical results of that theory
are invariant under gauge transformations so that, for example, many
different choices for the vector potential $\vec{A}$ give rise to the same

magnetic field $\vec{B}$. This situation gives rise to a variety of attitudes:

    (i) "The vector potential is meaningless!" (Philosophers)
    (ii) "Let's discuss all those elegant transformations which leave the theory invariant!" (Group-theory enthusiasts)
    (iii) "Why don't we simply pick a convenient gauge and get on with our calculations?" (Lowbrow physicists)

While I confess to being both an amateur philosopher and an amateur group-theorist, I strongly urge beginners in surface thermodynamics to adopt the lowbrow approach: choose a convention and get on with your calculations! For many purposes, though not all, the convention $V^S=0$, which I shall call the "Gibbs convention", is quite satisfactory.

Other excess quantities are defined in a similar way. Let U and S be the total energy and entropy of all the fluid in the box. We write

$$U = U^\alpha + U^S, \qquad S = S^\alpha + S^S, \qquad (2.5)$$

where U^α and S^α are bulk values <u>defined</u> by

$$U^\alpha = \bar{u}^\alpha V^\alpha, \qquad S^\alpha = \bar{s}^\alpha V^\alpha, \qquad (2.6)$$

where $\bar{u}^\alpha$ and $\bar{s}^\alpha$ are the energy and entropy per unit volume for the uniform bulk phase, and V^α is, of course, the same volume which appeared earlier in (2.1) and (2.3). Precisely because U^S and S^S are defined by (2.5) in terms of differences, their variations with a change in convention are easily computed from properties of the bulk phase:

$$\Delta U^S = \bar{u}^\alpha \Delta V^S, \qquad \Delta S^S = \bar{s}^\alpha \Delta V^S. \qquad (2.7)$$

A useful mnemonic for remembering (2.4), (2.7) and their analogs for other variables is: simply imagine that you have added an amount of bulk phase occupying a volume ΔV^S to the surface "phase".

In the case of a fluid mixture, there are separate mole numbers N_j, j=1,2,...c for each of the c distinct chemical components, and ρ_j^α stands for the number of moles of component j per unit volume in the bulk phase. For each component there is a surface excess amount N_j^S defined by the equation

$$N_j = N_j^\alpha + N_j^S, \qquad (2.8)$$

where

$$N_j^\alpha = \rho_j^\alpha V^\alpha. \qquad (2.9)$$

One should note that it is quite possible, given a particular convention (say $V^S=0$), for the $N_j{}^S$ to take on either positive or negative values. Likewise the excess surface entropy S^S can be negative, even though the bulk value is always positive.

It is also sometimes useful to define excess thermodynamic potentials. Recall that the Helmboltz (A), Gibbs (G), and grand (Ω) potentials are defined by:

$$A = U-TS, \qquad\qquad G = U-TS+pV,$$
$$\Omega = U-TS-\Sigma_j\mu_jN_j, \tag{2.10}$$

where T and p are the temperature and pressure, and μ_j is the chemical potential of the j'th component. Each of these can be written as a sum of a bulk and a surface part:

$$A = A^\alpha+A^S, \qquad G = G^\alpha+ G^S, \qquad \Omega = \Omega^\alpha+\Omega^S. \tag{2.11}$$

The equations (2.10) of course also hold for the bulk fluid, i.e., if one adds a superscript α to each extensive variable. Subtracting one from the other yields the results

$$A^S = U^S-TS^S, \qquad\qquad G^S = U^S-TS^S+pV^S,$$
$$\Omega^S = U^S-TS^S-\Sigma_j\mu_jN_j{}^S \tag{2.12}$$

for the surface excess quantities. Note that the variables T, p, and μ_j are identical in (2.10) and (2.12), i.e., we associate with the surface the same temperature, pressure, and chemical potentials as with the bulk. This is partly a matter of definition, and other definitions are certainly possible. Note that the stress tensor for the fluid near the wall will, in general, be anistropic, and thus has no simple connection with the bulk pressure.

In bulk thermodynamics it is often convenient to use "densities": ratios of two extensive variables, such as the energy per mole or the number of moles per unit volume. Analogous "surface densities" can be defined as ratios of surface-extensive quantities. One of the most useful is

$$\Gamma_j = N_j{}^S/a, \tag{2.13}$$

the excess amount per unit area of component j.

III. Fluid Against an Ideal Wall: Differential Relations

One of the standard equations for bulk thermodynamics is the differential relation

$$dU^{\alpha} = TdS^{\alpha} - pdV^{\alpha} + \Sigma_j \mu_j dN_j^{\alpha},\tag{3.1}$$

which can be interpreted as follows. There is a fundamental relation

$$U^{\alpha} = U^{\alpha}(S^{\alpha}, V^{\alpha}, N_1^{\alpha}, \ldots N_c^{\alpha})\tag{3.2}$$

which gives the energy as a function of the other extensive variables. Suppose that S^{α}, V^{α}, and the N_j^{α} are smooth functions of a parameter t. Then differentiating (3.2) by the chain rule yields

$$\frac{dU^{\alpha}}{dt} = \frac{\partial U^{\alpha}}{\partial S^{\alpha}}\frac{dS^{\alpha}}{dt} + \frac{\partial U^{\alpha}}{\partial V^{\alpha}}\frac{dV^{\alpha}}{dt} + \sum_j \frac{\partial U^{\alpha}}{\partial N_j^{\alpha}}\frac{dN_j^{\alpha}}{dt}.\tag{3.3}$$

Equation (3.1) is the same as (3.3), except for the omission, for compactness, of the dt's in the denominator of the latter, if one uses the definitions

$$T = \partial U^{\alpha}/\partial S^{\alpha}, \qquad p = -\partial U^{\alpha}/\partial V^{\alpha}, \qquad \mu_j = \partial U^{\alpha}/\partial N_j^{\alpha}\tag{3.4}$$

of the field variables. That is to say, (3.1) is a convenient "shorthand" for (3.3) and (3.4). (If you are ever in doubt as to the mathematical propriety of some manipulation involving differentials, it is helpful to go back to the "longhand" version, such as (3.3), and check it by the standard rules of calculus.)

From the fact that U^{α} is a homogeneous function of S^{α}, V^{α}, and the N_j^{α}, one can obtain the Euler equation:

$$U^{\alpha} = TS^{\alpha} - pV^{\alpha} + \sum_j \mu_j N_j^{\alpha}.\tag{3.5}$$

By differentiating this equation and combining the result with (3.1), one gets the Gibbs-Duhem equation:

$$S^{\alpha}dT - V^{\alpha}dp + \sum_j N_j^{\alpha}d\mu_j = 0.\tag{3.6}$$

Equations (3.1), (3.5), and (3.6) may be regarded as the basic relations of bulk thermodynamics, from which all other equations follow as applications of calculus[1]. The awkward superscript α has been retained to emphasize the fact that these hold <u>exactly</u> only for a <u>bulk</u> system, and are not satisfied for the total U, S, V, etc. (bulk plus surface) for a finite box. The generalization of these equations to surface excess quantities can be obtained from the following argument.

Let us suppose that one wall of our box, Fig. 3 is a movable piston, and its dimensions are now $\ell_1 \times \ell \times \ell$, where ℓ_1 is variable. If an external force F is applied to the piston and the system undergoes an adiabatic, reversible expansion, the first law of thermodynamics states that

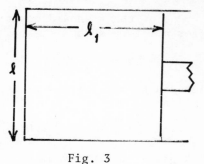

$$dU = -F d\ell_1, \qquad (3.7)$$

where F is positive if the force is pushing the piston to the left.

Fig. 3

If we would ignore effects at the perimeter of the piston, F would be equal to $\ell^2 p$, since the external force must balance the pressure of the molecules in the fluid, and the piston has an area ℓ^2. However, due to effects at the perimeter we might expect a correction term proportional to ℓ, and so we write

$$F = \ell^2 p - 4\ell\sigma. \qquad (3.8)$$

One should regard (3.8) in much the same spirit as (2.2): $4\ell\sigma$ is a "surface" contribution <u>defined</u> to be the difference between a physical quantity F and a "bulk" contribution, $\ell^2 p$, equal to the value which F would have <u>if</u> the force per unit area acting on each portion of the piston were precisely equal to p. The quantity σ is the "surface tension". You can visualize it as a force per unit length acting along the perimeter of the piston and tending to pull it to the left when σ is positive*. Another common notation for surface tension is γ.

[It is evident that the definition of σ makes use of a certain convention, through the appearance of $\ell^2 p$ rather than some other area times p on the right side of (3.8). However we shall adopt the definition (3.8) whatever the value of z_0, so that σ is independent of the convention which defines V^s.]

Upon noting that $\ell^2 d\ell_1$ is equal to dV and $4\ell d\ell_1$ is equal to d$\mathcal{a}$, we can rewrite (3.7), using (3.8), as

$$dU = -p dV + \sigma d\mathcal{a}. \qquad (3.9)$$

Since V, U, S, and the N_j are sums of bulk and surface contributions, (2.1), (2.5) and (2.8), and since S and the N_j are constant during the motion of the piston, we have:

* A stability condition requires $\sigma > 0$ for the interface between two bulk phases, but this is not required at an ideal wall.

$$dU = dU^\alpha + dU^S = -pdV^\alpha - pdV^S + \sigma d\mathbf{a}; \qquad (3.10)$$

$$0 = dS = dS^\alpha + dS^S; \qquad 0 = dN_j = dN_j^\alpha + dN_j^S. \qquad (3.11)$$

Next proceed as follows: Solve (3.10) for dU^S, eliminate dU^α using (3.1), and then eliminate dS^α and dN_j^α using (3.11). The result of this exercise is the equation

$$dU^S = TdS^S - pdV^S + \sum_j \mu_j dN_j^S + \sigma d\mathbf{a}, \qquad (3.12)$$

which resembles (3.1) except for the additional $\sigma d\mathbf{a}$ term at the end.

One expects that all surface extensive variables, for a given convention, will be proportional to $\mathbf{a}$, and this results in an Euler equation

$$U^S = TS^S - pV^S + \sum_j \mu_j N_j^S + \sigma \mathbf{a}, \qquad (3.13)$$

which is the surface counterpart of (3.5). Finally, by differentiating (3.13) and combining the result with (3.12), one obtains the Gibbs equation

$$S^S dT - V^S dp + \sum_j N_j^S d\mu_j + \mathbf{a} d\sigma = 0, \qquad (3.14)$$

which is the surface analog of (3.6).

Equations (3.12), (3.13), and (3.14) can be regarded as the basic equations of surface thermodynamics in much the same way as their bulk counterparts. We have derived them for the case of an ideal wall, but they are also formally identical (as we shall see) to the equations for a fluid-fluid interface, and apply at fluid-solid interfaces in certain special circumstances. Of course the full significance of such equations can only be appreciated by applying them to various situations and considering various special cases. This is particularly important because, in fact, the formal simplicity of the equations makes it easy for the beginning student to make serious blunders if he is not careful.

An example will help to show where the problems lie, and how to get around them. Let us compare the Gibbs-Duhem equation (3.6) and the Gibbs equation (3.14) in the case of a pure fluid, so that we can omit the summations and the subscript j:

$$S^\alpha dT - V^\alpha dp + N^\alpha d\mu = 0, \qquad (3.15)$$

$$S^S dT - V^S dp + N^S d\mu + \mathbf{a} d\sigma = 0. \qquad (3.16)$$

If we set $dT = 0$ in (3.15) and divide by V^α, the result is:

$$\rho^\alpha = N^\alpha/V^\alpha = (\partial p/\partial \mu)_T, \qquad (3.17)$$

a useful equation because it tells us how we can calculate the density
from a knowledge of $p(T,\mu)$, one of the "fundamental relations" for a
pure fluid. By analogy, we might set $dT = dp = 0$ in (3.16) and write

$$\Gamma = N^S/\boldsymbol{a} = -(\partial\sigma/\partial\mu)_{T,p}. \qquad \text{WRONG!} \qquad (3.18)$$

What is the problem with this equation? The problem is that the
right side makes no sense. For a pure fluid, μ is a well-defined
function of p and T, so that if T and p are held constant, there is
only one value of μ for which σ is defined, and thus its derivative
with respect to μ at fixed T and p is undefined. One should also be
suspicious of (3.18) because the left side depends on a convention,
whereas σ, as we have defined it, does not.

     The treacherous character of (3.16) from the viewpoint of some-
one familiar with bulk thermodynamics is that it seems to suggest
that three of the four variables T, p, μ, and σ can be varied
independently whereas, in fact, only two can be varied independently.
Note that there is nothing incorrect about (3.16), but one has to
employ the proper interpretation.

     There are two simple ways of getting rid of the extra differen-
tial in (3.16). One is to adopt a convention in which one of the
surface excess quantities vanishes. For example, if we adopt the
Gibbs convention, $V^S=0$, then we have:

$$S^S dT + N^S d\mu + \boldsymbol{a}\,d\sigma = 0 \qquad (V^S = 0) \qquad (3.19)$$

from which, by setting $dT=0$, we obtain the correct equation:

$$\Gamma = N^S/\boldsymbol{a} = -(\partial\sigma/\partial\mu)_T \qquad (V^S = 0) \qquad (3.20)$$

in place of (3.18). I have deliberately placed the convention in
parentheses following the equation to emphasize that the excess
quantities S^S, N^S and Γ have the values appropriate to this convention. Naturally such a notation, just like the vowel signs in
Hebrew, is for the benefit of beginners; experts don't need it
because they (should!) know what they're doing.

     The second, and more elegant, way to handle (3.16) is to solve
(3.15) for one of the differentials and insert the result in (3.16).
For example, if we solve (3.15) for dp and insert it in (3.16), we
have, after collecting terms:

$$(S^S - \bar{s}^\alpha V^S)dT + (N^S - \rho^\alpha V^S)d\mu + \boldsymbol{a}\,d\sigma = 0. \qquad (3.21)$$

If, once again, we set $dT=0$, the result is:

$$(N^S - \rho^\alpha V^S) = -\boldsymbol{a}(\partial\sigma/\partial\mu)_T. \qquad (3.22)$$

Note that by (2.4) the left hand side of (3.22) is <u>independent</u> of
the convention, and thus a physical quantity. Its <u>value is</u>, obviously,
just the value of N^S when $V^S=0$, so that (3.20) and <u>(3.22)</u> are
actually identical.

Of course, conventions other than the Gibbs convention can
certainly be employed, and are sometimes useful. One might, for
example, set $N^S=0$. This could result in a negative V^S, but so far
as the formalism is concerned, there is nothing wrong with that.

We have discussed an oddity associated with (3.14), and this
means, of course, that there must be a corresponding oddity connected
with (3.12). Indeed there is, but it is not quite as dangerous. So
far as <u>physical</u> changes are concerned, there is actually one less
independent surface-extensive variable than the number appearing on
the right hand side. The additional <u>formal</u> degree of freedom in
(3.12) arises from the fact that one can continuously change the
convention. We can, of course, get rid of it by the simple device
of fixing the convention. For example, the Gibbs convention yields:

$$dU^S = TdS^S + \sum_j \mu_j dN_j^S + \sigma d\mathbf{Q}, \qquad (V^S = 0) \qquad (3.23)$$

since if $V^S=0$, $dV^S=0$. However, (3.12) is a valid equation both for
changes in convention (pencil-pushing) and for physical changes
(piston-pushing), or for combinations. One can also rewrite (3.12)
in terms of invariant quantities, Problem 3.

IV.   Interfaces Between Fluid Phases: Surface Excess Quantities.

    Ideal walls of the sort previously discussed do not, of course,
exist in nature. Actual surfaces occur (typically) at a boundary
between two thermodynamic phases: imagine the meniscus between water
and water vapor, or the surface between metallic copper and a vacuum—
the latter to be thought of as a very dilute gas. The simplest of
these to describe in thermodynamic terms is a flat interface between
two fluids.

    In order to understand the concept of excess quantities in this
case it is helpful to imagine an $\ell \times \ell \times h$ cell, Fig. 4, with ideal walls
of the type previously discussed, whose height h is small compared
to ℓ. The cell contains
two phases, α and β,
separated by a meniscus
which is roughly half way
between the top and bottom
surfaces, indicated by a
solid line in Fig. 4. Of
course, we do not expect
the meniscus to be at a
precisely defined position.
A possible form for the

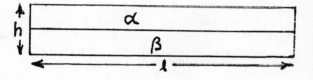

Fig. 4

density ρ as a function of height z in the case of a pure fluid
(for which α would be the vapor and β the liquid) is sketched in
Fig. 5. We shall assume that the
height λ over which the density
differs appreciably from ρ^{α}, the
bulk value in the upper phase,
or ρ^{β}, the bulk value in the lower
phase, is much less than h. [At
this point a comment should be
made about a slight technical
difficulty. It is necessary to
impose a weak gravitational field
in order to keep λ small if one
takes the limit of $\ell \to \infty$, $h \to \infty$. Since

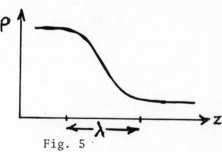

Fig. 5

the thermodynamic properties do not depend on details of the profile
shape, it seems unlikely that the analysis given below needs to be
altered because of this technical difficulty, but I am not aware of
anyone having worked out all the details.]

    The reason for using a flat cell h<<ℓ is that we shall wish to
take account of the top and bottom walls of the cell, in contact
with phases α and β, respectively, and it is convenient to be able
to ignore the effects of the side walls of the cell because of their
(relatively) small area. Note that the confining potential $\psi(\vec{r})$
could have different forms near the top and bottom walls.

It is convenient to use a slightly different notation from Sec. II and denote total values of extensive thermodynamic variables for the cell with a superscript t. Each of these will be written as a sum of five terms:

$$V^t = V^{\alpha w} + V^{\beta w} + V^{\alpha} + V^{\beta} + V^s, \qquad (4.1)$$

$$N_j^{\ t} = N_j^{\ \alpha w} + N_j^{\ \beta w} + N_j^{\ \alpha} + N_j^{\ \beta} + N_j^{\ s}, \qquad (4.2)$$

$$U^t = U^{\alpha w} + U^{\beta w} + U^{\alpha} + U^{\beta} + U^s, \qquad (4.3)$$

etc., with the following significance. We insert a set of four horizontal mathematical (i.e., no thickness) surfaces in the cell in a manner indicated schematically in Fig. 6, at heights z=a, a', b', and b, so that the cell is divided into five regions whose volumes are those indicated in the figure. The surface at z=a is sufficiently far from the top wall so that at this height the properties of the fluid differ insignificantly from those of bulk phase α. Similarly, at height z=b the properties are essentially those of bulk phase β, i.e., the influence of the bottom wall has disappeared. The heights z=a' and b' differ by a microscopic amount and are very close to the height of the meniscus.

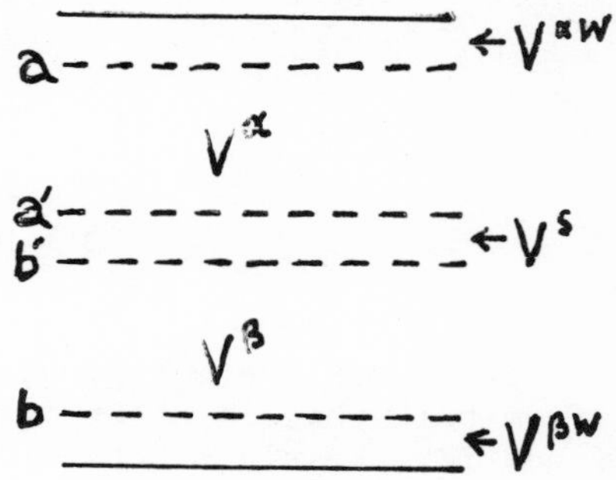

Fig. 6

Given the foregoing convention for $V^{\alpha w}$, the quantities $U^{\alpha w}$, etc. are to be determined by the difference procedure for obtaining surface quantities discussed previously in Sec. II, applied to a cell which contains only phase α (and has walls identical to the top wall of the cell in Fig. 4). An analogous comment applies to $U^{\beta w}$, etc. Hence, on the basis of what we already know, we can unambiguously define the extensive variables without a superscript by subtracting the αw and βw quantities from the totals:

$$V = V^t - V^{\alpha w} - V^{\beta w}; \qquad N_j = N_j^{\ t} - N_j^{\ \alpha w} - N_j^{\ \beta w};$$

$$U = U^t - U^{\alpha w} - U^{\beta w}; \qquad S = S^t - S^{\alpha w} - S^{\beta w}; \qquad (4.4)$$

etc.

Evidently V is just the total volume between the z=a and z=b surfaces in Fig. 6, and N_j is the total quantity of the j'th chemical species in this region. One can make a similar interpretation for U and S. Note, however, that there are possible problems if,

following Gibbs[5], one simply <u>defines</u> U as the total energy of the
fluid in the region a≤z≤b. The reason is that the energy includes
terms due to the interaction among groups of particles, some of
which are inside and some of which are outside this region. While
this problem can be settled by adopting an appropriate convention,
and likewise the problem of counting polyatomic molecules some of
whose constituent atoms are on one and some on the other side of
the dividing surfaces, there seems to be no similar way of handling
the entropy. Hence if one wants to know <u>precisely</u> how U and S, etc.,
are defined from the point of view of someone doing statistical
mechanics, it is useful to think about the top and bottom surfaces
of the cell, and then get rid of their effects through subtractions.
In the remainder of this section we shall assume that this has been
done, and we next focus attention on the equations which result from
combining (4.1) through (4.4):

$$V = V^\alpha + V^\beta + V^s,$$

$$N_j = N_j{}^\alpha + N_j{}^\beta + N_j{}^s, \qquad (4.5)$$

$$U = U^\alpha + U^\beta + U^s,$$

etc.

Here U^α is the energy associated through Eq. (2.6) with a
uniform bulk phase α occupying a volume V^α, and U^β is similarly
defined for phase β by changing α to β in (2.6). The surface excess
energy U^s is then the difference between the physical quantity U
and the sum $U^\alpha + U^\beta$. The latter clearly depends on two conventions,
the choice of V^α and V^β. Similar remarks apply to $\overline{N_j{}^s}$, S^s, etc.

The ambiguities introduced by the presence of these arbitrary
conventions are quite analogous to those considered earlier in
Sec. II, though of course two conventions are more confusing than
one convention. Suppose that V^α is altered by ΔV^α and V^β by ΔV^β.
Then a brief calculation, using the fact that V, N_j, etc. are fixed,
yields equations:

$$\Delta V^s = -\Delta V^\alpha - \Delta V^\beta,$$

$$\Delta N_j{}^s = -\rho_j{}^\alpha \Delta V^\alpha - \rho_j{}^\beta \Delta V^\beta, \qquad (4.6)$$

and similar expressions for other variables. On the right hand side
the densities ($\rho_j{}^\alpha$, etc.) refer to the bulk phases, and as the
properties of these phases are assumed to be known, changes due to
a shift in the conventions can always be calculated in a straightforward manner. (A useful mnemonic for (4.6) is that expanding the
region "occupied by" the surface phase can only be accomplished by
"eating up" the bulk phases.)

In the case in which the surfaces at z=a' and z=b' are

sufficiently far apart that at the former the properties of the
fluid are essentially those of bulk phase α and at the latter those
of bulk phase β, the surface excess quantities have a simple interpretation, as emphasized by Cahn[3]. Not only is V^S the <u>actual</u> volume
occupied by the surface region, but N_j^S is the actual <u>quantity</u> of
matter, U^S the "actual" energy and S^S the "actual" entropy of the
fluid in this region. Such a point of view is intuitively very
helpful, but the qualifications noted above Eq. (4.5) are again
applicable, which is why I have placed "actual" in quotation marks.

The <u>Gibbs convention</u> $V^S=0$ is often useful. If it is adopted,
a' and b' coincide in Fig. 6, and the single mathematical surface
which results is known as the <u>Gibbs dividing surface</u>. The vertical
position of this surface can be fixed by another convention, and a
common choice (also introduced by Gibbs) is to place it so that one
of the N_j^S, let us say N_1^S, is equal to zero. This is certainly
possible if $\rho_1^\alpha \neq \rho_1^\beta$. In terms of a density profile, the condition
$N_1^S=0$ is equivalent to putting the Gibbs surface at a height $z=z_1$
defined by the "equal area" condition (Problem 4):

$$\int_b^{z_1} [\rho_1{}^\beta - \rho_1(z)]\,dz = \int_{z_1}^a [\rho_1(z) - \rho_1{}^\alpha]\,dz \qquad (4.7)$$

illustrated in Fig. 7: the cross-
hatched areas to the left and right
of the vertical line are the two
sides of (4.7). In other words,
z_1 is chosen such that <u>if</u> phases
α and β were each <u>extended</u> with
uniform density right up to the
dividing surface, the total amount
of component 1 in the cell would
be the same as the actual amount.
You should beware of assuming that

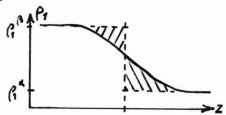

Fig. 7

the densities actually <u>are</u> uniform up to the dividing surface, or
that the presence of a <u>continuous</u> profile makes the concepts of
surface thermodynamics only approximate. (One occasionally finds
careless statements in the literature which might lead one to either
of these conclusions.)

The Gibbs, Helmholtz, and grand potentials are again defined
as in (2.10), and can be decomposed into bulk and surface contributions in the manner indicated in (4.5):

$$A = A^\alpha + A^\beta + A^S, \qquad G = G^\alpha + G^\beta + G^S,$$

$$\Omega = \Omega^\alpha + \Omega^\beta + \Omega^S. \qquad (4.8)$$

The field variables T, p, and the μ_j are identical in the two coexisting phases α and β. A subtraction procedure utilizing this fact yields (2.12), and thus expressions for A^S, etc., in terms of U^S, S^S and the N_i^S. Note that the values of T, p, etc. are the same in the interface as in the pure phases on either side of the interface (but see the comments following (2.12)).

V.   Interface Between Fluid Phases: Differential Relations.

        To find the differential relation for dU^s, we equip the cell
in Fig. 4 with a movable wall, Fig. 8.   Its dimensions are now
$\ell_1 \times \ell \times h$, where ℓ_1 is a variable,
and we shall assume that h is
very small compared to both ℓ and
ℓ_1 so that we can ignore effects
due to the side walls.

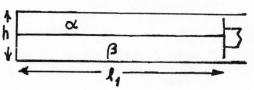

        In an adiabatic, reversible
process we have (compare with
(3.7)):

Fig. 8

$$dU^t = -Fd\ell_1, \qquad (5.1)$$

and we expect to be able to write the force F on the piston as:

$$F = h\ell p - \ell\sigma^{\alpha w} - \ell\sigma^{\beta w} - \ell\sigma, \qquad (5.2)$$

where the terms may be interpreted as follows: $h\ell p$ would be the
total force on the $\ell \times h$ piston produced by a uniform pressure p,
$\ell\sigma^{\alpha w}$ and $\ell\sigma^{\beta w}$ are the extra forces of tension acting on the top and
bottom edges of the piston, and $\ell\sigma$ is that due to the interface
separating α and β.   We assume that $\sigma^{\alpha w}$ and $\sigma^{\beta w}$ are known from
previous gedanken experiments carried out using cells containing
only phase α or only phase β.   Hence (5.2) serves to define $-\ell\sigma$
as a difference between a physical force F and other known quantities.
Note that the definition is independent of the choice of $V^{\alpha w}$, $V^{\beta w}$,
V^{α}, and V^{β}, and thus σ will not depend on the choice of the convention discussed in Sec. IV.

        Equation (5.1) may be rewritten with the help of (5.2) in the
form:

$$dU^t = -pdV^t + \sigma^{\alpha w}d\mathbf{a}^{\alpha w} + \sigma^{\beta w}d\mathbf{a}^{\beta w} + \sigma d\mathbf{a}, \qquad (5.3)$$

where $\mathbf{a}^{\alpha w}$, $\mathbf{a}^{\beta w}$, and $\mathbf{a}$ are the areas of the α-wall, β-wall, and
α-β interfaces, respectively.   We are now in a position to derive
the relationship for dU^s by means of the following steps, details
of which are left to the reader.   The first step is to eliminate
dU^t and dV^t from (5.3) using (4.4); i.e.,

$$dU^t = dU + dU^{\alpha w} + dU^{\beta w}, \qquad (5.4)$$

and the analogous relation for dV^t.   Next eliminate $dU^{\alpha w}$ and $dU^{\beta w}$
by using the equations obtained from (3.12) by replacing the
superscript s with αw (βw) and adding an αw (βw) superscript to σ
and $\mathbf{a}$.   The remaining quantities with αw and βw superscripts can
then be eliminated with the help of:

$$-dS = dS^{\alpha w} + dS^{\beta w}; \qquad -dN_j = dN_j^{\alpha w} + dN_j^{\beta w}, \qquad (5.5)$$

which are a consequence of (4.4) and the fact that N_j^t and S^t are
constants for the adiabatic process to which (5.1) applies. Thus
we have:

$$dU = TdS - pdV + \sum_j \mu_j dN_j + \sigma d\boldsymbol{a}. \qquad (5.6)$$

We now replace U, S, V, and N_j in this equation by the sums
in (4.5). The terms with superscript α can be eliminated using
(3.1). Of course (3.1), with a change in superscript, also applies
to the bulk phase β. What remains, then, is precisely Eq. (3.12):

$$dU^S = TdS^S - pdV^S + \sum_j \mu_j dN_j^S + \sigma d\boldsymbol{a}, \qquad (5.7)$$

and of course (3.13) and the Gibbs equation (3.14):

$$S^S dT - V^S dp + \sum_j N_j^S d\mu_j + \boldsymbol{a} d\sigma = 0, \qquad (5.8)$$

also apply.

The formal identity of (5.7) with (3.12) and (5.8) with (3.14)
does not imply a conceptual identity, for reasons which will become
obvious if we consider (5.8) for a pure substance:

$$S^S dT - V^S dp + N^S d\mu + \boldsymbol{a} d\sigma = 0. \qquad (5.9)$$

At a given temperature there is a unique vapor pressure at which
the liquid (β) and vapor (α) phases will coexist, and of course
this implies a unique chemical potential and a unique surface tension.
Hence only one of the four quantities T, p, μ, and σ in (5.9) can
be varied independently!

One way of (correctly!) utilizing (5.9) is to choose a convention by which two of the surface excess quantities vanish. For
example,

$$S^S = -\boldsymbol{a} d\sigma/dT, \qquad (V^S = N^S = 0) \qquad (5.10)$$

$$V^S = \boldsymbol{a} d\sigma/dp, \qquad (N^S = S^S = 0) \qquad (5.11)$$

where we have listed the conventions following the equations. Yet
another way of dealing with (5.9) is to use the Gibbs-Duhem equation
(3.6) for phase α and its counterpart for phase β in order to
eliminate two of the differentials in (5.9) and achieve a result in
which the coefficient of the remaining term (other than dσ) is
explicitly independent of convention[3]. See Problem 5.

In the case of a mixture of c components the Gibbs phase rule
states that if two phases are coexisting, the number of independent
fields is equal to c, or three less than the numbers of differentials

on the left side of (5.8). The simplest way of taking care of this is to adopt a convention in which two of the excess quantities vanish; e.g., $V^S = 0 = N_1^S$. A similar convention can them be employed in (5.7). The alternative approach using the Gibbs-Duhem equation for the pure phases has been discussed by Cahn[3].

A convention in which $N_j{}^S$ for some j is set equal to zero yields a situation with a very close formal analogy to that in which the remaining components form an interface at an ideal wall. Consider, for example, a 2-component mixture, and let $N_1^S = 0$. Then (5.8) becomes

$$S^S dT - V^S dp + N_2^S d\mu_2 + \mathbf{Q} d\sigma = 0, \qquad (5.12)$$

and apart from the subscript 2 it is identical to (3.16). Furthermore, in both cases one is able to set <u>one</u> of the (remaining) excess quantities equal to zero by convention. For example, if one is considering argon adsorbed on graphite, a two-component system, setting the surface excess of carbon equal to zero yields surface thermodynamic equations formally the same as those for argon against an ideal wall. (Of course, graphite is not a liquid, so the preceding assertion needs to be qualified by the remarks in Sec. VI, B below.)

Various relationships for the surface thermodynamic potentials A^S, G^S, and Ω^S can be worked out by combining (2.12) with (3.12) or (3.13) — note that all of these equations are valid for fluid interface as well as ideal walls. In particular one finds that:

$$A^S = -pV^S + \Sigma_j \mu_j N_j{}^S + \sigma \mathbf{Q},$$

$$G^S = \Sigma_j \mu_j N_j{}^S + \sigma \mathbf{Q} \qquad (5.13)$$

$$\Omega^S = -pV^S + \sigma \mathbf{Q}.$$

These equations are useful in understanding the assertion, which one often finds in the literature, that the surface tension is the "surface free energy per unit area". In bulk thermodynamics the term "free energy" is sometimes used for A and sometimes for G, and these quantities are not equal to each other. However, for surface thermodynamics the two <u>are</u> equal provided one adopts the Gibbs convention $V^S = 0$. Furthermore, in a pure substance one can adopt the additional convention $N^S = 0$, in which case σ is indeed equal to $A^S/\mathbf{Q}$ or $G^S/\mathbf{Q}$. But in a mixture σ is not (in general, at least) equal to $A^S/\mathbf{Q}$, though it is still equal to $G^S/\mathbf{Q}$ in a binary mixture if one uses the convention $N_1^S = N_2^S = 0$ (in which case one must expect $V^S \neq 0$). Provided $V^S = 0$, σ is equal to the $\Omega^S/\mathbf{Q}$ no matter how many components are present, and this is probably the best way of identifying the surface tension with an excess "free energy". Still another amusing possibility is pointed out in Problem 6.

VI.   Short Comments on Various Topics:

A.   Curved Interfaces

     A thermodynamic description of curved interfaces between fluid
phases is relatively simple provided the thickness λ of the interface,
Sec. IV, Fig. 5, is small compared to both the principal radii of
curvature, r_1 and r_2, of the surface.   The pressure difference
between the phases is given by the equation of Young and Laplace:

$$\Delta p = p^\beta - p^\alpha = \sigma(1/r_1 + 1/r_2), \qquad (6.1)$$

with r_1 and r_2 considered positive when phase β occupies a convex
region, such as a sphere.   In what follows we shall assume that the
interface is a sphere of radius r, so that

$$\Delta p = 2\sigma/r. \qquad (6.2)$$

Remember that the pressure is always higher inside the drop or
bubble than it is outside.   The temperature and the chemical
potentials, however, are the same in both phases, provided they are
in equilibrium.

     Gibbs[2] showed that if the dividing surface, with $V^S=0$, is placed
at the correct location, i.e., if one makes the proper choice for
the radius of the sphere, then (3.12), and consequently (3.13) and
(3.14), with $V^S=0$, are valid for a spherical droplet (or bubble).
In addition, since varying the radius of the droplet changes the
pressure difference (6.2), there is one more independent field
variable in the cases of phases coexisting across a spherical surface
of variable curvature than in the case of coexistence across a flat
interface.   For example, in the case of a pure fluid both dT and
dμ can be varied independently in the Gibbs equation

$$S^S dT + N^S d\mu + \boldsymbol{\mathcal{A}} \, d\sigma = 0, \qquad (6.3)$$

so that it makes sense to write

$$\Gamma = N^S/\boldsymbol{\mathcal{A}} = -(\partial\sigma/\partial\mu)_T . \qquad (6.4)$$

Here Γ is no longer subject to variation-by-convention because the
convention which fixes it is that mentioned at the beginning of the
paragraph.

     These remarks might lead one to suppose that the considerations
introduced in Secs. IV and V for flat interfaces are completely
inapplicable as soon as the surface acquires a slight curvature.
But such is not the case.   In most circumstances of practical interest
the radius r is very much larger than the thickness of the interface
λ.   The shift in the position of the dividing surface when going

from one convention to another, or the separation a'-b', Fig. 6, when $V^S \neq 0$, is on the order of λ or less, and the fractional errors caused by applying the formulas of Secs. IV and V to curved interfaces are of order λ/r. Since λ is, typically, of the order of at most a few nanometers, except near a critical point, these fractional errors are utterly negligible for droplets of macroscopic size. Hence as long as r is not too small, the convention of Gibbs mentioned in the previous paragraph may be regarded more as a way to produce a tidy formalism than as superior in a practical sense to other conventions. (And of course it remains the case that results in one convention can be easily translated into another.) The pressure difference between the phases, (6.1) or (6.2), is, however, an observable effect manifesting itself in phenomena such as capillary rise.

It is perhaps worthwhile adding a comment about the application of (6.2) in determining conditions of phase coexistence, since some of the treatments one finds in the literature are quite obscure. In Fig. 9 the chemical potential μ as a function of pressure p at a fixed temperature is shown as two separate curves, labeled α and β, for the vapor and liquid phases of a pure substance. For coexistence across a flat interface the vapor pressure is p_X and the corresponding chemical potential is μ_X. The two dots indicate the condition when phase β is a spherical droplet inside the phase α. Note that whereas the pressure difference is given by (6.2), the chemical potential is equal in the two phases, but higher than μ_X. The amount by which the pressure p^α in phase α exceeds p_X is denoted by δp, and is given by the formula (Problem 11):

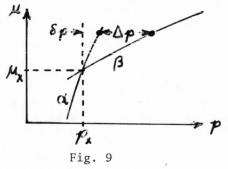

Fig. 9

$$\delta p = p^\alpha - p_X = \frac{2\sigma v^\beta/r}{(v^\alpha - v^\beta)} \qquad (6.5)$$

in the approximation in which one can treat the two curves in Fig. 9 as straight lines. Here v^α and v^β are the volumes occupied by one mole of fluid in phase α and phase β, respectively.

B.   Solid-Fluid Interface

The thermodynamic description of the interface between a
crystalline solid and a fluid, including the limiting case where
the fluid is a vacuum, involves a number of complications not
present in the case of fluid-fluid interfaces. (Solid-solid interfaces, such as grain boundaries, are even more complex, and I shall
say nothing about them.) For example, σ depends upon the crystal
face under consideration: the (111) surface of nickel is quite
distinct, both structurally and thermodynamically, from the (100)
surface. And in the case of a crystal lacking a center of symmetry
there is a certain ambiguity in defining surface excess quantities.

However, these are minor problems compared with the main head-
ache, which arises from the fact that solids can exist in states
with anistropic stress. In bulk thermodynamics this leads to the
necessity of introducing stress and strain tensors in place of the
scalars p and V which suffice for a fluid. And of course analogous
concepts appear in the corresponding surface thermodynamics. In
particular, the $\mathcal{a}d\sigma$ term in (5.7) must be replaced by one involving
the surface (excess) stress tensor. Equation (3.13) remains valid,
since one can think of it as defining the "surface tension" σ,
though the surface tension (or excess "free energy") must be distinguished from the surface stress, and "tension" is thus a somewhat
misleading terminology. Of course (5.8) must also be modified. I
shall not write down the modified equations, since to explain the
additional terms would require an extensive discussion[6].

There are two circumstances under which the equations previously
derived for fluids, in particular (5.7) and (5.8), are applicable to
solid-fluid interfaces. The first is that in which diffusion
processes in the solid are sufficiently rapid that they can relieve
the anisotropic stress in times which do not exceed the patience of
the experimenter. Unfortunately (or perhaps fortunately, given the
technological importance of solids!), this situation is unlikely to
arise except in small crystals close to their melting point.

The other circumstance is that in which the solid is sufficiently rigid that one can ignore changes in the dimensions of the surface
during the course of the experiment. (This may, for example, be a
reasonably good approximation when considering the adsorption of
small amounts of gas on a solid surface.) Then (5.7) is valid if
one sets $d\mathcal{a}=0$, and (5.8) is correct as written.

C.   Surface Phase Transitions

Since interfaces lie at the boundary between two bulk phases, one would expect that a phase transition in either of the bulk phases would be reflected in the properties of the surface and in the various surface excess quantities. However, there may be phase transitions in the surface layer in the absence of any transitions in the bulk phases; let us call these "intrinsic" surface phase transitions. The remarks below are all addressed to the subject of the thermodynamics of such intrinsic transitions, though they make use of certain analogies with bulk transitions.

One useful analogy is the following. A bulk phase transition manifests itself in thermodynamic terms through the fact that a thermodynamic potential is a continuous but unsmooth function of various field variables. Thus in Fig. 9 the first-order transition from vapor to liquid gives rise to a discontinuity in the slope of $\mu(p,T)$ as a function of p at fixed T. In fact both $\partial\mu/\partial p$ and $\partial\mu/\partial T$ are discontinuous along the phase transition line in the T, p plane. At a higher-order transition (also called a continuous transition or second-order transition or critical point), such as the onset of ferromagnetism, the first partial derivatives of the potential are continuous, but singular behavior (discontinuities or divergences or at least some non-analyticity) appears in the second- or higher-order partial derivatives.

In the case of intrinsic surface phase transitions, we expect the surface tension σ to be a continuous but unsmooth function of the independent thermodynamic fields. To take a specific example, an "insoluble" film of a suitable substance on the surface of water can exhibit a first-order phase transition which manifests itself through the fact that

$$\Gamma_2 = -(\partial\sigma/\partial\mu_2)_T \qquad (6.6)$$

is a discontinuous function of μ_2 at fixed T. (Here we are assuming $V^S=0=N_1^S$ with 1 referring to water and 2 to the substance on the surface.) It appears that higher-order transitions also occur in such films[7], and again it is helpful to think of the analogy between $\sigma(\mu_2,T)$ for the surface and $\mu(p,T)$ for the bulk phase of a pure substance.

There is one important respect in which the analogy between surface and bulk does not hold. The thermodynamic potential for a bulk system possesses certain stability properties which manifest themselves through the positivity of various heat capacities, compressibilities, etc. For a surface or interface there is a stability condition $\sigma>0$, but there is no requirement in general that the surface analogs of heat capacity, compressibility, etc. be positive. The reason is that the surface quantity is an excess quantity, the difference between an actual value and some conventional reference value, and thus can have either sign.

VII. PROBLEMS

No. 1. Show that when $V^S=0$, N^S in (2.2) is given by the expression

$$N^S = a \int_o^w [\rho(z) - \rho^\alpha] dz \qquad (P.1)$$

where a is the surface area $(6\ell^2)$ of the cubical box and w is some distance large compared to λ and small compared to ℓ.

No. 2. Find the equation analogous to (3.21) when (3.15) is solved for $d\mu$ and the answer inserted in (3.16). Use the notation $s^\alpha = S^\alpha/N^\alpha$, $v^\alpha = V^\alpha/N^\alpha$ for the entropy and volume per mole in the bulk phase.

No. 3. Rewrite (3.12) in terms of differentials of invariant quantities of the type which appear on the left side of (3.21). [Hint: Set $V^\alpha = 1$ in (3.1) and (3.5), and then combine these equations with (3.12).]

No. 4. Derive Eq. (4.7) from the condition $N_1{}^S = 0$ when $V^S = 0$.

No. 5. Solve the two Gibbs-Duhem equations for phases α and β for a pure substance, (3.15) and the equation obtained by replacing α by β, for $d\mu$ and dp , and substitute the expressions so obtained in (5.9) so as to obtain an equation in dT and dσ involving only quantities which are independent of the conventions which determine V^α and V^β.

No 6. Although it is seldom used in practice, one can formulate bulk thermodynamics in terms of the "null potential"

$$\Phi = U - TS + pV - \Sigma_j \mu_j N_j . \qquad (P.2)$$

Show that the surface tension for the flat interface between fluid phases is always equal to the surface excess null potential per unit area, independent of the conventions used to define V^α and V^β.

No. 7. For a pure substance against an ideal wall, derive the Maxwell relations ($V^S=0$):

$$\left(\frac{\partial S^S}{\partial \mu}\right)_{T,a} = \left(\frac{\partial N^S}{\partial T}\right)_{\mu,a} \qquad (P.3)$$

$$\left(\frac{\partial S^S}{\partial N^S}\right)_{T,a} = -\left(\frac{\partial \mu}{\partial T}\right)_\Gamma \qquad (P.4)$$

Show that the same relationships hold at an interface in a binary mixture if one makes a suitable convention and a suitable identification of the symbols by using appropriate subscripts.

No. 8. The isosteric heat of adsorption of a pure substance on an ideal wall is defined by:

$$q_{st} = T[s^{\alpha} - (\partial S^S/\partial N^S)_{T,\mathcal{Q}}] \qquad (P.5)$$

where s^{α} is the entropy per mole in the bulk phase. Show that if the bulk phase is an ideal gas, and $\Gamma(T,p)$ is known, q_{st} can be computed from the formula:

$$q_{st} = RT^2(\partial \ell np/\partial T)_{\Gamma} \qquad (P.6)$$

Hint: Recall that

$$s^{\alpha} = -(\partial \mu/\partial T)_p, \qquad (P.7)$$

and that for an ideal gas,

$$\mu(T,p) = \mu_{\circ}(T) + RT\ell np. \qquad (P.8)$$

One of the Maxwell relations derived earlier in Problem 7 will also be useful.

No. 9. Equation (P.6) has also been used in the case of krypton adsorbed on graphite. Does its validity depend on the approximation in which one replaces the graphite surface with an ideal wall? Why or why not?

No. 10. At an interface in a fluid mixture, let z_j be defined by replacing the subscript 1 by j everywhere in Eq. (4.7). Show that the value of Γ_j for the convention $V^S=0=N_1^S$ can be expressed in terms of $z_j - z_1$ and the properties of the bulk phases.

No. 11. (a) Derive (6.5) using the approximation that v^{α} and v^{β} are constant. Recall that $v=(\partial \mu/\partial p)_T$.

(b) If α is a dilute gas and r is very small, it may not be a good approximation to treat v^{α} as constant. Derive the result

$$RT\ell n(p^{\alpha}/p_x) = 2\sigma v^{\beta}/r, \qquad (P.9)$$

using the approximation $v^{\alpha} >> v^{\beta}$ and v^{β} a constant (Kelvin equation).

(c) What is the analog of (6.5) for the shift in the pressure of phase β when it coexists with a bubble of phase α?

No. 12. In certain dilute solutions it is found that the liquid-vapor surface tension decreases linearly with the mole fraction x of solute in the liquid phase if the temperature is fixed. Using the fact that under these circumstances the chemical potential of the solute is given to a first approximation by the formula

$$\mu = \mu° - RT\ell nx, \qquad\qquad (P.10)$$

where $\mu°$ depends only on T, derive the "surface equation of state" $\Gamma(\pi,T)$, where

$$\pi = \sigma_o - \sigma \qquad\qquad (P.11)$$

is the "surface" or "spreading" pressure, σ_o is the surface tension when x=0, and Γ refers to the solute with the convention that V^s=0 and the surface excess of solvent is also zero.

Acknowledgments

I am indebted to Professors J. C. Wheeler and F. A. Putnam for introducing me to the subject of surface thermodynamics and to Professors B. Widom and J. W. Cahn for a number of helpful comments and insights. My research work has been supported by the National Science Foundation through grant DMR 78-20394.

References

1. A very systematic approach to bulk thermodynamics will be found in H. B. Callen, "Thermodynamics", John Wiley & Sons, New York (1963).
2. J. W. Gibbs, "Collected Works", Yale University Press, New Haven, (1948), Vol. I, p. 219.
3. J. W. Cahn, Thermodynamics of Solid and Fluid Surfaces, in: "Interfacial Segregation", W. C. Johnson and J. M. Blakely (eds.), American Society for Metals, Metals Park, Ohio (1979).
4. For example, N. K. Adam, "The Physics and Chemistry of Surfaces", Oxford University Press (1941); A. W. Adamson, "Physical Chemistry of Surfaces", John Wiley & Sons, New York (1963); R. Aveyard and D. A. Haydon, "An Introduction to the Principles of Surface Chemistry", Cambridge University Press (1973); L. D. Landau and E. M. Lifshitz, "Statistical Physics" Addison-Wesley Publishing Co., Reading, Mass. (1969).
5. Gibbs was evidently concerned about this issue, see p. 220 of Reference 2. I am not satisfied with his explanation, perhaps because I don't understand it.
6. See Reference 3 and C. Herring, in: "Structure and Properties of Solid Surfaces", R. Gomer and C. S. Smith (eds.), University of Chicago Press (1953), p. 5 and in: "The Physics of Powder Metallurgy", W. E. Kingston (ed.), McGraw Hill Book Co., New York (1951), p. 143.
7. See the book by Adam, Ref. 4.

TWO DIMENSIONAL PHASE TRANSITIONS IN CLASSICAL VAN DER WAALS

FILMS ADSORBED ON GRAPHITE

Michel Bienfait

CRMC^2, Faculté des Sciences de Luminy
13288 Marseille Cédex 2 - France

ABSTRACT

Two-dimensional phase diagrams of adsorbed phases are usually determined by adsorption isotherm measurements that imply the existence of two-dimensional (2D) gases, liquids and solids. It is shown how the nature and the properties of these various phases, particularly 2D liquids and 2D solids, can be determined by different techniques like electron, neutron and X-ray scattering, calorimetry and ellipsometry. We pay here a special attention to the 2D gas-2D solid transition and to the phase transitions between 2D condensed phases, i.e. 2D melting and 2D polymorphism.

I - INTRODUCTION

One of the most significant developments of physics and physical chemistry over the past ten years deals with the two-dimensional (2D) adsorbed phases, that is, the phases in the monolayer range condensed on crystalline surfaces.

Some of these surface phases resemble three-dimensional (3D) gases, liquids or solids, but some of them have no analogues in the bulk matter.

Scientists have witnessed a real explosion of new results in this field. Every day now brings a bit more of comprehension about the microscopic and macroscopic properties of these 2D phases. These results are mainly due to the improvement of the traditional calorimetry, to the appearance of accurate techniques for surface

analysis, like low energy electron diffraction (LEED) and Auger
electron spectroscopy (AES), to the adjustement of tools for bulk
characterization like NMR, neutron scattering and Mössbauer spectroscopy, to surface studies and also to the activity of theoreticians who have transposed to surfaces their techniques previously
developed for bulk matter. But, a few results, probably among the
most beautiful and significant, have been obtained by using very
carefully an old technique, I mean adsorption volumetry.

The purpose of these lectures is to illustrate recent
findings in this field using examples of the phase transitions
of classical van der Waals films adsorbed on graphite. This review
is far from being exhaustive : a few typical results are analyzed
in order to give the flavour of work on surface phases. A detailed
review of the work published before 1975 can be found in the book
of J.G. Dash[1] and a critical analysis of the results obtained on
helium films has been published by the same author[2].

II - THERMODYNAMICS

II.1 - Adsorption Isotherms

The most spectacular and easily comprehensible results on
macroscopic thermodynamic properties of 2D adsorbed phases have
been obtained by measuring adsorption isotherms.

In order to draw an adsorption isotherm, one measures, for a
given temperature, the coverage θ on the substrate surface as a
function of the pressure of the adsorbing gas surrounding it.
According to the pressure range, one can use adsorption volumetry
($10^{-5} < p < 100$ torr) or Auger electron spectroscopy
($10^{-10} < p < 10^{-4}$ torr). Adsorption volumetry[3-5] is the most
precise method but operates with crystalline powders. Let us
recall that a thermostated glass container is filled with this
powder. The adsorbent, whose pressure is measured before and after
adsorption, is enclosed in a calibrated volume. The amount of
adsorbed gas is determined by the perfect gas law. On the other
hand, AES works with a single crystal whose cleaness can be monitored during adsorption[6]. Here the coverage is obtained after
calibration from the intensity of an Auger transition. The method
is powerful but suffers some disadvantages[6] such as electron
stimulated desorption, but reliable results have been obtained
from both techniques for Xe and Kr adsorbed on (0001) graphite.
The most valuable results have been obtained about ten years ago
by Prof. Duval and his school[3-5,7-12] by volumetry, on exfoliated
graphite, a special, well-defined, uniform, homogeneous graphite
powder. Their results were extended and completed by AES a few
years later[6,13,14]. All these findings are drawn schematically in
figures 1 and 2.

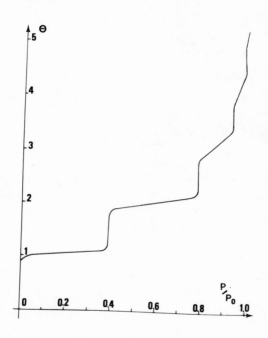

Fig. 1. Krypton condenses on (0001) graphite in a layer by layer mode[9].

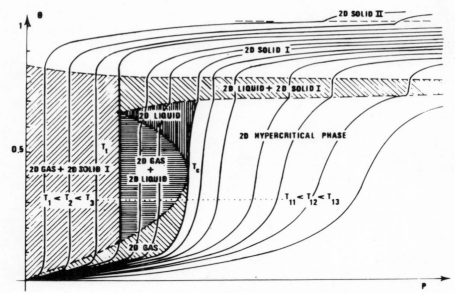

Fig. 2. A sketch of the adsorption isotherms of Xe, Kr, CH_4 or NO
on (0001) graphite. Various two dimensional phases are
stable[5].

    The adsorption isotherm represented in figure 1 exhibits five
height steps, obtained at undersaturation ; that is, at a pressure
lower than the 3D equilibrium pressure. This figure shows obviously
that the condensation occurs in a layer by layer mode. Now, let us
focus our attention on the first step (the first layer). At low
temperature and pressure (fig. 2), we can observe a first order
transition between a dilute and a condensed phase. At intermediate
temperature and pressure, the step is split into two substeps, the
first one with a landing at a density of about 0.7, and the second
one with a plateau near $\theta = 0.9$. On the right part of the figure,
the first substep disappears which means that the first transition
becomes continuous ; it is always followed by a second clear-cut
substep. By analogy with bulk matter, these successive transitions
resemble the transitions between gas, liquid and solid. The surface phases have been called 2D gas, 2D liquid and 2D solid by Thomy
and Duval[5]. Their stability regions are shown in figure 2. Below
a triple point T_t, only the transition between a 2D gas and a 2D

solid occurs. The successive transitions 2D gas $\rightleftarrows$ 2D liquid $\rightleftarrows$ 2D solid are observed between the triple point and a critical temperature T_c. Above T_c, the adsorbed monolayer undergoes a transition between a hypercritical phase and a solid. The values of T_c and T_t for Xe are listed in table I.

Table I

	T_t	T_c	$H_g - H_{2D}^{sol}$ / ΔH_{2D}^{sol} Kcal mole^{-1}		ΔH_{2D}^{liq}	ΔH_{3D}^{sol}	ΔS_{2D}^{sol} cal mole^{-1}K^{-1}
			AES	Calorimetry	Volumetry		AES
Xe	99 K (5)	117 K (5)	5.5 ± 0.1 (13)	5.7 ± 0.3 (30)	5.10 (5)	3.76 (5)	38.4 ± 1 (13)(86)

If we look at figure 2 a little more carefully, we can observe another slight substep at high pressure near $\theta = 1$. This transition between two condensed phases is a transition between two 2D solids. The structural analysis (§ IV) will specify the properties of these new two dimensional phases.

Hence, several phases, looking analogous to classical phases of bulk matter, occur in the first layer of simple molecules adsorbed on graphite. The characterization of these phases by various techniques will allow us to say in the next paragraphs how the 2D phases are close to the 3D ones.

But prior to leaving the thermodynamics, we have to analyze the preceding raw data. First, it is interesting to define clearly the equilibrium between the different coexisting phases. The answer to this question is given by the phase rule[1]. Let us consider the simple case of a monolayer (or a submonolayer) adsorbed on a homogeneous crystalline surface. The system is described by four pairs of conjugate variables : (T,S), (P,V), (μ,N) and (Φ,A) where Φ and A are the spreading pressure and the surface area respectively, and where the other symbols have their usual signification. If there are c constituents distributed amongst ϕ phases, the variance of the system is

$$v = c + 3 - \phi \qquad (1)$$

In the case of Xe or Kr adsorbed on graphite, c = 1. Hence
for v = 0, one has ϕ = 4. This means that the 2D triple point is
in fact a quadruple point where a 2D gas, a 2D liquid, a 2D solid
and a 3D vapor are in equilibrium. If the variance v = 1, as along
the vertical part of a step (fig. 2), there are three coexisting
phases ; for instance, in the low temperature regime, the equili-
brium can be written : 2D gas $\rightleftarrows$ 2D solid $\rightleftarrows$ 3D gas. As a matter of
fact, relation (1) is valid as far as the first adsorbed layer only
is concerned. If we consider the second, the third... layers, that
is, if we take account of the gradient field caused by the
substrate, we have to introduce one more pair of conjugate variables per new layer. Hence for a double layer, (1) becomes

$$v = c + 4 - \phi \tag{2}$$

Along the vertical part of a step (v = 1), one has ϕ = 4 and the
preceding equilibrium may be rewritten : 2D gas(in the first layer)
$\rightleftarrows$ 2D solid(in the first layer) $\rightleftarrows$ 2D gas(in the second layer) $\rightleftarrows$ 3D
gas. So, it is important to define carefully the system before
using the phase rule. As indicated in fig. 1, every adjacent layer
can behave with thermodynamic individuality. This discussion shows
that the surface phases are not purely twodimensional because they
are not only attracted by the substrate but they continuously
exchange molecules with the 3D surrounding vapor.

More quantitative thermodynamic information can be taken out
from an adsorption isotherm network. The transitions (the steps)
are characterized by pairs of variables (p,T). These pairs define
the coexisting region of two surface phases. We can plot them on
a log p, T^{-1} diagram which is close to a 2D phase diagram since
log p is proportional to the chemical potential variation of the
3D vapour and since the chemical potentials of the surface phases
and of the 3D vapour are equal in the coexisting region. Such a
phase diagram is drawn schematically in figure 3. One can
recognize the stability domain of the 2D gas, 2D liquid, 2D solid
in the first layer, the transition lines between the 2D solids and
between the first and the second layer, the 3D sublimation curve
and the critical and triple points of the monolayer as well. The
coexistence lines are straight lines because the transitions obey
the relation

$$\left(\frac{\partial \mathcal{L}n\ p}{\partial\ 1/T}\right)_{\theta,A} = -\frac{q_{st}}{R} \tag{3}$$

where the isoteric heat of adsorption q_{st}[15-18] is constant for a
given transition. This isosteric heat is simply related to the
latent heat of transformation when one phase is diluted and the
other one condensed (for instance, 2D gas-2D solid or 2D gas-2D
liquid) ; then q_{st} yields $h_g - h_{2D}$[19], the difference between the
enthalpy of the 3D vapour phase and the enthalpy of the 2D

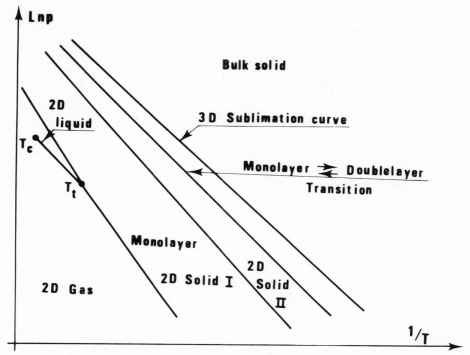

Fig. 3. Schematic phase diagram of simple molecules adsorbed on
(0001) graphite[5],[6],[13].

condensed phase. The values of $h_g - h_{2D}$ for 2D liquid and solid Xe
are given in table I. However, the meaning of q_{st} is more complicated when considering a transition between two condensed phases
like liquid-solid and the obtained relation is not of practical
use[19].

Finally, in the favourable cases of 2D gas–2D liquid and
2D gas–2D solid transitions, the intersection of the straight line
with the ℓnp axis yields $S_g - S_{2D}$ the entropy of vaporization or
sublimation. The values obtained for Xe on graphite are listed in
table I.

A microscopic parameter can be deduced from the latent heat of
sublimation of the 2D crystal ; the energy of the 2D crystal
$H_{2D}^{cryst.}$ can be approximated by $u_\perp + u_{/\!/} + u^{vib}$ where $u_\perp$ is the perpendicular binding energy at 0 K of a rare gas atom on the basal plane
of graphite, $u_{/\!/}$ the lateral binding energy and u^{vib} the vibrational

energy of atoms in the 2D phase. u is assumed equal to 0.5 u of
the 3D crystal, and u^{vib} is obtained by integrating experimental
film heat capacities[13,14]. In this way $u_\perp$ can be estimated.

The microscopic mechanism of 2D phase transitions can be
schematically understood by using thermostatistical models. At first,
the early stages of condensation on a substrate can be quite different depending on the relative strength of the adsorbate-adsorbate and adsorbate-substrate binding energies (see for instance
20). If the lateral energy between pairs of atoms adsorbed on a
crystal surface is appreciably stronger than the adsorption energy
of one individual atom, three dimensional clusters are initially
formed on the bare substrate. If the ratio between these energies
is reversed, condensation occurs in a layer by layer fashion. Of
course, this argument is fairly rough and a rigorous discussion of
the condensation mechanisms must turn on the relative value of the
free energy of the 2D and 3D systems.

Let us see now the transitions inside the 2D layer. Several
statistical models permit us to obtain some clues regarding which
features of interparticle interactions are important for determining the various transitions. The simplest are the Fowler and the
Ising models[18,21]. In both cases one considers a lattice gas with
nearest neighbor lateral attraction. The Fowler model which uses
the mean field approximation is unrealistic near the critical
temperature but has the advantage of analytical solutions. Both
models give the same qualitative results : there is a first order
transition below a critical temperature between a dilute phase and
a dense one ; the lateral interaction is responsible for this
cooperative phenomenon.

Increasing the number of parameters of the models obliges us
to use numerical methods and computers. Numerous works have been
achieved in this field. All have enlarged the range of the interparticle interaction at least to the second neighbor sites and
sometimes up to the fifth neighbors or more[22-27]. These papers
have one result in common : a nearest neighbor repulsion or
exclusion, and more distant attractions, are necessary to obtain
a phase diagram with a 2D gas, liquid and solid. For instance,
one can get a diagram looking like the one drawn in fig. 3[24].
Sometimes, the phase diagram can be extremely rich and include
several solid-solid transitions[27] depending on the range, the
strength and the sign of pairs of neighbor interaction. Other
models better fitted to the comprehension of such or such
particular transition will be analysed later. Anyway let us
retain that models yield the main qualitative features of the
experimental phase diagrams.

II.2 - <u>Calorimetry</u>

An adsorption isotherm network should give every thermodynamic information on surface phases, at least in principle. As explained in II.1, we can obtain the enthalpy and the entropy of phase transitions. From classical thermodynamic relations, we should also determine the specific heat. But the determination of the entropy is not very precise and its variation with temperature is not measurable. The only way to obtain the specific heat is to measure it directly, for instance by calorimetry. This technique is also used to determine the heat of adsorption.

II.2.1. <u>Heat of adsorption.</u> By introducing continuously the adsorbate into an isothermal calorimeter, Rouquerol[28] is able to record directly the isosteric heat of adsorption. The introduction is slow enough to be always in quasi-equilibrium[29]. The results are illustrated in fig. 4 for Kr adsorbed on (0001) graphite below the 2D triple point[30]. To help comprehension, one has also reported the adsorption isotherm at the same temperature (77.3 K).

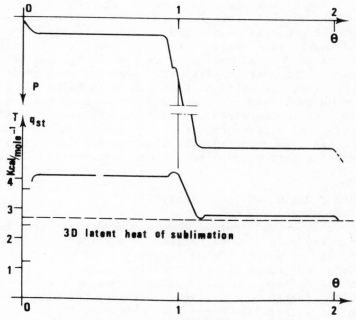

Fig. 4. Heat of adsorption of Kr adsorbed on graphite versus coverage. The upper curve gives the adsorption isotherm at the same temperature[30].

The constant value of the adsorption isosteric heat for $\theta < 0.9$ corresponds to the 2D gas-solid transition ; the little rounded peak near $\theta = 1$ is due to the solid-solid transition. One can also notice a second plateau, in the second layer region, at an energy a little higher than the 3D latent heat of sublimation. These plateaus are characteristic of first order transitions. In this case, q_{st} gives the 2D latent heat of transformation (see II.1).

II.2.2. <u>Specific heat</u>. Specific heat is a valuable technique to determine phase boundaries occurring at nearly constant temperature. The heat capacity at constant coverage exhibits peaks at boundaries when the temperature is changed. This technique has been extensively used to determine the phase diagram of ^{4}He and ^{3}He adsorbed on graphite[2]. The aim of this paper being to analyze the phase transitions of van der Waals films, I just want to illustrate the possibilities of this technique on the example of Kr adsorbed on graphite.

The gas-solid and solid-solid transitions in the first layer of krypton are fairly well known (II.1, II.2.1 and IV.1). However the gas-liquid coexistence domain has never been observed. From adsorption isotherm measurements Larher[11] claimed that the 2D triple point (84.8 K) was very close to the 2D critical temperature (85.3 K). Very recently Butler et al[31] explored this temperature region by heat capacity measurements. They succeeded to observe a strong and sharp peak centered at 85.5 K for $\theta \simeq 0.5$. Of course, they expected two peaks, one at T_t and one at T_c. Furthermore, the maximum of the peak shifts with coverage. From their results they suggest that the critical temperature is below the triple point. The unusual shape of this 2D phase diagram must be considered cautiously because the width of heat capacity peak (~ 2 K) is larger than the expected difference between T_c and T_t and one knows that heterogeneities can smear an otherwise first order transition[32]. Measurements on better graphite powders could settle the question.

III - SURFACE PHASE   CHARACTERIZATION

As seen above, adsorption isotherm and calorimetry measurements imply the existence of various 2D adsorbed phases called 2D gases, liquids or solids. Other techniques are needed to characterize these so-called 2D phases. The next table gives the possibilities of the different surface probes. Some of them are very popular like electron and neutron scattering but some others are developing very fast like X-ray scattering and NMR.

Table II

		Thermo dynamics	Long range order	Vibra-tions	Electron surface states	Fluid mobility	Short range order	Inter-action energy
Electrons	Elastic		Struc-ture	Debye temp.				
	Inelas-tic AES ELS	Adsorp. isoth.			Surf. transi-tions		Configu-ration	
Neutrons	Elastic cohe-rent	Phase diagram	Struc-ture	Debye temp.			Corre-lation range	
	Elastic inco-herent			Debye temp.				
	Inelas-tic cohe-rent			Phonons				
	Inelas-tic inco-herent	Phase **diagram**		Localized modes		Diffusion coeffi-cient		
R X	Elastic		Struc-ture	Debye temp.			Corre-lation range	
	EXAFS						Configu-ration	
Adsorp-tion isotherms		Phase diagram						Enthalpy entropy
Calori-metry		Phase diagram						Enthalpy specific heat
I R				Localized modes				
NMR						Mobility		
Mössbauer			x	Debye temp.		Mobility	x	
Neutral beams (He)				x				Inter-action states
Photo-emission					Surface states		Configu-ration	

The best techniques to study the crystallography of a solid
are neutron, X-ray, and electron diffraction. The first two
can in principle yield the position of atoms on the surface because
they obey the kinematic theory of the diffraction. Furthermore,
they can operate whatever the value of the pressure surrounding
the adsorbate. However, the low neutron or X-ray cross sections
obliges us to use powders as a substrate. This shortcoming disappears with the various forms of electron diffraction which work
on monocrystals. But the operating pressure of the 3D gas must be
weaker than 10^{-3} torr and last but not least multidiffraction
prevents from using a simple theory to interpret the LEED patterns.
One can directly obtain only the adsorbate mesh and its epitaxial
relation with the substrate. Fortunately, in the simple case of
monoatomic rare gases adsorbed on graphite, it is possible, using
symmetry arguments, to determine the position of atoms on the
substrate.

The surface vibrations can be determined by neutron or X-ray
scattering and by infra-red spectroscopy. The Debye temperature T_D
of 2D solids can be obtained by diffraction measurements but the
interest of this average value of thermal vibrations is rather
limited. More valuable information about the collective movements
of atoms (phonons) can be given by neutron inelastic coherent
scattering and information about the localized vibrations by
neutron inelastic incoherent scattering or infra-red spectroscopy.

The short range order in disorder surface phases is obtained
from neutron and X-ray diffraction measurements and the local
configuration of adsorbed molecules by electron loss spec⁺roscopy
(ELS) and EXAFS.

The surface electron states are not very well known although
two techniques, ELS and photoemission, can yield the peaks of
density of states. One can expect a strong development in this area
in the near future.

Finally the mobility of 2D fluids can be measured by quasi-
elastic neutron scattering or NMR.

None of these tools can give a complete characterization of
the various surface phases. All of them are complementary, and the
nicest and more reliable results usually combine several techniques.

IV - STRUCTURAL PROPERTIES AND THE SOLID-SOLID TRANSITION

One can imagine two extreme situations for a solid adsorbed on
a crystalline surface. If the variation of the attraction potential
of the substrate is very smooth, the structure is imposed by the
adsorbate-adsorbate binding energy : the monolayer is incommensurate with the substrate. However, strong periodic potential wells

can be present on the surface ; if these wells are stronger than
the interactions between atoms at the distances between nearest
neighbor sites, a registry structure may appear. In fact, the
actual situation is often between the two extremes, especially for
simple molecules adsorbed on graphite. The competition and interference between the depth, the symmetry, the spacing of the sites
and the size, the coverage, the interaction energy of adatoms will
produce some complex patterns, a few examples of which are given
below.

IV.1 - The Commensurate-Incommensurate Transition

Two 2D solids have been inferred in the Kr and Xe monolayers
from the Thomy and Duval adsorption isotherm studies[5](II.1). Still,
the earliest diffraction measurements have pointed out one 2D
crystal only. LEED experiments for Xe[33,13] and Kr[19], and neutron
scattering studies for Kr[34,35] have revealed a $\sqrt{3}$ solid in epitaxy
on graphite. The deregistered transition has been only confirmed
very recently by transmission high energy electron diffraction
for Xe[36] and by high-resolution LEED[37-39] and X-ray diffraction[40,41]
for Kr. All diffraction experiments show an apparently second-
(or higher-) order phase transition from a $\sqrt{3}$ in-registry 2D solid
to an out-of-registry structure.

Let us focus our attention on the Fain and Chinn[37-39] results
obtained on krypton.

The diameter of krypton atoms is smaller than the distance
between the $\sqrt{3}$ sites. Hence, at a coverage slightly below unity, a
commensurate $\sqrt{3}$ structure is expected (fig. 5). As soon as the
coverage goes beyond unity (by increasing pressure above the
substrate), krypton atoms are expected to be forced out of their
site position to form an incommensurate lattice. LEED measurements[37-39] and recent THEED measurements have confirmed these
previsions[42]. Superstructure peaks are observed. They are caused
by an hexagonal closed packed structure either in-registry or
out-of-registry with the substrate depending on temperature and
pressure. The results are presented in figure 6a in a shape
directly comparable to the adsorption isotherm findings of Thomy
and Duval[5] (fig. 6b) ; $(d_0/d)^2$ is the square of the ratio of the
in-registry Kr-Kr distance d_0 and the actual mean distance of
atoms d ; θ_0 is the onset of the deregistry transition. The
comparison would be rigorous only for a perfect monolayer without
vacancies or adatoms in the second layer. In spite of this restriction, it is obvious that fig. 6a and 6b represent the same transition observed with two different techniques. One sees that, at a
given temperature, below an onset pressure, the Kr-Kr distance is
constant within the limit of the experimental uncertainty. Increa-

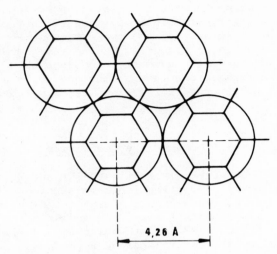

Fig. 5. The $\sqrt{3}$ x $\sqrt{3}$ structure of epitaxial in-registry Xe, Kr and
CH$_4$ monolayers on basal planes of graphite.

sing the pressure results in a deregistration of the overlayer and
in a gradual decrease of the distance between atoms which form a
triangular structure covering the whole area available for
adsorption with a monolayer.

These striking features owing to the competition between the
adsorbate-adsorbate and adsorbate-substrate forces provide a unique
opportunity to test the various theories of epitaxy[43-50].
Theoreticians have used the dislocation formalism by introducing
misfit dislocations[43,44,46,49,50] in the adsorbed layers in order
to accomodate the difference between the lattice parameters of the
substrate and the overlayer. This approach has been successful in
interpreting the main features of the heavier rare gases deregistry
transition[38]. The misfit dislocations introduce modulations into
the overlayer which have been called static distortion waves[47,48].
The study illuminates the behaviour of incommensurate structures
having a preferred relative orientation. From lattice dynamics
arguments, it is shown that the equilibrium configuration of a
monolayer in epitaxy on a crystalline substrate has a definite

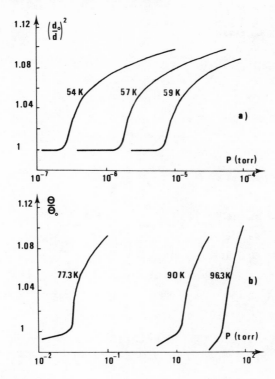

Fig. 6. Comparison of LEED measurements and vapour pressure isotherms near the transition from an in-registry solid to a
compressed, out-of-registry solid.
a) d_0 is the registered layer Kr-Kr distance and d is the
mean observed Kr-Kr distance.
b) θ is the coverage measured by Thomy, Régnier and Duval[9]
at 77.3 K, 90.0 K and 96.3 K ; θ_0 = 0.930, 0.940 and
0.945, respectively.

orientation with a fixed rotation angle with respect to the graphite lattice. This angle is due to the competition between vibrational energy terms caused by transverse or longitudinal phonons.
The film partially accomodates to the substrate structure by taking advantage of its shear modulus, which is weaker than its com-
pressive modulus. Numerical predictions indicate a 2° rotation
angle for Kr on (0001) graphite which is consistent with spot
elongations of the LEED pattern[39]. For argon, calculations predict
an angular deviation that depends on the overlayer density.
Figure 7 shows the observed relative orientation of the Ar layer
with respect to the graphite substrate as a function of the mean
nearest neighbor distance[51]. This low energy electron diffraction
measurement is a good verification of the modeling of the overlayer

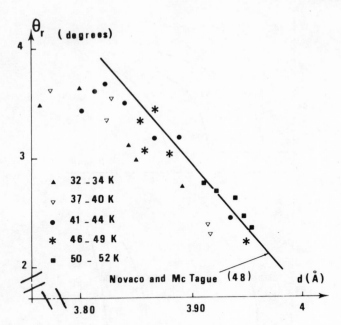

Fig. 7. Relative orientation θ_r between Ar monolayers and
a($\sqrt{3}\times\sqrt{3}$) structure versus argon mean nearest-neighbor
distance d for various temperatures[51].

distortion by lateral argon-graphite forces.

IV.2 - The Incommensurate-incommensurate Transitions

As explained above, the deregistry transition results from
the competition between, on the one hand, the depth and the width
of the substrate potential wells and on the other hand, the size
and the lateral interaction energy of the adsorbed atoms. For non
"spherical" molecules, other types of transitions can occur. If
the size of the molecule is clearly larger or smaller than the
distance between potential wells, the structure of the layer is
incommensurate with respect to the substrate lattice. One can
imagine that the "anisotropic" molecule can either lie down or
stand up on the surface according to the coverage and the temperature. Then, a transition between two incommensurate structures can
occur.

These ideas can be illustrated with the example of nitric oxide adsorbed on the graphite basal plane[52]. In fact, this molecule is dimerized and is almost rectangular. Neutron diffraction experiments showed that at low temperature (10 K) and concentration (coverage 0.40), the scattering pattern is drastically different from that measured at the same temperature but near layer completion ($\theta \simeq 1$). This high density structure is still stable at 74 K. Figures 8 and 9 present the neutron diffraction patterns of the two incommensurate nitric oxide solids. The saw-tooth shape of peaks clearly visible in figure 9 is characteristic of the scattering by "powders" of 2D crystals. These spectra have been interpreted and the resulting structures are given above the spectra.

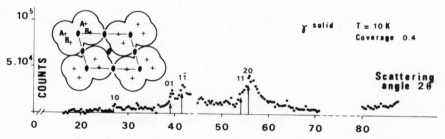

Fig. 8. Neutron diffraction pattern from NO adsorbed on graphite (0001). Background scattering from the substrate has been subtracted. 2D crystal at 10 K for coverage 0.40. Molecules are lying down according to the inset model. The arrows indicate the position and the relative intensities of Bragg peaks calculated from this structure.

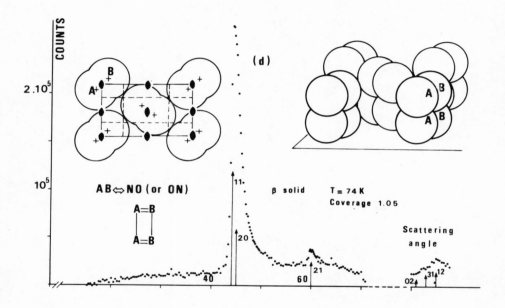

Fig. 9. Same as fig. 5. 2D crystal at 74 K for coverage 1.05.
N_2O_2 molecules are standing up on the graphite substrate.

Both are close-packed structures but with quite different packings. The high density 2D crystal has N_2O_2 dimers standing up, whereas the low density solid is formed of molecules lying down on the surface. The "standing up" structure is very close to the packing of the densest plane of the bulk nitric oxide structure. However, the "lying down" packing has no equivalent in three dimensions, indicating that the substrate field is the most important parameter in the stabilization of this 2D solid.

Another transition between two incommensurate structures presents a great interest in 2D magnetism. An antiferromagnetic ordering has been discovered in the first layer of oxygen adsorbed on graphite. One knows that bulk oxygen has two solid phases α and β. In both phases O_2 molecules are perpendicular to the densest packed planes. There is an antiferromagnetic order in the α phase whose basal plane structure is a skew triangular lattice. The long range magnetic order disappears at the $\alpha - \beta$ transition and the structure becomes equilateral triangular.

Oxygen adsorbed on graphite (0001) exhibits two analogous phases. This result comes from neutron diffraction studies of O_2 overlayers[53]. For a dense layer, the diffraction pattern shows at low temperature a diffraction peak consistent with a skew triangular structure very close to the densest packed lattice of the bulk α-phase. Furthermore, a superstructure peak is observed and attributed to a magnetic superlattice reflection which is a signature of the antiferromagnetic ordering. Above 10 K, this superstructure peak disappears and the 2D lattice becomes equilateral triangular. This magnetic transition is probably continuous. Investigation is continuing to say to what extent the 2D magnetic long range order and its disappearance are comparable to that observed in three dimensions.

V - THERMAL VIBRATIONS

Surface thermal vibrations in 2D phases are associated with the loss of rotational or translational degrees of freedom of adsorbed molecules. The measurement of vibrational properties of 2D adsorbed phases is interesting because it gives microscopic information about the bonding and the molecular conformation and location on the surface. Experimentalists can measure either average vibrational properties like the Debye-Waller factor or the eigenfrequencies and eigenvectors of isolated molecules or the dispersion relation and the density of states of molecular collective excitations (phonons).

V.1 - Mean Squared Displacement of Adsorbed Atoms

The Debye-Waller factor[54] gives us information about the mean squared amplitude of the adsorbed molecule vibrations. Three methods have been used to measure this thermal property of simple molecules adsorbed on graphite : LEED, neutron scattering and Mössbauer spectroscopy. The 2D crystal of xenon has been investigated by LEED[55,56]. A value of about 0.015 $\mathring{A}^2$ is found at 80 K for the mean squared displacement perpendicular to the (0001) graphite surface. Its variation with temperature is also measured. If compared to the atomic vibrations normal to a (111) face of a 3D xenon crystal, the amplitude of the Xe 2D solid is five times lower. This indicates a partial hindering of the surface vibrations on graphite with respect to vibrations in the (111) plane of 3D xenon ; the modeling of this effect can determine the force constant of xenon atoms adsorbed on graphite. It is, of course, stronger than the corresponding parameters associated to the xenon-xenon bonding. The study also shows that the eigenfrequencies of the xenon atoms are significantly lower than the equivalent modes of the graphite substrate. This is a general effect which is used in neutron scattering to discriminate the adsorbate and substrate inelastic processes[57,58].

In Mössbauer spectroscopy, one has to use molecules containing a Mössbauer resonant atom. Among other things, this technique provides a direct measurement of the nuclear mean squared displacement. For adsorption studies, two resonant nuclei have been chosen : ^{119}Sn [59] and ^{57}Fe [60]. The compounds are relatively simple : $Sn(CH_3)_4$, $SnCl_4$, SnI_4 and $FeCl_2$. The Debye-Waller factor is used to estimate the binding energy of the adsorbed molecules.

V.2 - Intramolecular and Localized Surface Modes

Loss spectroscopy is of common use for the study of molecule excitations. Up to now, three techniques have succeeded in measuring surface modes. The first two, infrared spectroscopy and energy loss electron spectroscopy, are limited by selection rules. Still they have proved powerful for the study of surface modes of molecules chemisorbed on monocrystals[61-63]. But, as yet, the only method used for physisorption studies is inelastic neutron scattering. Although it needs substrate powders and adsorbate with large cross section to obtain detectable signals, it has several advantages with respect to the former methods. At first, the scattering laws are well known, which enables to compare quantitatively the intensities. Secondly, the energy resolution is excellent and offers the opportunity to measure low energy excitations.

Whereas neutrons couple to the single particle correlations, via the spin incoherent scattering cross section, the coherent cross-section relates to the collective correlations of matter. In this paragraph, we consider the surface modes of individual molecules. In V.3 the results on surface phonons obtained with coherent inelastic scattering will be reviewed. Butane loss incoherent spectra[64] exhibit a very rich structure as shown in figure 10. We observe four well-defined peaks at energy transfers of 259 cm^{-1} (32.1 meV), 217 cm^{-1} (26.9 meV), 157 cm^{-1} (19.5 meV), 112 cm^{-1} (13.9 meV) and a broader band centered at 50 cm^{-1} (6.2 meV).

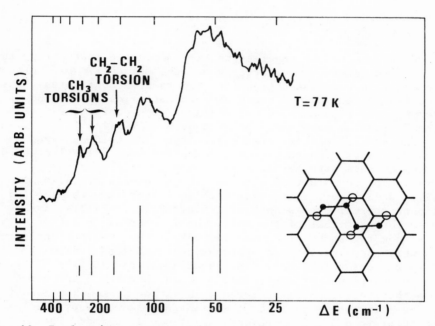

Fig. 10. Inelastic spectra of butane adsorbed on graphite at 77 K.
Comparison of the monolayer and calculated spectra. The
background has been subtracted from the observed spectrum,
and the error bars represent the statistical uncertainty
since multiple-scattering effects are believed negligible.
The arrows indicate the energy of the three lowest-lying
modes of the bulk solid. The calculated spectrum is for
the two-parameter model described in[64]. The inset shows
the proposed orientation of butane with respect to the
graphite basal plane. Only the four coplanar hydrogen
atoms (0) on one side of the carbon skeleton have been
included for clarity.

The first three peaks correspond to the intramolecular CH_3 and CH_2 torsions of the bulk solid. The fourth peak (112 cm^{-1}) is interpreted as a rocking surface mode about an axis in the film plane parallel to the chain direction. Furthermore, the model calculations provide two other modes consistent with the band centered at 50 cm^{-1}. The first one is a rocking mode with an axis perpendicular to the chain direction and the second one is a uniform oscillatory motion of the entire molecule normal to the substrate. Finally, the authors are able to infer the orientation of the butane molecule with respect to the graphite basal plane (fig. 10).

V.3 – Surface Phonons

Turning now to the coherent neutron inelastic scattering, the recorded spectra give information about the collective excitations of the overlayer. One uses graphite powders with considerable basal plane orientational order. Hence one can partly decompose the scattering into in-plane and out-of-plane components. The most remarkable results have been obtained with ^{36}Ar films[57,58,65]. First, the crystallography of the monolayer has been analyzed by neutron diffraction. Argon forms a dense triangular lattice which is incommensurate with the substrate. The lattice constant is always smaller than that of a registered √3 overlayer (4.26 Å) (see IV.1 and fig. 5) ; its value at low temperature is 3.86 Å ; the linear thermal expansion coefficient (2.10^{-3} K^{-1}) is five times as large as that of the bulk. The incoherent inelastic spectra show two peaks at 3 and 5.5. meV in the in-plane configuration which are interpreted as transverse and longitudinal zone boundary phonons polarized and propagating in the 2D film planes. In the out-of-plane configuration, one peak between 5 and 6 meV is also observed. It is associated with either a resonant coupling of the out-of-plane motions of the film to bulk modes of the substrate or to a localized Einstein out-of-plane mode. These results provide interesting insights into the dynamical properties of the overlayer. Modeling the findings shows that in-plane vibrations are little influenced by the substrate and can be, to some extent, described by an ideal 2D like crystal.

As temperature increases, the neutron spectrum broadens and the transverse peak softens. The broadening and softening are not clear-cut, which indicates a continuous transition in the 40-60 K temperature range. It is interpreted as a melting region.

VI - MELTING AND 2D LIQUIDS

VI.1 - The 2D Liquid Mobility

        As seen in II.1, 2D liquids have been inferred from adsorption
isotherm measurements but no direct evidence of the existence of
these new liquids has been presented until recently. The occurrence
of 2D liquids came from mobility measurements by NMR[66,67],
Mössbauer spectroscopy[68,69] and quasi-elastic neutron scattering[70,71].

        This last technique probably brought the most significant
results in this field. It enables us to distinguish two kinds of
adsorbed fluids : the 2D hypercritical fluid (see II.1) with a
high compressibility, occupying the whole surface available for it,
and the 2D liquid whose domains are limited by a well-defined
boundary. The results have been obtained in the case of a submonolayer of methane adsorbed on the basal plane of graphite and the
mobility was measured by quasi-elastic neutron scattering[70,71].
The principle of the experiment is very simple and is currently
used to measure the diffusion coefficient of bulk liquids. A monochromatic neutron beam interacts with molecules executing brownian
movements and gains or loses energy, resulting in a broadening of
the energy neutron distribution. According to theory, this broadening is Lorentzian and for small scattering vector Q the width ΔE
of the peak is proportional to the translational diffusion coefficient D

$$\Delta E = 2\hbar DQ^2 \qquad\qquad (3)$$

        The phase diagram of methane adsorbed on graphite is fairly
well known[5] and resembles the one drawn in figure 2. The 2D critical temperature and the 2D triple point are 72 and 56 K respectively.
Let us consider the neutron quasi-elastic spectra measured at two
temperatures (61 and 90 K) for different coverage. They are drawn
in figures 11 and 12. The widths of the spectra are larger than
the instrumental resolution indicating a mobility of the CH_4
molecules. At 90 K, the broadening clearly varies with coverage
whereas at 61 K it does not, which means that the diffusion coefficient is coverage dependent above the 2D critical temperature and
coverage independent below. A detailed analysis of all experimental
results yields the various values of D listed in table III and IV.
The experimental conditions at 61 K have been chosen in the
coexistence domain of the so-called "2D liquid" and "2D gas". In
our neutron scattering experiment, we cannot observe the 2D dilute
phase (2D gas) because of its high mobility and small density which
yield a very flat and broad spectrum. Our peaks are mainly due to
the most dense phase. This is confirmed by the observation of a
proportionality between the amplitude of the Lorentzian curve and

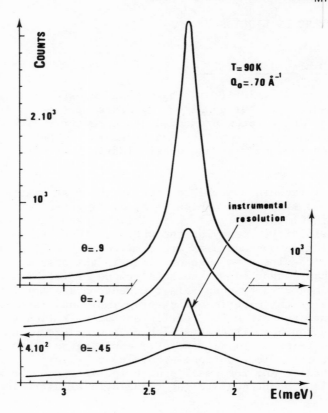

the coverage at constant $\vec{Q}$ and T. Hence, the constant value of the diffusion coefficient is proof that the coverage change only produces an increase of the surface occupied by the densest phase whose mobility is an intensive property. This is precisely the property of a liquid. Furthermore, the measured value of D is the same order of magnitude as the bulk liquid within the temperature range for which this 3D liquid is stable, namely above 90 K.

This can be illustrated by a set of 2D liquid islands bathing in a sea of very low density 2D gas. Our experimental results show that the diffusion coefficient does not depend on the fraction of the graphite surface covered with the 2D liquid. This behavior is quite different from that of the 2D hypercritical fluid, i.e. the fluid stable above the 2D critical temperature. In this case, molecules tend to occupy all the available surface and the mobility is strongly coverage dependent (see table IV). A simple model fitting the diffusion coefficient variation suggests that the mobility of the hypercritical fluid is proportional to the free space between molecules.

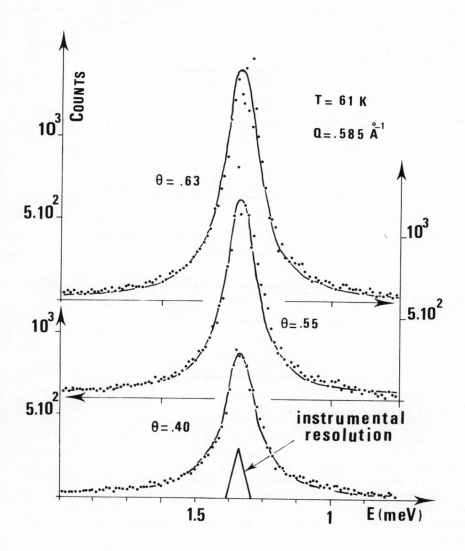

Fig. 12. Difference of the CH_4 covered and bare graphite incoherent
quasi-elastic spectra at constant temperature and scattering vector for three coverages. The broadening is coverage
independent. The solid curves represent the best fits with
a Lorentzian curve convoluted by the instrumental resolution. Two-dimensional liquid.

Table III. Diffusion coefficient D of the saturated 2D liquid for
           temperature 61 K and different coverages θ. D is
           coverage independent.

θ	$D(10^{-5}\ cm^2\ s^{-1})$ 61 K
0.30	2.2 ± 0.4
0.40	2.5 ± 0.3
0.55	2.6 ± 0.3
0.63	2.7 ± 0.3

Table IV. Diffusion coefficient of the 2D hypercritical fluid at
          90 K. D is coverage dependent.

θ	$D(10^{-5}\ cm^2\ s^{-1})$
0.45	13 ± 1.3
0.7	7 ± 0.7
0.9	2.2± 0.3

VI.2 - 2D Melting

Referring to figure 2 which exhibits a well-defined triple
point, the melting of a 2D solid must be a first order transition.
Still, all measurements of the triple point either by calorimetry[32] or neutron scattering[58] have shown that melting is more
or less abrupt depending on the gas adsorbed on graphite. Usually,
it is difficult to say if the transition is first or second order
because results obtained from different techniques can be contra-
dictory.

In the case of Ar adsorbed on graphite, the decrease of long
range order on melting can be observed by neutron diffraction.
The intensity and the shape of the diffracted peaks are function
of the range of order, or of the size of the diffraction arrays.
The spatial correlation range has been measured for adsorbed Ar
monolayers. The decrease at melting in correlation range is gradual, indicating at least apparently, a continuous transition.
However, adsorption isotherm measurements exhibit a vertical rise
in the low coverage range between the "2D triple point" and the

2D critical temperature[72]. This seems to show that a 2D gas and a
2D liquid coexist between these two temperatures and that melting
is a first order transition. The neutron diffraction and isotherm
adsorption experiments have been performed on two kinds of graphite substrate, recompressed and exfoliated powders respectively.
The quality of the latter is better and the isotherm adsorption
measurements are more reliable, more especially as experiments[73]
have shown that the abruptness of the melting transition for argon
strongly depends on the substrate quality. Here the width of the
melting transition can be explained by size effects due to
substrate imperfections, which broaden an otherwise sharp first
order transition[32]. Thus the question of the order of the melting
transition for Ar is still open.

The best way to decide if melting is first order is to
observe the coexistence of a 2D liquid and a 2D solid. In a
diffraction experiment, a solid gives sharp diffracted peaks
whereas a liquid exhibits only broad bumps. If the two phases
coexist, the resulting diffracted pattern is the superposition
of a broad bump and a sharp peak. This has been observed in the
case of NO adsorbed on graphite. For this system, melting is a
first order transition[52].

All these findings can be tested against theories of 2D melting, but very few theories of the 2D solid-fluid transition have
been proposed up to now ; the most detailed one is based on the
dislocation model of melting[74,75,90]. It is assumed that a 2D solid
has pairs of dislocations which can move under the influence of a
local shear stress. As the temperature is raised, the dislocation
pair mean separation increases. At a given temperature T_m, the
average distance diverges ; there is a complete loss of elastic
restoring force and the 2D crystal melts. According to this model,
the melting temperature is given by

$$T_m = (2\pi)^{-2}(mk/8\hbar^2)a \; \theta_D^{\;2}$$

where m is the molecular mass, a the area per molecule, θ_D the 2D
Debye temperature and the other symbols have their usual meaning.
This expression has the same form ($T_m \sim \theta_D^2$) as the empirical
Lindemann melting law of bulk matter[76] but, here, the proportionality coefficient is specified. Numerical applications of the above
equation agree with experiment results concerning rare gases adsorbed on graphite. However, the theory does not tell us anything
about the nature of the transition and takes no account of the
influence of the substrate structure. Hence, it seems important
to reconsider now the theory of 2D melting, in the light of the
new experimental findings.

VI.3 – Critical Temperature

About fifty 2D critical temperatures have been measured for different systems[77]. They can be compared to the bulk critical temperatures $T_c(3D)$ of the adsorbate. The ratio $T_c(2D)/T_c(3D)$ is always close to 0.4 (extrema values : 0.36 and 0.56) and lies between the values obtained from the mean field theory (0.5) and from the 2D Ising model (0.37). No clear quantitative analysis of the variation of this ratio with the strength of the adsorbate-adsorbate or adsorbate-substrate interactions have been published so far.

VII – GAS-SOLID TRANSITION

The mechanism of the 2D gas-solid transition in the first adsorbed layer has been throughly analyzed by Auger electron spectroscopy for the Xe/(0001) graphite system. Let us recall that the intensity of an Auger peak of the adsorbed atoms gives, after calibration, the surface coverage θ. The kinetics of adsorption or desorption can be monitored by following, at constant T, the variation of this intensity with a pressure change Δp. From the 2D phase diagram (fig. 3), the supersaturation for the gas-solid transition is defined by

$$\frac{\Delta p}{p} = \frac{p' - p}{p}$$

where p is the pressure of this transition at a given temperature and p' is the actual pressure.

VII.1 – Adsorption Kinetics[78-80]

The adsorption kinetics is measured by suddenly establishing a supersaturation (p' is chosen so that the 2D crystal is stable). Typical adsorption kinetics curves are plotted in figure 13 at 79 K for a few supersaturations. More than 30 such curves have been recorded for xenon between 74 and 76 K for different supersaturations. Three conclusions can be drawn out from the results.

i. Not any induction time before condensation is measured which means there is little or no activation energy of 2D nucleation.

ii. The coverage versus time curves shows three regimes : the first one(I) corresponds to the adsorption of the 2D gas, the second one(II) to the growth of 2D crystal in the presence

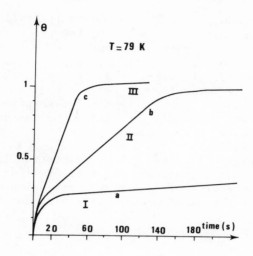

Fig. 13. Adsorption kinetics of the first monolayer of xenon on (0001) graphite for different supersaturations.

p'(10^{-7} torr)   a   b   c
                      1.2   1.6   2.1
p = 1.15 10^{-7} torr

of the 2D gas, and the third one(III) to the completion of the 2D crystal.

    iii. In regime II (coexistence of the 2D gas and the 2D crystal), the $\theta(t)$ curves are always straight lines ; the rate of adsorption dθ(II)/dt is independent of θ. Thus the condensation follows a zero order kinetics. Moreover, this rate of adsorption is proportional to p'/$(2\pi mkT)^{1/2}$ where m is the mass of xenon atom, k the Boltzmann constant and T the gas temperature : the condensation kinetics is determined by the incident flux only. This means that in regime II the xenon atoms condensing on the part of the surface covered with the 2D gas or on the other part covered with the 2D crystal, move on this surface and manage to settle in a growth site of a 2D crystal edge. In other words, atoms condense in a two step process : 3D gas → 2D gas (in the first or in the second layer) → 2D crystal.

VII.2 - Desorption Kinetics[80]

In these experiments, the 2D crystal is first formed at a given temperature and pressure. At t = 0, the pressure is suddenly decreased and a desorption curve is recorded. The results for different T are plotted in figure 14 for xenon. Once again, we recognize the three regimes I,II,III already described in thermodynamics and adsorption kinetics. We also see that in regime II the rate of desorption is independent of the coverage : desorption has, like adsorption, a zero order kinetics. Hence the mechanism of 2D crystal desorption is also a two step process.

According to figure 14, the rate of desorption in regime II depends strongly on the temperature. Plotting $d\theta^{II}/dt$ versus $1/T$ gives the activation energy of desorption. Its value 6 ± 0.4 Kcal $mole^{-1}$ agrees within the experimental error with the thermodynamics values of table I.

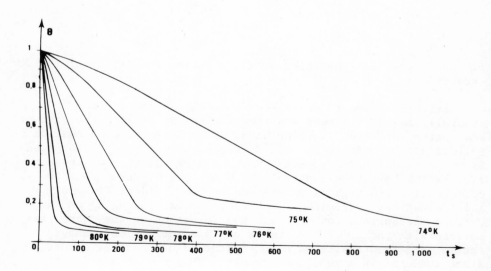

Fig. 14. Isothermal desorption curves of Xe at different temperatures.

In all the processes where a zero order reaction is detected, an infinite reservoir supplies particles at a constant rate, independent of the reaction progress. Here, for both desorption and adsorption kinetics, the reservoir is the 2D gas in the first or in the second layer. This means that the concentration of these gases is constant, i.e. the 2D gas $\rightleftarrows$ 2D solid reaction is much faster than the 2D gas $\rightleftarrows$ 3D gas reaction[79]. If there is no reservoir, the reaction no longer obeys a zero order kinetics. Other systems have to be experimentally studied to check whether the two step process for the 2D gas-solid transition is a general rule in the adsorption of simple molecules on graphite.

VIII - OPTICAL PROPERTIES

Optical properties of simple molecules adsorbed on graphite have never been thoroughly studied. The only detailed papers dealing with this subject report an ellipsometric measurement of xenon condensed on graphite and its interpretation[81,82]. Results are spectacular enough to be summarized here. In ellipsometry, a plane polarized wave is reflected on a surface and the ratio of the amplitudes ψ and the phase shift Δ of the two orthogonal (out and in-plane) components of the reflected wave are measured. For xenon on graphite, it is easy to measure the variation of Δ and ψ as a function of the gas pressure at a given temperature. The curves obtained are comparable with the adsorption isotherm already drawn in figures 1 and 2. As a matter of fact, multistep functions $\Delta(p)$ or $\psi(p)$, very analogous to figure 1 are recorded as p varies from $\sim$ 0 to p_o, the sublimation pressure of bulk xenon, checking again the layer-by-layer growth mode of this system. In the submonolayer regime, the ellipsometric signals are proportional to the coverage[81] yielding adsorption curves identical to those obtained by volumetric measurements (see II.1). The method is very interesting because it operates with a monocrystalline substrate and high or low gas pressure.

The optical parameters Δ and ψ are interpreted with the Fresnel-Drude framework[83-85] of light reflection. They depend on the refractive indexes of the adsorbate and substrate. One finds that, unlike 3D xenon, the adsorbed layers are appreciably anisotropic. Some of the refractive indexes are listed in table V. The index $n_{//}$ parallel to the graphite basal plane is not very sensitive to the graphite surface field whereas the perpendicular index $n_{\perp}$ is strongly influenced by the substrate. This important birefringence is related to the dipolar moment induced by the graphite crystal[82].

Table V - Refractive indexes of xenon for $\lambda \to \infty$

	Bulk xenon	Adsorbed monolayer·	Adsorbed double layer
$n_{\parallel}$	1.533	1.53	1.53 first layer 1.53 second layer
$n_{\perp}$	1.533	1.64	1.64 first layer 1.64 second layer

Owing to these interesting results, the ellipsometry is worth being developed studying the optical properties of the various 2D adsorbed phases. The interpretation is very simple for physisorbed molecules because there is no charge transfer from the overlayer to the substrate keeping therefore its bulk optical properties.

IX - DISCUSSION AND CONCLUSION

All these new exciting findings about the 2D adsorbed phases and their transitions have been obtained thanks to the exceptional properties of the (0001) graphite surface. Some of the used techniques like adsorption volumetry, calorimetry, NMR, neutron scattering, Mössbauer spectroscopy, need graphite powder in order to increase the surface/bulk ratio. Exfoliated graphite has very clean, uniform, homogeneous basal plane surfaces[3,4]. Still the crystal size is so small that it can induce some intergranular condensation or some broadening of a first order transition[32]. The best way is to use cleaved graphite monocrystal when the observation technique (LEED, ellipsometry) allows it. Then it is possible to check and monitor the structure and the cleaness of the surface during adsorption[13].

The exceptional qualities of graphite surfaces have encouraged the numerous studies reviewed here. They enable the measurement of very interesting properties of the adsorbed films, bringing out the macroscopic and microscopic features of the 2D adsorbed phases. As explained in these lectures, the thermodynamic, structural, vibrational and optical properties of the adsorbed phases on graphite have been determined for several systems. Furthermore, the mobility of the 2D fluids can be measured and the mechanism of the first order transitions begins to be understood.

But nothing is known about the electronic properties of the surface phases, except very rare dipole moment determinations[87] and a few results on orbital calculations of rare gases[88] or atomic hydrogen[89]. This is due to the small modification of the electron levels during the physisorption of molecules on graphite. Hence the perturbation is not easily measurable.

Some other properties are still to be thoroughly studied like the 2D superfluidity, the magnetic ordering, the solid-solid transitions and the structure and dynamics of the 2D liquid. Finally, one will have to pay special attention to the mechanism of the continuous transitions.

References

1. J.G. Dash, "Films on Solid Surfaces", Acad. Press, N.Y. (1975).
2. J.G. Dash, Physics Reports 38C (March 1978).
3. A. Thomy and X. Duval, J. Chim. Physique 66:1966 (1969).
4. A. Thomy and X. Duval, J. Chim. Physique 67:286 (1970).
5. A. Thomy and X. Duval, J. Chim. Physique 67:1101 (1970).
6. J. Suzanne and M. Bienfait, J. Physique, Colloq. C4, suppl.
     n° 10, t. 38, p. 31 (1977).
7. X. Duval and A. Thomy, C.R. Acad. Sc. Paris 259:4007 (1964).
8. A. Thomy and X. Duval, "Adsorption et Croissance Cristalline",
     ed. CNRS, Paris, p. 81 (1965).
9. A. Thomy, J. Régnier and X. Duval, Colloq. Intern. CNRS n° 201,
     ed. CNRS, Paris, p. 1 (1971).
10. A. Thomy, J. Régnier, J. Menaucourt and X. Duval, J. Cryst.
     Growth 13/14:159 (1972).
11. Y. Larher, J. Chem. Soc. Faraday Trans I, 70:320 (1974).
12. X. Duval and A. Thomy, Carbon 13:242 (1975).
13. J. Suzanne, J.P. Coulomb and M. Bienfait, Surf. Sci.40:414
     (1973) - 44:141 (1974) - 47:204 (1975).
14. H.M. Kramer and J. Suzanne, Surf. Sci. 54:659 (1976).
15. T.L. Hill, J. Chem. Phys. 13:520 (1949).
16. T.L. Hill, Advan. Catalysis 4:211 (1952).
17. T.L. Hill, J. Chem. Phys. 18:246 (1950).
18. A. Clark, "The theory of adsorption and catalysis", Acad. Press,
     N.Y. (1970).
19. Y. Larher, J. Colloid Interface Sci.37:836 (1971).
20. J.A. Venables and G.L. Price, "Nucleation of Thin Films in
     Epitaxial Growth" part B, p. 381, Acad. Press, N.Y. (1975).
21. T.L. Hill, "Introduction to Statistical Thermodynamics",
     Addison-Wesley Reading, Mass. (1960).
22. B.J. Alder and T.E. Wainwright, J. Chem. Phys.33:1439 (1960)-
     Phys. Rev.127:359 (1962).
23. J. Orban and A. Bellemans, J. Chem. Phys. 49:363 (1968).
24. J. Orban, J. Van Craen and A. Bellemans, J. Chem. Phys. 49:1778
     (1968).
25. C.E. Campbell and M. Schick, Phys. Rev. A5:1919 (1972).
26. H.P. Neumann, Phys. Rev.A11:1043 (1975).
27. K. Binder and D.P. Landau, Surf. Sci. 61:577 (1976).
28. J. Rouquerol, "Thermochimie" ed. CNRS, Paris, p. 537 (1972).
29. C. Letoquart, F. Rouquerol and J. Rouquerol, J. Chim. Phys.
     3:559 (1973).
30. J. Régnier, J. Rouquerol and A. Thomy, J. Chim. Physique
     3:327 (1975).
31. D.M. Butler, J.A. Litzinger, G.A. Stewart and R.B. Griffiths,
     Phys. Rev. Lett. 42:1289 (1979).
32. T.T. Chung and J.G. Dash, Surf. Sci.66:559 (1977).

33. J.J. Lander and J. Morrison, Surf. Sci. 6:1 (1967).
34. P. Thorel, B. Croset, C. Marti and J.P. Coulomb in "Proc.
    Conf. on Neutron Scattering", Gatlinburg (1976).
35. C. Marti, B. Croset, P. Thorel and J.P. Coulomb, Surf. Sci.
    65:532 (1977).
36. J.A. Venables, H.M. Kramer and G.L. Price, Surf. Sci.
    55:373 (1976 ; 57:782 (1976).
37. M.D. Chinn and S.C. Fain Jr., J. Vac. Sci. Technol. 14:314
    (1977).
38. M.D. Chinn and S.C. Fain Jr., Phys. Rev. Lett. 39:146 (1977).
39. S.C. Fain Jr. and M.D. Chinn, J. Physique Colloq. C4,
    suppl. n° 10, t. 38, p. 99 (1977).
40. T. Ceva and C. Marti, J. Physique Lettres 39:L 221 (1978).
41. P.M. Horn, R.J. Birgenau, P. Heiney and E.M. Hammonds,
    Phys. Rev. Lett. 41:961 (1978).
42. J.A. Venables and P.S. Schabes-Retchkiman, private communica-
    tion.
43. F.C. Frank and J.H. van der Merwe, Proc. Roy. Soc. London,
    Ser. A 198:205,216 (1949) ; 200:125 (1949).
44. J.H.van der Merwe, J. Appl. Phys. 41:4725 (1970).
45. S.C. Ying, Phys. Rev. B12:4160 (1971).
46. J.A. Snyman and J.H. van der Merwe, Surf. Sci. 42:190 (1974);
    45:619 (1974).
47. A.D. Novaco and J.P. McTague, Phys. Rev. Lett. 38:1286 (1977).
48. A.D. Novaco and J.P. McTague, J. Physique Colloq. C4,
    suppl. n° 10, t. 38, p. 116 (1977).
49. J.A. Venables and P.S. Schabes-Retchkiman, J. Physique
    Colloq. C4, suppl. n° 10, t. 38, p. 105 (1977).
50. J.A. Venables and P.S. Schabes-Retchkiman, Surf. Sci.
    71:27 (1978).
51. C.G. Shaw, S.C. Fain and M.D. Chinn, Phys. Rev. Lett. 41:955
    (1978).
52. J. Suzanne, J.P. Coulomb, M. Bienfait, M. Matecki, A. Thomy,
    B. Croset and C. Marti, Phys. Rev. Lett. 41:760 (1978).
53. J.P. McTague and M. Nielsen, Phys. Rev. Lett. 37:596 (1976).
54. See for instance : International tables for X-ray Crystallo-
    graphy, Vol. III, p. 232, Kynoch Press, Birmingham (1968).
55. J.P. Coulomb, J. Suzanne, M. Bienfait and P. Masri,
    Sol. State Commun. 15:1585 (1974).
56. J.P. Coulomb and P. Masri, Sol. State Commun. 15:1623 (1974).
57. H. Taub, L. Passell, J.K. Kjems, K. Carneiro, J.P. McTague
    and J.G. Dash, Phys. Rev. Lett. 34:654 (1975).
58. H. Taub, K. Carneiro, J.K. Kjems, L. Passell and J.P. McTague,
    Phys. Rev. B16:4551 (1977).
59. S. Bukshpan, T. Sonnino and J.G. Dash, Surf. Sci. 52:466 (1975).
60. H. Shechter, J.G. Dash, M. Mor, R. Ingalls and S. Bukshpan,
    Phys. Rev. B14:1876 (1976).

61. K. Horn and J. Pritchard, J. Physique Colloq. C4, suppl. n° 10, t. 38, p. 164 (1977).

62. H. Froitzheim, H. Ibach and S. Lehwald, Phys. Rev. B14:1362 (1976).

63. H. Ibach, Proc. 7th Intern. Vac. Congr. & 3rd Intern. Conf. Solid Surfaces, ed. R. Dobrozemsky et al. Berger and Söhne, p. 743, Vienna (1977).

64. H. Taub, H.R. Danner, Y.P. Sharma, H.L. Murry and R.M. Brugger, Phys. Rev. Lett. 39:215 (1977).

65. K. Carneiro, J. Physique Colloq. C4, suppl. N° 10, t. 38, p. 1 (1977).

66. R.J. Rollefson, Phys. Rev. Lett. 29:410 (1972).

67. B.P. Cowan, M.G. Richards, A.L. Thomson and W.J. Mullin, Phys. Rev. Lett. 38:165 (1977).

68. H. Shechter, J. Suzanne and J.G. Dash, Phys. Rev. Lett. 37:706 (1976).

69. H. Shechter, J. Physique Colloq. C4 suppl. n° 10, t. 38, p. 38 (1977).

70. J.P. Coulomb, M. Bienfait and P. Thorel, J. Physique Colloq. C4, suppl. n° 10, t. 38, p. 31 (1977).

71. J.P. Coulomb, M. Bienfait and P. Thorel, Phys. Rev. Lett. 42:733 (1979).

72. F. Millot, J. Physique Lettres 40:L9 (1979).

73. Y. Grillet, F. Rouquerol and J. Rouquerol, J. Physique Colloq. C4, suppl. n° 10, t. 38, p. 57 (1977).

74. J.M. Kosterlitz and D.J. Thouless, J. Phys. C5:124 (1972); 6:1181 (1973).

75. R.P. Feynman, private communication to R.L. Elgin and D.L. Goodstein.

76. F.A. Lindemann, Z. Phys. 11:609 (1910).

77. F. Millot, thesis Nancy (1976) n° 12185.

78. J. Suzanne, J.P. Coulomb and M. Bienfait, J. Cryst. Growth 31:87 (1975).

79. J.A. Venables and M. Bienfait, Surf. Sci. 61:667 (1976).

80. M. Bienfait and J.A. Venables, Surf. Sci. 64:425 (1977).

81. G. Quentel , J.M. Rickard and R. Kern, Surf. Sci. 50:343 (1975).

82. G. Quentel and R. Kern, Surf. Sci. 55:545 (1976).

83. M. Born and E. Wolf,"Principles of Optics", Pergamon Press (1959).

84. P. Drude and Wiedemans, Ann. Physik. Chem. 32:584 (1887).

85. P. Drude, Ann. Physik. Chem. 36:532 (1889).

86. J. Suzanne, P. Masri and M. Bienfait, Jap. J. Appl. Phys. suppl. 2 part 2 p. 295 (1974).

87. G. Honoré and J.P. Beaufils, Le Vide 163-165:50 (1973).

88. E. Giamello, C. Pisani, F. Ricca and C. Roetti, Surf. Sci. 49:401 (1975).

89. F. Ricca, J. Physique, Colloq. C4, suppl. n° 10, t. 38, p.173 (1977).

90. A. Holz and J.T.N. Medeiros, Phys. Rev. B17:1161 (1978).

THEORY OF HELIUM MONOLAYERS

M. Schick

Department of Physics
University of Washington
Seattle, WA 98195

ORGANIZATION OF LECTURES

These lectures are somewhat more general than the title
indicates. I shall use the adsorbed helium system to illustrate
several of the transitions that can be observed in adsorbed systems.
In the first lecture I shall review the phase diagram of adsorbed
helium and describe the nature of the observed phases and of the
transitions between them. Particular attention is paid to the
highly quantum-mechanical low density fluid phase. Order-disorder
transitions in adsorbed systems are the subject of the remaining
two lectures. In the first of these I derive the universality
classes which these transitions can represent. The particular
order-disorder transition which occurs in adsorbed helium and
krypton systems is the subject of the final lecture in which renormalization group methods for the calculation of phase diagrams
is reviewed.

PHASES OF ADSORBED HELIUM

Consider a monolayer of structureless atoms adsorbed upon a
substrate. At temperatures which are small compared to the characteristic energy of excitations perpendicular to the substrate, the
system behaves as if it were two-dimensional. The adsorbate can
exist in several phases, some of which have their counterpart in
bulk and some of which do not. A bulk system exists within a
background which is translationally and rotationally invariant,
but the adsorbed system is bound to a substrate which is invariant
only under certain rotations, translations and reflections which
define a space group G_0. The various phases which are observed in
adsorbed monolayers are most easily described by comparing the

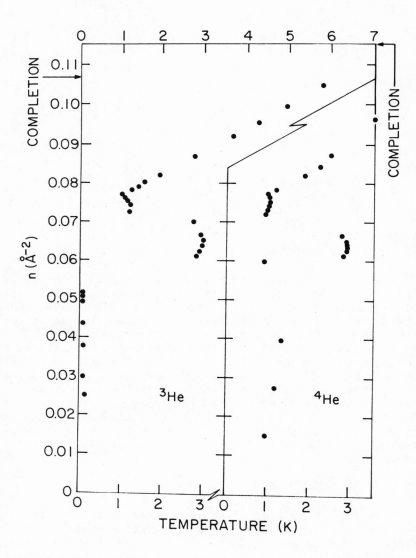

Fig. 1 Loci of specific heat maxima for helium adsorbed on
        Grafoil.

space group G which leaves the density $n(\bar{r})$ of the phase invariant
to the space group of the substrate. There are three possibilities:
(1) G is the same as G_0 so that $n(\bar{r})$ is invariant under all operations which leave the substrate invariant. This is the case in
the gas and liquid, or more generally, fluid phases. (2) G is a
subgroup of G_0. This corresponds to a commensurate or registered
solid phase. (3) G is not a subgroup of G_0. This is the case of
an incommensurate solid. The classification of phases according
to their types of symmetry makes clear the kind of phase transitions
which can be observed. There are transitions between phases of the
same type, the most notable example being that between liquid and
gas. This transition is discussed by Michel Bienfait in this
School. There are transitions between phases of different types.
Transitions between phases of types 1 and 2 are denoted order-
disorder transitions. I shall discuss them at length in my second
and third lectures. Transitions between phases of types 1 and 3,
denoted melting or sublimation transitions, are discussed by
Michael Kosterlitz. Lastly, the transitions between phases of
types 2 and 3 are denoted commensurate-incommensurate transitions.
They are the subject of the lectures of Alan Luther.

    With the above ideas in mind I shall turn to the experimental
observations of helium monolayers adsorbed on the basal planes of
graphite. The data points shown in Fig. 1 are, for the most part,
the loci of specific heat maxima taken by the University of
Washington groups.[1,2] They are supplemented by adsorption isotherms[3] above 4 K and neutron scattering[4] in the case of He[3]. My
interpretation of the observations is shown in the phase diagrams
of Fig. 2. There are a few features worthy of note. First, there
is a single fluid phase, not liquid and gas phases. As I shall
show below, the low density fluid phase is extremely quantum
mechanical. Second, there is a registered solid. Its existence,
in helium, is also due to large quantum effects, a point amplified
below. The registered phase is also noteworthy in that it exists
as a single phase over an appreciable range of densities and is
well separated from other phases, making it well suited for studying critical phenomena. Third, there is an incommensurate solid
at high densities. It is sufficiently compressible that it exists
over a wide range of coverages. This enables the melting transition to be studied far from the complications introduced when the
monolayer nears completion.

    Let us now look more closely at these phases in turn, begin-
ning with the low density phase. By low densities I mean those
less than about 0.06 $\overset{\circ}{A}{}^{-2}$. Specific heat measurements on He[4] at
densities of 0.027 (plusses), 0.028 (circles), 0.040 (crosses),
and 0.042 (triangles) are shown in Fig. 3. Note that at 4 K,
C/Nk is less than unity, rises to a peak at about 1 K then falls
to zero. The signals from He[3] shown in Fig. 4 are somewhat
different. The densities shown there are 0.0279 (circles) and

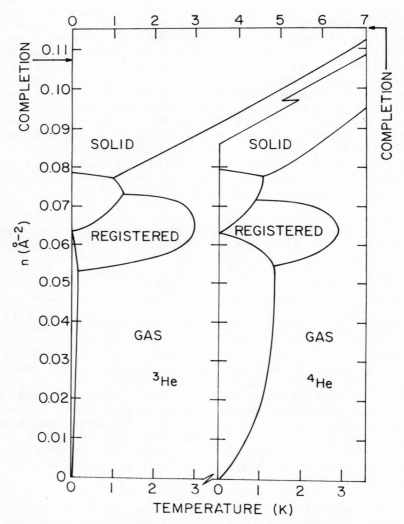

Fig. 2. Phase diagram of helium adsorbed on Grafoil which arises
from my interpretation of the data of Fig. 1.

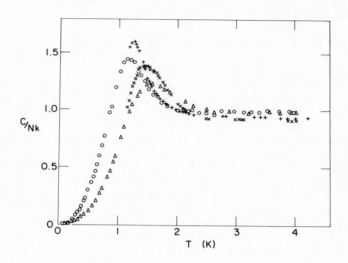

Fig. 3.   Specific heats of adsorbed ^{4}He at low densities.

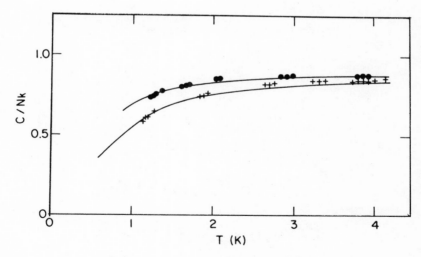

Fig. 4.   Specific heats of adsorbed ^{3}He at low densities.

0.0415 (plusses). C/Nk is again less than unity at 4 K but they
do not rise at lower temperature. However a small peak or shoulder
is seen at about 0.1 K as shown in Fig. 5.

These are the experimental facts. How are they to be interpreted? The first step the theorist must take is to examine the
one-particle problem, that is, the analog of the free-electron
model in Solid State theory. For this one must calculate the
potential $V(\bar{r})$ due to the substrate which is seen by a single
helium atom. If it is assumed that this is just a sum of the pair
interactions between He and the C atom of the graphite substrate,
one obtains the result of Fig. 6 taken from Hagen, Novaco and
Milford.[5] The potential is shown as a function of the distance perpendicular to the substrate measured in units of a = 2.46 Å the
side of a unit cell within the substrate plane. The difference
between the well depth over the adsorption site and directly over
a carbon atom is about 20 K. The ground state energies of He[3],
He[4] in the potential are indicated by the horizontal lines and are
some 60 to 70 K above the absolute minimum of $V(\bar{r})$. Recently the
actual substrate potential has been reconstructed from the results
of experiments in which a beam of atomic helium is scattered from
graphite.[6] The potential obtained is shown in Fig. 7. While it
has the same qualitative features of the Novaco-Milford potential,
it is not as attractive. The well depth is 18 meV or 209 K. The
He[4] and He[3] ground states are 69 K and 74 K respectively above this.

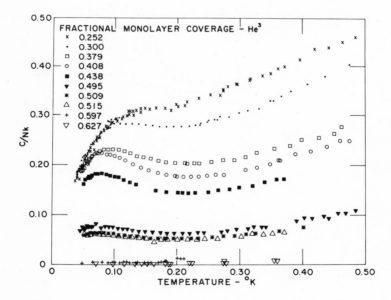

Fig. 5. Low-temperature specific heats of adsorbed [3]He.

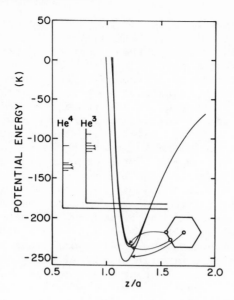

Fig. 6.   Potential profiles calculated in Ref. 5 showing the sub-
          strate potential as a function of z for three fixed points
          in the x-y plane.   Also shown are energy levels diagrams
          for the lowest eight states (zero crystal momentum) of
          ^{4}He and ^{3}He.

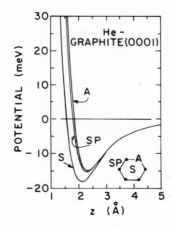

Fig. 7. Potential profiles reconstructed from scattering experiments.

    Because $V(\bar{r})$ is periodic in the plane of the substrate one
must, in order to obtain the energies of the single particle states,
perform a band calculation. The free-particle band (i.e., $V(\bar{r})$
= 0) is shown in the upper part of Fig. 8. It is unrecognizable,
of course, because the simple parabolas of the free-particles have
been folded back onto the reduced zone of the substrate. Points
of high symmetry are indicated. Below it is the band obtained by
Milford and Novaco. (As of this writing, the band obtained from
the measured potential is not available.) The effect of the potential on the lower bands, the most important ones for low densities
and temperatures, is very small. Some degeneracies are lifted and
a small band gap is introduced. The first state involving an
excitation perpendicular to the substrate has an excitation energy
of 80-85 K. From this information we conclude that at the temperatures of interest, the helium can be well-approximated by a truly
two-dimensional system and that any effect of the substrate other
than making the helium behave two-dimensionally can be ignored.
These assumptions are of course in accord with the experimental
observation of Figs. 3 and 4 that at temperatures above 4 K the
specific heat appears to approach the value unity which is characteristic of a two-dimensional ideal gas (no interactions within
the gas or between it and the substrate). In order to determine

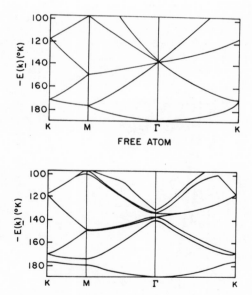

Fig. 8.   a) Free particle band structure plotted in the Brilloiun
          zone of a triangular lattice. b) Band structure of He[4]
          calculated from the potential of Fig. 6.

what information the deviations from unity contain we have recourse
to a theoretical framework which is appropriate to systems at low
densities and high temperatures. This is, of course, the virial
expansion. As derivations of the virial expansion for two-dimensional systems can be found elsewhere[7,8] I shall be brief.

The ℓth virial coefficient B_ℓ is defined in terms of the
expansion of the (two-dimensional) pressure in terms of the average
density, n:

$$\phi = \beta^{-1}n + \beta^{-1} \sum_{\ell=2}^{\infty} B_\ell(\beta)n^\ell \tag{1.1}$$

All other thermodynamic functions have a similar expansion. Two
which are of interest are the inverse isothermal compressibility

$$K_T^{-1} = \beta^{-1}n + \beta^{-1} \sum_{\ell=2}^{\infty} \ell B_\ell(\beta)n^\ell \tag{1.2}$$

and the specific heat at constant area

$$C/Nk = -\beta^2 \frac{d^2}{d\beta^2} \left| \ell n\left(\frac{nA}{Z_1}\right) + \sum_{\ell=1}^{\infty} \ell^{-1} B_{\ell+1}(\beta)n^\ell \right| \tag{1.3a}$$

where A is the area and Z_1 the one-particle partition function.
For helium on graphite where we have seen that the potential of
the substrate can be ignored,

$$Z_1 = (2s + 1)A/\lambda^2$$

where s is the spin of the particle and λ its de Broglie wavelength.
The specific heat then takes the form

$$C/Nk = 1 - \beta^2 \frac{d^2}{d\beta^2} \sum_{\ell=1}^{\infty} \ell^{-1} B_{\ell+1}(\beta)n^\ell \tag{1.3b}$$

The first correction to ideal gas behavior in Eqs. (1.1)-(1.3)
comes from the second virial coefficient B_2 which is determined by
the one and two particle partition functions Z_1, Z_2 according to

$$B_2(\beta) = -\lim_{A\to\infty} A \left| \frac{Z_2(A,\beta) - \frac{1}{2} Z_1^2(A,\beta)}{Z_1^2(A,\beta)} \right| \tag{1.4}$$

From Eq. (1.1) we intuit that for repulsive potentials B_2 is positive whereas for attractive potentials it is negative. It is instructive to evaluate Eq. (1.4) <u>classically</u> for the case of a two-particle interaction which is spherically symmetric. One obtains

$$B_2^{CL} = \frac{1}{2} \int_0^\infty (1 - e^{-\beta v(\rho)}) \, 2\pi\rho \, d\rho$$

which exhibits the expected behavior (i.e., positive for repulsive potentials and negative for attractive ones). If we now substitute this expression into Eq. (1.3b) retaining only the second virial coefficient

$$C/Nk = 1 - n\beta^2 \frac{d^2 B_2}{d\beta^2} + O(n^2) \tag{1.5}$$

we obtain

$$C/Nk = 1 + n\beta^2 \int_0^\infty v^2(\rho) \, e^{-\beta v(\rho)} 2\pi\rho \, d\rho + O(n^2)$$

which states that the specific heat is always greater than unity. As this is contrary to the experimental data, it follows that the helium system cannot be treated classically. The quantum mechanical evaluation of Eq. (1.4) leads to the following result for spinless Bose or Fermi systems[8]

$$B_2(\beta) = \mp \frac{\lambda^2}{4} - 2\lambda^2 {\sum_m}' \left(\sum_n (e^{-\beta\varepsilon_{nm}} - 1) + \frac{\lambda^2}{\pi^2} \int_0^\infty dk k \delta_m(k) e^{-\beta \frac{\hbar^2 k^2}{2\mu}} \right) \tag{1.6}$$

The upper sign in the first term is for the Bose system and the lower sign for the Fermi. The prime on the sum over the azimuthal quantum number m indicates that it is over all even integers, both positive and negative for bosons and all odd integers for fermions. The energy of the nth bound state of azimuthal quantum number m is ε_{nm} and is negative. The reduced mass μ is simply one half the usual particle mass and $\delta_m(k)$ is the phase shift of the mth partial wave. The second virial coefficient of a fermion with spin 1/2, $B_{2F}^{(1/2)}$ is simply

$$B_{2F}^{(1/2)} = \frac{3}{4} B_{2F} + \frac{1}{4} B_{2B} \tag{1.7}$$

Let us check that the terms of Eq. (1.6) conform to our

intuition. The first term is the virial coefficient of the ideal
gas and the sign correctly reflects the statistical attraction of
bosons and repulsion of fermions. The second term arises from
the possibility of bound states. This term is always negative in
conformity with the intuition that bound states reduce the pressure.
When it is recalled that the phase shifts of repulsive potentials
are negative and those of attractive potentials positive, it is
seen that the last term, too, is in accord with our expectations.

It is straightforward to apply these results to helium by
using the standard Lennard-Jones interaction. The results are
shown in Fig. 9. It is noteworthy that the classical value of B_2
differs from its quantum mechanical value by 50% at 30 K! Even
at 60 K it is in error by about 15%. The difference between the
results for ^{3}He and ^{4}He is completely due to the mass difference
above 4 K. We are now in a position to compare theory with experiment. According to Eq. (1.5), a plot of $(C/Nk - 1)n^{-1}$ should be
a universal function of temperature. The experimental data for
He^3 plotted in this way are shown in Fig. 10 where they are compared with the theoretical results. The agreement is excellent.
The results for He^4 are shown in Fig. 11. The agreement is not
as impressive but is still adequate. It is interesting to understand where the difference in behavior of the two systems origi-
nates. It is due to the fact that at low temperatures, the dominant

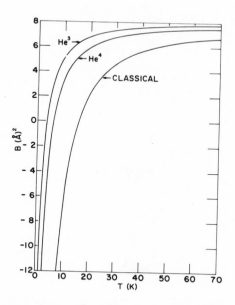

Fig. 9.   Second virial coefficients for two-dimensional systems
of ^{3}He and ^{4}He.

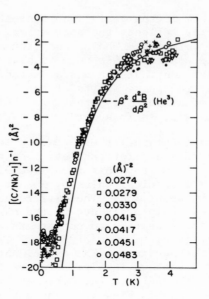

Fig. 10.   Comparisons of calculated and observed specific-heat
           deviation per unit density for ^{3}He.

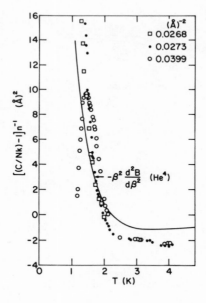

Fig. 11.   Comparison of calculated and observed specific-heat
           deviation per unit density for ^{4}He.

contributions to B_2 come from the phase shifts of lowest azimuthal quantum number. For He^4 this is the m=0 phase shift. For He^3 it is the m=0 phase shift weighted by 1/4 and the two identical m = ±1 phase shifts weighted by 3/4 each. The m=0 phase shift primarily reflects the hard core of the potential while the m = ±1 phase shifts are dominated by collisions with an impact parameter of the order of the thermal wavelength. At 4 K this is about 5 Å so that these collisions reflect the attractive part of the interaction. Thus, due to the statistics, the two isotopes at low temperatures "see" very different parts of the same Lennard-Jones potential which results in different specific heats.

A direct measure of the effect of the statistics is the magnetic susceptibility of He^3. How is it that a spin-independent interaction affects the magnetic susceptibility? If all scattering processes were allowed as in a gas of distinguishable particles the interaction would have no effect at all on the susceptibility. This quantity would be equal to its value in a non interacting classical gas, the Curie susceptibility. But all scattering processes are not allowed. Those in the singlet state must also be in states of even angular momentum. At low temperatures this is predominantly m=0 scattering which is from the hard core. Similarly that in the triplet state must be in states of odd m and is predominantly m = ±1 at low temperatures, reflecting the attractive interaction. Thus the effect of the potential is not the same for triplet and singlet scattering and the potential makes its prescence known in the susceptibility. This quantity has been measured[9] and the results are shown in Fig. 12. The agreement between theory and experiment is very good. Note that the statistics are responsible for the deviation of the susceptibility from the Curie value but that the interaction tends to suppress the effect of the statistics. This is easily understood. Statistical effects are

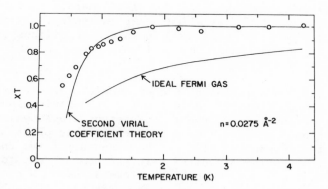

Fig. 12   Comparison of calculated and observed magnetic susceptibility (multiplied by temperature).

generally important when two particles are within a thermal wavelength of one another. When this wavelength is smaller than the
hard-core radius, which occurs for T $\gtrsim$ 4 K in helium, the particles
can never be so close and the effects of statistics are suppressed.

As we have seen, at high temperatures and low densities the
helium fluid behaves like a quantum mechanical imperfect gas. But
what is its state at low temperatures? In particular what do the
peaks in the specific heat signify? I believe that these peaks
certainly indicate a condensation, at least in the He^4. Numerous
calculations indicate that the He^4 gas is unstable at T=0. The
virial expansion of the inverse compressibility, Eq. (1.2), indicates
an instability if B_2 is sufficiently negative and the virial coefficients of both isotopes do become strongly negative. Lastly
the measured total heat capacity at low temperatures increases
linearly with the number of adsorbate atoms which is a necessary,
but not sufficient, condition for a two-phase coexistence region.
Given that there is a condensation, the question remains as to the
nature of the condensed phase. I shall postpone further discussion
of this point until later in this lecture.

Before leaving the low density system I would like to remark
briefly on some amusing aspects of the helium system adsorbed on
graphite pre-plated with rare gases such as Kr, Ar, etc. Rehr and
Tejwani[10] have approached this problem from a tight-binding formalism arguing that the potential provided by the rare gas atoms is
much deeper than that provided by the graphite alone. Hence the
helium atoms will tend to be more localized. This view is supported
by the band calculations of such a system which show appreciable
band gaps. The virial expansion may now be reworked in this tight-
binding framework and the results are in rather good agreement with
experiment. They show the specific heat peak in He^4 which occurred
at 1 K on bare graphite to be shifted to higher temperatures. The
shift depends systematically on the preplating atom. Thus the
bare graphite system is analogous to the alkali metals in which
the free electron model is an excellent starting point. By preplating it, one produces a system more analogous to silicon or
germanium where the tight-binding model is more appropriate. A
nice versatility this!

Let me now pass to a region of higher density, densities near
n = 0.064 Å^{-2}. As the temperature is lowered, the fluid undergoes
an order-disorder transition to the $\sqrt{3}$ x $\sqrt{3}$ structure shown in
Fig. 13. As mentioned earlier, the existence of this phase is
also due to quantum mechanics. To see this consider first the
ground state of the heavier classical rare gas atoms. Due to the
attractive interaction, these adsorbed systems all form a close-
packed solid. Whether this solid is in registry with the substrate
in the $\sqrt{3}$ x $\sqrt{3}$ structure or is incommensurate depends upon the
relative strengths of the atom-substrate potential and the atom-atom

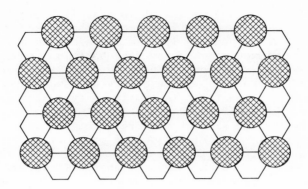

Fig. 13.   Arrangement of adsorbed atoms in $\sqrt{3}$ x $\sqrt{3}$ structure.
Shaded circles represent adsorbed helium atoms and are
drawn to the relative size of the bare interaction
potential parameter σ = 2.56Å.

potential. Consider Xe first. In the three-dimensional solid at
80 K the nearest-neighbor distance d between atoms is 4.4 Å compared to the 4.26 Å of the $\sqrt{3}$ x $\sqrt{3}$ structure on graphite. It is
not surprising then that Xe forms the commensurate structure only
under pressure. The nearest-neighbor distance in solid Kr at 60°
is 4.1 Å. If this structure is pulled apart slightly until it
registers with the graphite, more energy is gained from the substrate than is lost by the Kr-Kr interaction, or so the experiments
tell us for the Kr does form a registered $\sqrt{3}$ x $\sqrt{3}$ structure. The
nearest-neighbor distance in Ar at 40 K is 3.8 Å, far from the
4.26 Å of the $\sqrt{3}$ x $\sqrt{3}$ structure. Thus Ar forms only an incommensurate solid. This is also true of Ne with d = 3.2 Å at 115 K.
So why should He form this structure? The reason is that the incommensurate solid is not stable at zero pressure for the same
quantum mechanical reasons which make the bulk solid unstable at
zero pressure. Thus the incommensurate solid is simply not a
candidate for the ground state. The competition is between the
$\sqrt{3}$ x $\sqrt{3}$ structure and a liquid in the case of He[4]. It is the more
stable of these two phases which will be the high density phase
in the two-phase coexistence region discussed above. Novaco[11] has
argued that the liquid is the more stable state. However the uncertainty in the model substrate potential which was used is of
the order of magnitude of the energy difference between the two
states so the result is not reliable. Now that the substrate
potential is known from experiment, the above calculation should
be repeated with this potential. I once believed that the liquid

would be more stable. Now my belief is that it will be the $\sqrt{3}$ x $\sqrt{3}$
structure. Thus the low density region of the phase diagram will
be the same as observed in Kr, i.e., a commensurate phase with a
multicritical point and a two-phase region at low temperatures and
densities. This is just a prejudice however. Experiment will
shortly tell us the result. I can easily envisage two experiments
which will do this. The first is simply to systematically map out
the low density region of the He[4] system using specific heat as a
probe. This is now the kind of experiment a graduate student
learns his craft on. Either the loci of peaks will, with increasing
density, increase in temperature and march toward the region in
which the commensurate phase exists or they will decrease in
temperature and plunge toward zero temperature. The former behavior would indicate a commensurate ground state, the latter a
liquid one. A second experiment would entail a neutron diffraction
measurement at a density in the two phase region. If the high
density phase is commensurate, the unmistakable superlattice peaks
will be seen. If they are not, the ground state is liquid. In
He[3] the ground state will either be registered or a gas, for there
is excellent reason[12] to believe that He[3] does not liquefy on
graphite. My prejudice is that the commensurate state is again
favored and that the low temperature peaks of Fig. 5 indicate a
condensation to that phase. Experiments can again provide the
answer but they are more difficult because of the low temperature
involved.

As order-disorder transitions are the subject of lectures 2
and 3 I will say no more about the ordered state here. Instead I
shall pass to higher densities where the helium is found in an
incommensurate solid phase at low temperatures. The specific
heat signals[1,3] are shown in Fig. 14. At low temperatures they
can be fit by a Debye model as they decrease like T^2. The Debye
temperatures behave with density exactly as their three-dimensional
counterparts. As the temperature is raised a small peak is observed. The sharp peak at higher densities actually reflects
second-layer promotion. This large and uninteresting peak sits
on top of the small one characteristic of the first layer as shown
in the inset to Fig. 14. If this peak is identified with the
melting of the two-dimensional solid, then this melting is continuous, at least above 1 K or so, as there is no evidence of a
latent heat. At lower temperatures, the melting is first order
and there is a two phase region as shown in the phase diagram. A
continuous melting at high temperatures and a discontinuous one
at low temperatures seems to be characteristic of all the adsorbed
rare gases. The continuous melting is, of course, unique to two
dimensional systems. A beautiful and successful dislocation (and
disclination) theory of this phenomenon based on the ideas of
Kosterlitz and Thouless[13] has been developed by Halperin and
Nelson.[14] Unfortunately for thermodynamic measurements, the theory

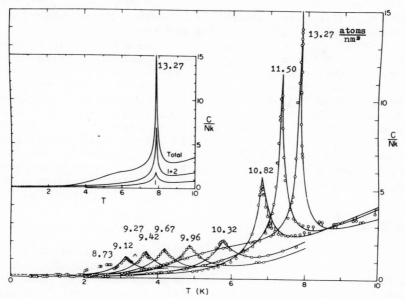

Fig. 14. Heat capacity of high density ^{4}He films near melting.
Coverages in (nanometers)$^{-2}$ are shown for each peak. In
the inset the peak for 13.27 atom/mm^2 is decomposed into
contributions from melting in the first layer, promotion
into the second layer, and desorption into the gas phase.

predicts only an essential singularity in thermodynamic functions
so it is impossible to verify by such means. However discontinuities in elastic constants are predicted so that the theory is
certainly verifiable. As it is such a beautiful theory it deserves
a detailed comparison with experiment. The lectures of Michael
Kosterlitz can be consulted for further details.

The only transition which I have not yet discussed is the
commensurate-incommensurate transition between the $\sqrt{3}$ x $\sqrt{3}$ structure and the incommensurate solid. This is first order in helium.
The densities of the two structures differ considerably. In Kr
this same transition appears to be continuous. An enormous amount
of attention has been devoted to transitions of this kind as they
appear in many different kinds of systems, for example, those that
exhibit charge density waves. Nevertheless at the time of this
conference the theoretical dust has yet to settle--at least my
vision is still obscured--so it is with pleasure that I leave the
task of clarifying the situation in his lectures to Alan Luther.

This concludes my survey of the phase diagram of helium adsorbed on graphite. In summary: there exist three phases, a fluid phase which, at low densities and temperatures exhibits easily observed quantum effects, an ordered phase which is commensurate with the substrate, and a close-packed solid which is not in registry with the substrate. The three transitions between these three phases are all of current interest. They are the solid-fluid, or melting transition, the commensurate-incommensurate transition, and the order-disorder transition. It is the object of my next two lectures to demonstrate that transitions of the latter kind in adsorbed systems provide physical realizations of theoretical models of current interest and that their phase diagrams can now be reliably calculated with a consequent agreement between theory and experiment unusual in Surface Physics.

ORDER-DISORDER TRANSITIONS

I shall consider in this lecture transition from the fluid phase to an ordered phase which is commensurate with the substrate over a finite temperature range. To have a definite transition in mind you can think of the transition helium undergoes to the $\sqrt{3} \times \sqrt{3}$ structure of Fig. 13. Such transitions are extremely common, particularly in chemisorbed systems. Furthermore they are easily observed either directly by the diffraction of neutrons, x-rays, or low-energy electrons, or indirectly by adsorption isotherms, specific heats, or nuclear magnetic resonance.

While numerous observations of ordered structures have been reported, it is only recently that the transitions to these structures have been studied. This is due, of course, to the fact that our ideas about phase transitions have changed greatly in the last decade. It is useful to recall two concepts which have been introduced. First is the characterization of transitions by critical exponents. For example, as the transition temperature T_c is approached from below, the intensity of a superlattice spot in a diffraction experiment will vanish like

$$I \propto \left| T - T_c \right|^{2\beta}$$

which defines the exponent β. Similarly the specific heat diverges as T_c is approached from either side according to

$$C \propto \left| T - T_c \right|^{-\alpha}$$

which defines the critical index α. For the two dimensional Ising model, $\beta = 1/8$ and $\alpha = 0$ (i.e. the divergence is logarithmic). The second idea is that of the universality class. It is believed that, in general, the values of the critical indices depend only on very few things such as the dimensionality of the system, the number of components of the order parameter and some basic symmetries. Thus many systems which appear superficially to be quite different will undergo transitions with the same values of the critical indices. All such transitions are said to be in a particular universality class which is characterized by certain values of the critical indices.

My interest in this subject arose from the transition to the $\sqrt{3} \times \sqrt{3}$ structure which helium undergoes. This transition was first observed by specific heat measurements.[1] The results are shown in Fig. 15. They were originally compared to a logarithmic divergence which they fit fairly well. Most physicists who thought about this experiment believed that the specific heat should be logarithmic. If I were asked at this time why this should be so I would have probably responded that the system can clearly be modeled by a lattice gas which, as is well known, can be mapped to an Ising model. Onsager solved the Ising model and a logarithmic specific heat resulted. If my questioner pointed out that the lattice gas with repulsive nearest-neighbor interaction, used to describe the helium system, mapped to an <u>antiferromagnetic</u> Ising model, not the ferromagnetic Ising model that Onsager solved, I would have noted that the transitions of the two models on the square lattice have the same logarithmic singularity. If my

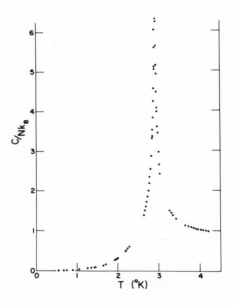

Fig. 15. Specific heat of ^{4}He adsorbed on Grafoil at a density of $0.064\mathring{A}^{-2}$.

questioner were so rude as to note that the lattice of adsorption
sites provided by the graphite was <u>triangular</u> and not square I
would have responded to the effect that the lattice geometry was
irrelevant and muttered something about universality.

It was Alexander[15] who pointed out that this line of reasoning
was false because this particular transition could not be in the
universality class of the ferromagnetic Ising model. The $\sqrt{3} \times \sqrt{3}$
state has three degenerate ground states while the ferromagnetic
Ising model has only two. Thus the symmetries of the two systems
are completely different. With that, Bretz[16] remeasured the
specific heat singularity using a somewhat more homogeneous substrate than the Grafoil used previously and found a value of α of
0.36, the largest value of any α ever reported and certainly not
zero! This established conclusively that the transition to the
$\sqrt{3} \times \sqrt{3}$ was in a different universality class than that of the
ferromagnetic Ising model.

With that particular cat out of the bag one would like to
know what other felines might be hidden therein. In particular
one would like to know what order-disorder transitions can be
continuous, what universality classes do they represent and, if
possible, what are the exponents of these classes. In effect a
guide or table is desired to indicate to the experimentalist what
transitions to what ordered states are in what class. It is the
object of the remainder of this lecture to prepare such a table.

The way this is done is to make use of the phenomenological
theory of Landau and Lifshitz.[17] Consider the density of the adsorbed system which, in this lecture, I shall denote $\rho(\bar{r}_j)$ where
$\bar{r}_j$ is the position of an adsorption site. Above T_c, $\rho(\bar{r}_j)$ has
the symmetry G_o of the substrate. Below T_c, however, it can be
written

$$\rho(\bar{r}_j) = \rho_o(\bar{r}_j) + \delta\rho(\bar{r}_j)$$

where $\rho_o(\bar{r}_j)$ has the symmetry G_o and $\delta\rho(\bar{r}_j)$ transforms like G, a
subgroup of G_o. This additional piece can always be expanded in
terms of the complete set of basis functions $\phi_{\ell i}(\bar{r}_j)$ of the irreducible representations of G_o,

$$\delta\rho(\bar{r}_j) = \Sigma' \Sigma \; C_{\ell i} \phi_{\ell i}(\bar{r}_j) \qquad\qquad (2.1)$$
$$\phantom{\delta\rho(\bar{r}_j) = } {\scriptstyle \ell \; i}$$

where $C_{\ell i}$ are coefficients, ℓ labels the representation and i the
function of a given representation. The prime on the sum indicates
that the unit representation is excluded. For the simple Bravais
lattices which I will be discussing for the most part, a given
representation is completely described by the m independent wave-
vectors $\bar{k}_1$, $\bar{k}_2$, ...$\bar{k}_m$ of the "star" of $\bar{k}$. By independent I mean

that the wavevectors are not related to one another by a reciprocal
lattice vector of the substrate. The integer m is the dimensionality
of the representation. The functions $\phi_{\ell i}(\bar{r}_j)$ can be taken to be
$\sin \bar{k}_i \bar{r}_j$ or $\cos \bar{k}_i \bar{r}_j$. According to the Landau-Lifshitz theory a
transition can be continuous if certain conditions are satisfied.
The first of these is that $\delta\rho(\bar{r}_j)$ transform like a single irreducible representation of G_o so that

$$\delta\rho(\bar{r}_j) = \sum_{i=1}^{m} C_{\bar{k}_i} \phi_{\bar{k}_i}(\bar{r}_j) \tag{2.2}$$

An order parameter of m components can be defined by

$$\psi_{\bar{k}_i} = \sum_r \phi_{\bar{k}_i}(\bar{r}) n(\bar{r}_j) \tag{2.3}$$

where $\rho(\bar{r}_j)$ is the thermal expectation value of $n(\bar{r}_j) = 0$ or 1.
The thermal expectation value of all components of the order parameter vanish above T_c.

There are still an infinite number of irreducible representations to be considered. The second and third rules limit the
allowable representations drastically. The second rule states
that the representation should be one for which it is not possible
to construct an invariant which is of third order in the $\psi_{\bar{k}_i}$.
While apparently correct in three dimensions, this rule is known
to be violated in some cases in two dimensions and so we do not
employ it. That leaves only the third rule which states that if
T is the representation in question, then the antisymmetric part
of T^2 should not contain the vector representation. The essence
of this statement is that if the transition has anything to do
with the symmetry of the substrate then the vectors $\bar{k}_i$ character-
izing the representation whould occur at points of high symmetry
in the Brillouin zone of the substrate. Only such representations
can satisfy this rule, the so-called Lifshitz condition. There
are continuous transitions which violate this rule[18] but the
ordered state in this case is characterized by a k vector which
changes continuously with temperature. This would show up by the
change with temperature of the position of the superlattice spots
in a scattering experiment. Such a state would not be commensurate
with the substrate over a finite range of temperatures. We exclude
all such transitions from consideration.

Classification proceeds as follows. We consider a given
symmetry G_o of the adsorption sites provided by the substrate.
There are only five such symmetries as there are only five Bravais
lattices in two dimensions. For the given space group we look at
the $\bar{k}$ values of high symmetry and construct the basis functions
$\phi_{\bar{k}_i}$ of the representations. As noted earlier, they are either

sin or cos functions. Substituting these into (2.2) we obtain the
form of the ordered state. Next, in order to determine the universality class of the transition, we construct from the order-para-
meter components $\psi_{\bar{k}_i}$ the Landau-Ginzburg-Wilson (LGW) Hamiltonian.
In practice this reduces to constructing all products of three or
four $\psi_{\bar{k}_i}$ which are invariant under the operations of G_o.
Although there is no systematic way of doing this, invariance
under translations requires a third-order invariant to be of the
form $\psi_{\bar{k}_1} \psi_{\bar{k}_2} \psi_{\bar{k}_3}$ where $\bar{k}_1 + \bar{k}_2 + \bar{k}_3$ equals a reciprocal lattice
vector. A similar restriction holds for higher-order invariants
too. This simplification permits such invariants to be obtained
by inspection. We compare the resulting LGW Hamiltonian with those
describing known models. If the Hamiltonian of the system of
interest is identical to that of one of these models, we identify
the universality class of the transition of interest with that of
the known model. If we are fortunate the exponents of the model
are known. As an example of the procedure consider the symmetry
group P2mm, the symmetry group of a rectangular array of adsorption
sites. The Brillouin zone is shown in Fig. 16a. There are three
one dimensional representations whose k values occur at points of
high symmetry. They are labeled p, q, s. (To see that the representations are one-dimensional apply to any of them the point group
operations of the rectangle. One obtains no independent vectors,
only ones related to the original vector by a reciprocal lattice
vector.) Consider the representation labeled $\bar{p}$. The one basis

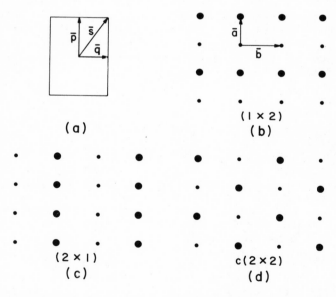

Fig. 16. Adsorption sites form a lattice with space group P2mm;
(a) the Brillouin zone and the vectors $\bar{p},\bar{q},\bar{s}$ that correspond to the structures of (b), (c), and (d), respectively.

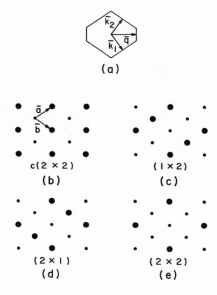

Fig. 17.   (a) Brillouin zone of C2mm lattice: $\bar{q}$ corresponds to
the structure of (b); the structures of (c) and (d) are
generated by the two-dimensional representation that
belongs to $\bar{k}_1$ and $\bar{k}_2$. The structure in (c) can be
reached continuously even though in general it belongs
to two irreducible representations. See Ref. 19.

function is $A \cos \bar{p} \cdot \bar{r}$ so that

$$\rho = \rho_o + A \cos \bar{p} \cdot \bar{r} .$$

As $\bar{p}$ is one-half a reciprocal lattice vector, the state looks like that of Fig. 16b. It's easy to see that there are no odd-order invariants and that the even order invariants of order ℓ are simply $(\phi_{\bar{p}})^{\ell}$. The LGW Hamiltonian is that of the Ising model. Thus a transition to the (1x2) state can be continuous and is predicted to be in the class of the ferromagnetic Ising model. The exponents of the transition will be $\alpha=0$ $\beta=1/8$. A more detailed analysis is presented in the papers by Domany, Schick, Walker, and Griffiths[19] and Domany and Schick[20] so I will only summarize the results here.

For the space group P2mm we find that transitions to the three ordered states of Fig. 16 can be continuous and that all are Ising-like. That these three transitions should all have the identical critical behavior is an example of universality.

Results for the simplest space group, that of the skew lattice P2ll, are identical to those for P2mm. The analogous three ordered states should be obvious.

The space group C2mm is that of the centered rectangular lattice, an example of which is provided by the (110) plane of an fcc lattice. The results of this space group are identical to those of P4mm, the group of a square array, so I will discuss the latter. The results for C2mm are shown in Fig. 17 and for P4mm in Fig. 18. There is a one-dimensional representation characterized by the vector $\bar{q}$ leading to the c(2x2) state shown in (b). The transition to this state is also in the universality class of the Ising model. There is also a two-dimensional representation spanned by $\bar{k}_1$ and $\bar{k}_2$. Ordered states which they give rise to generally have a (2x2) symmetry; i.e. one with an arbitrary unit cell of four atoms. The two primitive lattice vectors of this structure are each twice as large as those of the substrate. Three special cases of ordered structures obtained from these k vectors are shown in (c), (d), and (e). The LGW Hamiltonian is that of the x-y model with cubic (square?) anisotropy, that is, two perpendicular directions are singled out. This is an extremely interesting model because it is believed that its transitions are non-universal.[21] The exponents α and β depend on all those details which universal values ignore. It would be extremely interesting then to compare the exponents from, say, two (2x2) transitions undergone by different adatoms or to monitor the exponents of a given transition while the coverage and thus the spreading pressure was changed. Currently the system of O on W(110) which undergoes a transition to a (2x1) structure is being studied by Lagally[22] and co-workers. It would be useful to have a physisorbed system that could be studied by specific heat measurements which would undergo such a transition. But first a

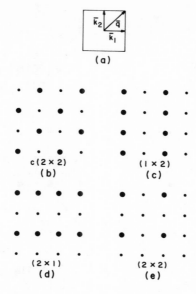

Fig. 18.   Brillouin zone for P4mm (a); the structure (b) corre-
           sponds to q; (c) and (d) to k_1 and k_2 respectively.
           The structure in (e) can be reached continuously even
           though in general it belongs to two irreducible repre-
           sentations. See Ref. 19

substrate with C2mm or P4mm symmetry is needed. Perhaps MgO will
prove to be sufficient.

      Lastly there is the symmetry group of the triangular lattice
P6mm. There is a two-dimensional and a three-dimensional representation to investigate as shown in Fig. 19a. The ordered state
corresponding to the two dimensional representation is the $\sqrt{3}$ x $\sqrt{3}$
observed in adsorbed helium. The LGW Hamiltonian is that of the
three-state Potts model which was first pointed out by Alexander.[15]
This is a very interesting model. I believe from some recent work
of den Nijs[23] and of Nienhuis, Berker, Riedel and myself[24] that the
exponents of this model are known. They are $\alpha = 1/3$ and $\beta = 5/48$
$= 0.104....$ As noted earlier Bretz[16] measured $\alpha = 0.36$. Recently
Horn, Bergeneau, Heiney and Hammond[25] determined β from x-ray
scattering and found $\beta = 0.09 \pm .03$. These values tend to corrob-
orate the underlying assumptions of the classification scheme
presented here.

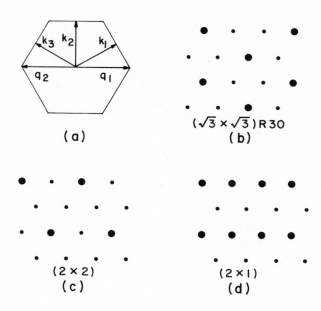

Fig. 19.   (a) Brillouin zone for P6mm.   (b) The structure that
belongs to the two-dimensional representation with $\bar{q}_1$,
$\bar{q}_2$.   (c) and (d) belong to the three-dimensional
representation with k_1, k_2, k_3.

The three dimensional representation can give rise to the
structures illustrated in Fig. 19c and d.   Examination of
the LGW Hamiltonian indicates that just below T_c the ordered
structure will have, in general, a (2x2) structure and not the
special case of the (2x1).   The LGW Hamiltonian is that of the
four-state Potts model which has been the subject of considerable
theoretical interest.[24]   Its exponents are believed to be $\alpha = 2/3$,
$\beta = 1/12$.   It would be exciting to measure the specific heat of a
system which would undergo such a transition.   Clearly one needs
an atom or molecule sufficiently large that there are nearest-
and second-nearest neighbor repulsions.   Xenon is the largest
rare gas atom one is likely to work with and it does not undergo
a transition to a (2x2) state.   Perhaps a large molecule should
be tried.   Another scheme for preparing a four-state Potts transition is presented below.   First let me present in Table I the
transitions which can be continuous.   The table is given in
Fig. 20.   All transitions not in this table are expected to be
first order.   By comparing the values of α and β in each class
the advantage of working with specific heat is easily seen.

TABLE 1

Substrates (first column) of various symmetries. Universality classes that can be realized and the expected critical behavior are listed in first row. The entries identify the corresponding ordered superlattice structures.

	Ising $\alpha=0\,(\log)$ $\beta=1/8$	X-Y with cubic anisotropy nonuniversal	Three-state Potts $\alpha=1/3$ $\beta=5/48$	Four-state Potts $\alpha=2/3$ $\beta=1/12$
P2mm Ex fcc(110)	(2×1) (1×2) $c(2\times2)$			
C2mm Ex bcc(110)	$c(2\times2)$	(2×1) (2×2)		
P4mm Ex fcc(100) bcc(100)	$c(2\times2)$	(2×1) (2×2)		
P6mm Ex bcc(111) fcc(111) Graphite			$(\sqrt{3}\times\sqrt{3})\,R30°$	(2×2)

   Thus far I have considered substrates which present simple
Bravais lattices of adsorption sites. Eytan Domany and I have
recently extended the above analysis to other cases of interest.[20]
One of these is the honeycomb lattice, a triangular Bravais lattice
with a basis of two sites per lattice point. The states which can
be reached by continuous transitions are shown in Fig. 20. The
transitions are different for honeycomb lattices encountered in
chemisorption by cleaving an fcc or bcc along a (111) plane than
for those encountered in physisorption by preplating a substrate
with a close-packed layer. I will review the physisorption results
here. Consider Fig. 20a first. The transition to this (1x1) structure is predicted to be Ising-like. This suggests an interesting
experiment[26] recently performed very beautifully by Tejwani and
Vilches.[27] They first adsorbed helium on graphite and observed
the √3 x √3 transition. The specific heat results shown in Fig. 21
yielded $\alpha = 0.28$, not so far from the value of 1/3 expected for the
three state Potts model. They then removed the helium and preplated
the graphite with Kr, which presents a honeycomb array of adsorption
sites. They then reintroduced the same amount of helium as before

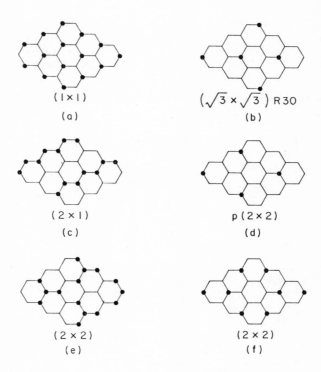

Fig. 20. Superlattice structures on the honeycomb lattice corre-
          sponding to k vectors of Fig. 19(a).

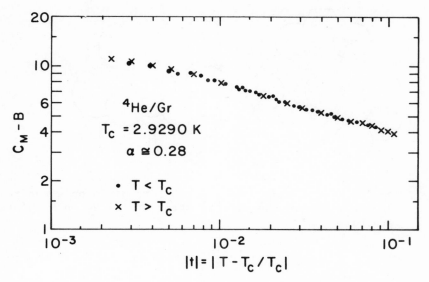

Fig. 21.   Specific heat of ^{4}He adsorbed on graphite foam at the
           transition to the $\sqrt{3}$ x $\sqrt{3}$ structure which is shown in
           the left side of Fig. 22. The specific heat (minus a
           constant background) is displayed versus reduced tem-
           perature on a log–log plot.

and observed the (1x1) transition (see Fig. 22). The specific
heat results are shown in Fig. 23 and are well fit by a logarithmic
singularity! This beautiful experiment convincingly demonstrates
the crucial role of symmetry in determining the universality class.
Next consider Fig. 20f. Transitions to this (2x2) structure are
again in the class of the four-state Potts model.[26] This (2x2)
transition could be expected to occur in a system in which atoms
A were used to preplate graphite and then atoms B, slightly larger
than A, were adsorbed on top of them. For thermodynamic stability
A must be more strongly bound to the graphite than B. One possible
candidate[26] is N_2 adsorbed on Kr plated graphite as N_2 is slightly
larger than Kr and it is much more weakly bound to graphite. The
slight asymmetry of N_2 is of no concern as long as ordering of the
direction of the molecules takes place at a lower temperature than
that to the (2x2) structure. The above experiment may not work.
The N_2 might ignore the Kr sites entirely and form an incommensurate solid. On the other hand, the results if it worked would
be sufficiently exciting that a quick-and-dirty experiment to see
whether the ordering does occur seems justified.

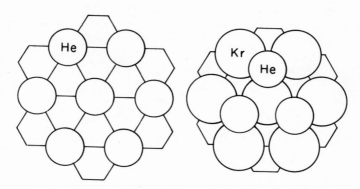

Fig. 22.   On the left is the $\sqrt{3} \times \sqrt{3}$ structure of He on graphite.
On the right is the $\sqrt{1} \times \sqrt{1}$ structure of He on Kr-plated
graphite.

        Structures of Fig. 20c, d, and e can be reached by a continuous
transition in the universality class of the Heisenberg model with
cubic anisotropy, a model about which very little[28] is known at
least in the way of exponents.   Structure d can also be reached by
two transitions: the first to a structure like f via a four-state
Potts transition followed by an Ising transition to d.   Finally the
$\sqrt{3} \times \sqrt{3}$ structure of b is in the class of a model studied by
Griffiths and myself[29] for reasons only a theorist could love.   It
was amusing to find it arise here.   I should note that the Lifshitz
condition is <u>not</u> obeyed here and the transition is expected to be

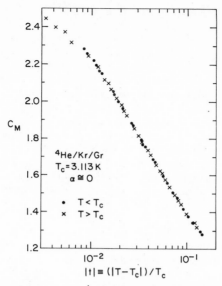

Fig. 23.   Specific heat of ^{4}He adsorbed on Krypton-plated graphite
foam at the transition to the $\sqrt{1}$ x $\sqrt{1}$ structure shown in
the right side of Fig. 22. The specific heat is displayed
versus reduced temperature on a semi-log plot.

first order. Of course all transitions other than those discussed
above are again expected to be first order. This concludes results
for the honeycomb lattice. Domany and I also considered the transitions of molecules like Br_2 and the transitions from one ordered
structure to another. The interested reader is referred to the
paper[20] for details.

In conclusion we have found that continuous order-disorder transitions of simple adsorbed atoms can be in the universality classes
of the following models: Ising, x-y and Heisenberg both with cubic
anisotropies and three- and four-state Potts model. All of these
models are the objects of much current study and their realization
by easily studied adsorbed systems is of great interest.

CALCULATION OF ORDER-DISORDER PHASE DIAGRAMS

In the last lecture I indicated how one could determine the universality class of an order-disorder transition and therefore the value of the critical exponents. However, one often wants far more information than just this set of numbers. For example the value of the transition temperature and its dependence on coverage may be desired or any other quantity which is embodied is the phase diagram of a particular system. Unfortunately for the theorist this means real work (unlike the unreal work of the Landau theory of the last lecture in which one puts in very little and gets out a great deal). However, the path to be taken is well known in principle. All information must emerge from a calculation containing two ingredients; first a Hamiltonian that describes the system accurately and, second, Statistical Mechanics.

Adsorbed systems have the great advantage that they can be rather well described by a lattice gas Hamiltonian due to the presence of adsorption sites provided by the substrate. At least this description should be a good one in the density and temperature regime in which the commensurate ordered phase is experimentally observed.

The basic variables of the lattice gas model are the site occupation numbers n_i which take the values zero and unity. The lattice-gas Hamiltonian has the form

$$\mathcal{H} = \beta\mu\sum_i n_i - \beta\Sigma' v_{ij} n_i n_j - \beta\Sigma' v_{ijk} n_i n_j n_k + \ldots \tag{3.1}$$

where μ is the chemical potential, v_{ij} is the two-particle interaction, v_{ijk} the three particle interaction and so on. The prime on the second and third sums indicates summation over distinct pairs and triples respectively. Note that I am following what has become common practice in employing the reduced Hamiltonian which is equal to the usual Hamiltonian of classical mechanics multiplied by $-\beta$, the negative of the inverse temperature. This system can be mapped onto an Ising model by introducing at each site a spin σ_i which takes the values ± 1 and is related to n_i by

$$\sigma_i = 1 - 2n_i$$

The Ising Hamiltonian takes the form

$$\mathcal{H}_N = NC + H\Sigma\sigma_i + \Sigma' K_{ij}\sigma_i\sigma_j + \Sigma' K_{ijk}\sigma_i\sigma_j\sigma_k + . \tag{3.2}$$

where
$$C = \frac{\beta}{2}(\mu - \frac{1}{4}\sum_j v_{ij} - \frac{1}{24}\sum_{jk} v_{ijk} + \cdot) \quad,$$

$$H = \frac{\beta}{2}(\mu - \frac{1}{2}\sum_j v_{ij} - \frac{1}{8}\sum_{ij} v_{ijk} + \cdots) \quad,$$

$$K_{ij} = -\frac{\beta}{4}(v_{ij} + \frac{1}{2}\sum_k v_{ijk} + \cdot\cdot) \quad,$$

$$K_{ijk} = -\frac{\beta}{8}(v_{ijk} + \cdot\cdot) \quad.$$

The number of spins is N. In the case in which there are two-body interactions only, the phase diagram is symmetric about zero mag-netic field H which corresponds to a value of the chemical poten-tial $\mu_0 = \frac{1}{2}\sum_j v_{ij}$. This chemical potential corresponds to a lattice-gas density of 1/2.

With the Hamiltonian defined, we need only calculate the free energy per spin according to

$$e^{Nf(K)} = Tr\ e^{\mathcal{H}_N(K;\{\sigma\})} \tag{3.3}$$

where K here stands for the set of interaction constants in (3.2) and f is the usual free energy per spin multiplied by $-\beta$. Eq. (3.3) is of course the great expression of statistical mechanics linking thermodynamics on the left which depends on but a few quantities K with classical mechanics on the right depending on a multitude of uninteresting spins σ as well as on the K. To carry out the trace in (3.3) is very much like conducting an election: the individual votes are of little interest, all that matters is the simple outcome of the final count. Traditional methods of statistical mechanics focus on carrying out this trace or election.

Renormalization group (RG) procedures approach the problem slightly differently.[30] Let me continue with the election metaphor. There are other methods of determining the will of the people than a direct one-step election. For example, the people can elect a congress. If the election procedure is sufficiently good, the will of the people will be carried out in the congress. This system has a nice advantage. It is not merely that there are fewer congressmen than people so that it is easier to carry out a vote among them. One need not even do that. If one has a really good election procedure, it can be used by the congress to elect a committee and by the committee to elect a chairman and then all we need do is to ask him!

The analogue of the election in Statistical Mechanics is a

mapping from N spins σ to N' spins σ' and the election procedure is defined by a projection operator $P(\{\sigma\},\{\sigma'\})$ such that

$$\text{Tr}' \ P(\{\sigma\},\{\sigma'\}) = 1 \tag{3.4}$$

Returning now to Eq. (3.3) we proceed as follows

$$e^{Nf(K)} = \text{Tr}\, e^{\mathcal{H}_N(K,\{\sigma\})} \ ,$$

$$= \text{Tr}\,\text{Tr}' \ P(\{\sigma\},\{\sigma'\}) e^{\mathcal{H}_N(K,\{\sigma\})} \ ,$$

$$= \text{Tr}'\,\text{Tr}\, P(\{\sigma\},\{\sigma'\}) e^{\mathcal{H}_N(K,\{\sigma\})} \tag{3.5}$$

where I have used Eq. (3.4) in the second line. The central statement of the renormalization group is that

$$\text{Tr}\, P(\{\sigma\},\{\sigma'\}) e^{\mathcal{H}_N(K,\{\sigma\})} = e^{\mathcal{H}_{N'}(K',\{\sigma'\}) + Ng(K)} \tag{3.6}$$

where the Hamiltonian on the left and right are of exactly the same form. The values of the interaction constants in the new Hamiltonian are different however. Clearly if Eq. (3.6) is true the new set of interaction constants must depend upon the old. Symbolically

$$K' = K'(K) \tag{3.7}$$

a set of equations called the recursion relations.

The function g(K) is a part of the free energy one obtains in return for reducing the degrees of freedom from N to N'. Clearly if one had carried out the trace in one step, Eq. (3.6) would still be correct but all K' would vanish and g would be the entire free energy. In general then g is just a piece of the free energy. The total free energy is easily obtained from g however. Upon substituting Eq. (3.6) into (3.5) we obtain

$$e^{Nf(K)} = e^{Ng(K)}\,\text{Tr}'\, e^{\mathcal{H}_{N'}(K',\{\sigma'\})} \ ,$$

$$= e^{Ng(K) + N'f(K')} \ .$$

This follows from the fact that $\mathcal{H}_{N'}$ is of precisely the same form as $\mathcal{H}_N$ and the definition of the free energy Eq. (3.3). We have

$$f(K) = g(K) + \frac{N'}{N} f(K') \tag{3.8}$$

We now iterate the procedure (Congress elects a committee). From
K we have gone to K'. The recursion relation (3.7) tells us that
from K' we can go to K" and so on.

Iteration of Eq. (3.8) yields

$$f(K) = \sum_{\ell=0}^{\infty} \left(\frac{N'}{N}\right)^{\ell} g\left(K^{(\ell)}\right)$$
(3.9)

where $K^{(\ell)}$ is the ℓ'th iterate of the initial value $K^{(0)} \equiv K$.
Note that Eq. (3.9) expresses the non-analytic function $f(K)$ as a
sum of analytic functions $g(K^{(\ell)})$. The non-analyticy arises because
the number of terms in the sum is infinite.

Having seen that RG methods work in principal, we might argue
that in practice they are no improvement because carrying out the
trace in Eq. (3.6) is just as formidable as in the original problem,
Eq. (3.3). This is correct and approximation methods must be
employed. The crucial difference is that in older methods one
tried to approximate the free energy itself which is a non-analytic
function of its arguments. In RG methods one approximates the
recursion relations Eq. (3.7) which are analytic. It is far easier
to systematically approximate an analytic function than a non-
analytic one and therein resides the power of the renormalization
group. Typically one proceeds by considering a small finite
system governed by the Hamiltonian of interest. As the system is
finite, the trace can actually be carried out and the recursion
relations and functions $g(K)$ can be obtained. They are approximations to those which would be obtained from the infinite system.
The approximate function $g(K)$ and the approximate recursion relations are then used in Eq. (3.9) to obtain a free energy which,
like the desired exact one, will be a non-analytic function of
its argument.

Let us now apply these methods to the order-disorder transi-
tion to the $\sqrt{3} \times \sqrt{3}$ structure in the helium system. It should be
clear that in order for such a structure to form, the nearest-
neighbor interaction between adsorbed atoms must be repulsive.
Further attractive interactions contribute nothing to the formation
of this structure but they do alter the phase diagrams. I shall
ignore these attractions for the helium case where they are weak,
although not ignorable. This effect can be added later. In the
Kr system where these interactions are quite strong, they will be
dealt with ab initio.

The lattice-gas Hamiltonian then is

$$\mathcal{H} = \beta\mu \sum_i n_i - \beta v \sum' n_i n_j$$

where v is positive. This maps to the Ising Hamiltonian

$$\mathcal{H} = H \sum_i \sigma_i + K \sum{}' \sigma_i \sigma_j$$

where $H = \frac{\beta}{2} (\mu - 3v)$

and $K = -\frac{\beta}{4} v$

As v is positive, K is negative or antiferromagnetic. We thus
deal with a triangular Ising antiferromagnet. This can be solved
exactly in zero field which corresponds to the physically uninteresting density of 1/2. (It is uninteresting because long
before this high density is reached, the lattice-gas model will
fail to describe the physical system. Rather, some adatoms will
be in a second layer sitting on top of an incommensurate solid.)
Nonetheless the result is interesting theoretically. The exact
result[31] is that the transition temperature vanishes at this
density. The reason for this is clear. Given equal numbers of
up and down spins there is simply no unique ground state. This
is reflected in the finite entropy per spin[32] at zero temperature,
S/N = 0.324. The system is an example of one which, in current
jargon, is "totally frustrated" - a concept employed in descriptions
of spin glasses. For positive values of the field H which are not
too large, corresponding to densities which are not too low, the
ground state will be that of Fig. 13.

     A most important consideration in choosing an RG projection
operation is that the ground states of the renormalized system
look just like those of the original   system. It is a simple
matter to preserve the symmetry of ferromagnetic ground states
but not such an easy one to preserve the three sublattice structures
of Fig. 13. There are at least two ways to proceed. One can map
this system onto a ferromagnetic one in a one-step transformation,
and then proceed as usual. This is the method used in the Kr
system described below. The other method is to build the sublattice structure into the projection operation itself. This is
the procedure followed for the helium case. Mentally divide the
triangular lattice into three sublattices A, B, and C, where three
lattice sites which are all nearest neighbors of one another belong
to different sublattices. The problem of finding an appropriate
projection operator would be phrased in the political analogy
somewhat as follows. A minority of one third of the citizens feel
very differently about some issue than the two-thirds majority.
How are voting districts to be drawn so that their voice in congress
will be heard proportionately to their members? One must "gerry-
mander", that is, draw the districts so as to exclude the members
of the majority from some districts which will represent the
minority. It surprises me that this solution was introduced into
the RG by J. M. J. van Leeuwen[33] who, I believe, was not raised in

Boston but in Holland where I thought the politics to be less
Byzantine. In any event, the districting method chosen is shown
in Fig. 24. The approximate recursion relations and functions g
were those of a system of three cells, one from each sublattice or
nine spins in all. The system was treated as it it were periodically continued.

    The results of this very simple approximation are as follows.[34]
The phase diagram in the temperature density plane is shown in
Fig. 25. Note that the exact result that $T_c(n=1/2) = 0$ is repro-
duced. The entropy we obtain at zero temperature at this density ·
differs by about 0.3% from the exact result. The transitions
temperature rises as the density is decreased from 1/2, peaks
extremely near the density n = 1/3 as expected and falls to zero
again at n = 0.24. This value is close to the series results[35]
which place this density of minimum ordering at 0.27. This latter
number is in satisfactory agreement with neutrons scattering experiments[4] on He[3] where the ordered phase is seen at densities greater
than 0.283 but is not seen for densities less than 0.25, at least
at 1K. The value of the one adjustable parameter $|J| = K/\beta$ is
determined by matching the maximum transition temperature to its
experimentally observed value. This yields J of minus two degrees
Kelvin or a nearest neighbor repulsion of 8 degrees, a reasonable
value. The satisfactory agreement between the theoretically and
experimentally determined phase boundary is shown in Fig. 26. One
can now calculate any thermodynamic function desired. Theoretical
values of the specific heat at constant density are compared to

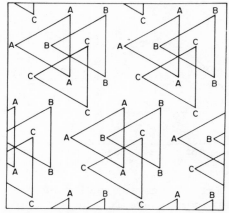

Fig. 24.   Sites of the triangular lattice divided into three
           sublattices A, B, and C and grouped into inter-
           penetrating three-cell clusters.

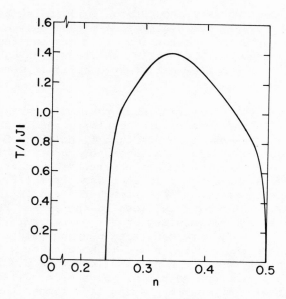

Fig. 25.   Phase diagram of the triangular lattice gas with nearest-
neighbor repulsive interaction of 4J.   Diagram is in the
temperature-density plane.

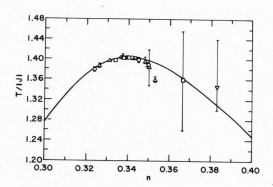

Fig. 26.   Detail of Fig. 25 showing region near the peak n_{max} =
0.337.   Data points are for ^{4}He on Grafoil: circles,
cell A; triangles, cell B; both from Bretz et al.,
Ref. 1; squares, Bretz Ref. 16; inverted triangles,
Herring et al., Ref. 2.

experimental values taken on Grafoil[1] at a series of densities in
Figs. 27 to 30. Agreement is particularly good on the low side of
the transition, less so on the high side. It is interesting to
note that the amplitude of the singularity decreases as n departs
from 1/3 in accordance with experimental observation. The only
major failing of the approximation lies in the approximate critical
exponent α = -0.09 as opposed to the theoretically expected value
of 1/3 or the experimentally observed value of 0.36. Thus,
sufficiently close to T_c, the approximate results must become poor.
This is seen in Fig. 31 where the experimental data is that of
Bretz[16] using the ZYX substrate.

Adsorption isotherms can be calculated[36] and one is shown in
Fig. 32. The compressibility has a singularity of the form $|t|^{-\alpha}$
as does the specific heat. If our value of α were positive as it
should be, the compressibility, shown in the lower part of the
figure, would diverge at the densities of the phase boundary at
this temperature. These densities are indicated by the two small
vertical marks on the adsorption isotherm. The only effect on the
isotherm if α were positive in the calculation is that the slope
of the isotherm would be horizontal at those two densities. That
is all, and is the reason why continuous phase transitions are
difficult to see by adsorption isotherms. This concludes my
discussion of the helium calculation. The sublattice RG method
employed alone can be used for other cases of interest as in the
formation of the (2x2) structure on the triangular and honeycomb
lattices. This particular case has been worked out by J. S. Walker
and myself.[37]

Let us now consider the effect of the further neighbor
attractive interactions which simply cannot be ignored for the Kr

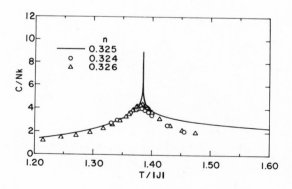

Fig. 27. Heat capacity per particle at constant coverage. Data
          from Bretz et al., Ref. 1 for densities somewhat below
          the maximum n_{max} = 0.337.

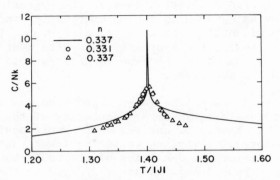

Fig. 28. Heat capacity per particle at constant coverage at n_{max}. Data are from Bretz et al, Ref. 1.

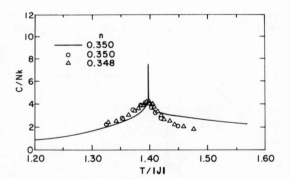

Fig. 29. Heat capacity per particle at constant coverage. Data from Bretz et al., Ref. 1 for densities somewhat above the maximum.

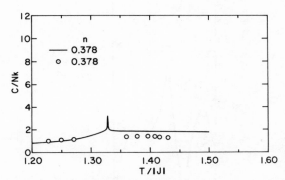

Fig. 30. Heat capacity per particle at constant coverage. Data
          from M. Bretz, Ph.D. thesis (University of Washington
          1971) (unpublished).

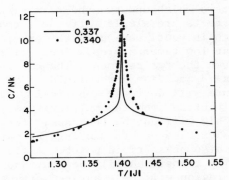

Fig. 31. Heat capacity per particle at constant coverage. Data
          for ^{4}He on ZYX from Ref. 16 at n_{max}. Theoretical curve
          shown is same as Fig. 28.

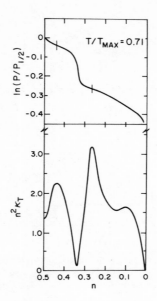

Fig. 32.   Upper part: calculated adsorption isotherm of the
           triangular lattice gas at T/T_{max}=0.71.   Lower part:
           calculated isothermal compressibility (multiplied by
           the square of the density) of the triangular lattice
           gas versus density N.

and N_2 adsorbed systems which undergo the $\sqrt{3}$ x $\sqrt{3}$ transition.   The
expected physical effects are simple.   If the attraction is weak
the transition, which is continuous at high temperatures, will
become first order at low temperatures.   There will be a multicritical point between the two regimes.   The coexisting phases
will be ordered $\sqrt{3}$ x $\sqrt{3}$ and disordered fluid.   The first-order
transition temperature will decrease monotically as the density
changes in either direction from the multicritical density.   For
strong attractions this will not be the case.   There will be a
critical point which enables one to distinguish gas and liquid
phases.   There will also be a triple point.   Again at higher temperatures the ordering transition will be continuous.   This region
and the first order region will be separated either by a multicritical point or a critical end-point.   There has been debate[38]
as to which of these two possibilities, i.e., weak or strong
attractions, is exhibited by Kr and N_2.

     This system has been approach via RG methods by Ostlund and
Berker.[39]   It would be tedious to use the sublattice method employed
in the helium problem and include the attractive interactions.

They therefore employ a one-step prefacing transformation which maps the Ising spins onto more complicated objects but which has the virtue that the $\sqrt{3} \times \sqrt{3}$ ordered state corresponds to the ferromagnetic state of these objects. As noted earlier, it is an easy matter to preserve a ferromagnetic symmetry for any kind of object. They divide the triangular system into cells of three spins which are all nearest-neighbors of one another. They further consider the case of infinite nearest-neighbor repulsions, a good first approximation for Kr and N_2. Therefore in lattice gas language the cell of three spins can be in four states. In the first, there are no atoms in the cell. In the remainder, there is one atom at one of the three positions. The first case is mapped to an empty state, or vacancy. The other three are represented by states of a vector which can point in three positions. Thus the n_i of the lattice gas has been replaced by variables $t_i = 0,1$ which reflect whether the cell is empty or full and, if the latter, a vector s_i that takes on three values. If the reader draws the possible configurations of two neighboring cells he will see that the ordered $\sqrt{3} \times \sqrt{3}$ structure corresponds to one in which the vectors in each cell are pointing in the same direction, i.e., a ferromagnetic ground state. Further the interaction between two vectors which point in the same direciton is that of the second neighbor attraction in the lattice gas. Interaction between vectors in different orientations however depends on the directions of the two vectors and can be either the nearest neighbor repulsion or the third or fourth neighbor attractions. Because of this dependence of the interaction on relative vector position the model Hamiltonian is not that of a dilute three-state Potts model as in an earlier exploratory calculation.[40] For the interaction, Ostlund and Berker take the experimental values of the Lennard-Jones interaction at the appropriate distances. They employ the Migdal approximation which is a particularly simple one. The kind of results they obtain for Kr and N_2 are shown in Fig. 33. A multicritical point is obtained but critical and triple points are not. They went on to include finite size effects by assuming that for a given chemical potential there would be a smearing ΔT of the temperature among the graphite platelets which would yield a variation Δn of the density about the phase boundary of the infinite system. These smeared boundaries are shown by dotted lines in Fig. 33. Note that in Figs. 33-35 a full monolayer is defined as one atom per three adsorption sites. The specific results obtained for Kr and N_2 including this effect are shown in Figs. 34 and 35. I believe that the fit to experiment is extremely good. The conclusion that Kr does not exhibit a triple point has been buttressed by recent data of Butler, Stewart and Griffiths.[41] It is my belief that the calculations for helium, Kr and N_2 indicate that the phase boundaries of order-disorder transiton can now be calculated with an accuracy that has been unusual heretofore in surface physics.

Finally let me briefly address the subject of finite size

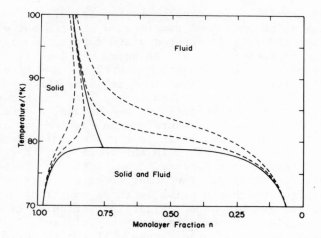

Fig. 33.   Temperature variation and finite size effects on the
           multicritical region of the phase diagram for krypton
           submonolayers on graphite.   Full curves correspond to
           an ideal thermodynamic system ($\Delta T=0$).   Dashed curves
           were calculated with temperature smearings of $\Delta T=0.1$
           and 0.8K.

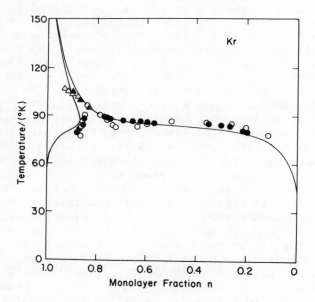

Fig. 34.   Comparison of experimental phase diagram with that
           predicted with $\Delta T=0.8$K.   System is Kr on graphite.

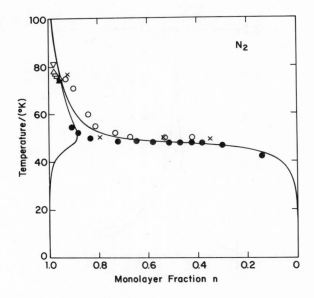

Fig. 35. Comparison of experimental phase diagram with that
          predicted with ΔT=0.8K. System in N_2 on graphite.

effects briefly introduced above. There are several ways one can
approach the calculation of thermodynamic quantities of finite
systems. If the properties of the infinite system can be calculated exactly, most likely those of the finite system can be also.
This is the case of the Ising model. The exact results for the
specific heat of the finite square Ising model with periodic
boundary conditions as obtained by Ferdinand and Fisher[42] are
shown in Fig. 36. The rounding of the specific heat peak and
shift in the temperature at which the maximum specific heat appears
are evident. The direction of this shift depends on the kind of
boundary conditions employed.

The extreme suitability of RG methods to the calculation of
such effects seems to have been recognized by many workers. The
idea is quite simple. One begins with a finite system of N_{INT}
spins and applies the RG transformation until only a few spins are
left. The free energy of this system is then calculated exactly.
The effect is to truncate the sum of Eq. (3.9) after a finite
number of terms of order $\ell \sim \log N_{INT}/\log(N/N')$. All thermodynamic
quantities are now analytic.

Berker and Ostlund[43] and Callaway and Schick[44] have carried

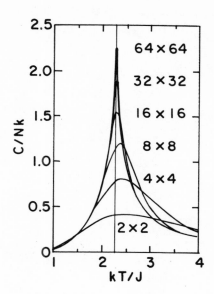

Fig. 36.   The specific heat per spin for small Ising lattices;
exact results for the nxn square lattice with periodic
boundary conditions are displayed for n=2, 4, 8, 16,
32 and 64. N=n^2. The limiting critical point is
marked by a vertical line.

out this procedure on respectively the $\sqrt{3}$ x $\sqrt{3}$ transition observed
by Bretz[16] and the (1x1) transition observed by Tejwani and Vilches[27]
for the helium on Kr plated graphite system. The results of both
calculations are qualitatively similar. If one starts with the
size system one believes one has experimentally, the calculated
specific heat maximum is about twice the size of the experimentally
observed one. Either the uniform regions are smaller than expected
or the fact that there are an ensemble of systems all with different
specific heats tends to reduce the maximum. It would be nice if
the latter were not the case for then comparison of experimental
and theoretical results would yield the effective sizes of the
experimental system.

     This is a good place to stop and summarize very briefly. The
phases of adsorbed systems are understood rather well today and
the general outlines of the transitions between these phases is
also clear. Adequate understanding as evidenced by the agreement
between theory and experiment is present only for the order-disorder
transition (in my unbiased opinion). Experiments to confirm the
dislocation theory of melting have yet to be performed on adsorbed

systems. The situation is worse for the commensurate-incommensurate transition where there seems to be positive disagreement between theory and experiment.[45] Many interesting experiments remain to be performed. These include the detection of the discontinuities in the elastic constants of the incommensurate solid on melting and the detection of the transition to the "hexatic" phase[14] which should exist when substrates with other than triangular symmetry are employed. There is also my pet, the detection of a transition in the class of the four-state Potts model. Multicritical points have yet to be explored. On the theoretical side it can be noted that the patchwork of theories, one for each kind of transition, is most unsatisfying. One would like a single framework describing on a microscopic level the statistical mechanics of commensurate as well as incommensurate structures. Then all transitions could be dealt with uniformly and the complete phase diagrams of individual systems could be calculated. Of course there is also the application of all the techniques which have proved so successful in physisorbed systems to the more difficult chemisorbed systems. But this is subject matter for another set of lectures which I gladly leave to Professor Bauer.

ACKNOWLEDGMENTS

I am pleased to thank my colleagues of the University of Washington, Greg Dash, Eytan Domany, Sam Fain, Bob Puff, John Rehr, Eberhard Riedel, Ed Stern, and Oscar Vilches, for generously sharing their knowledge of Surface Physics with me over the past several years. Much of the content of these lectures is due to Bob Siddon and Jim Walker whom I had the pleasure of supervising. Particular thanks are due Eytan Domany and Michael Wortis for patient instruction and joyous collaboration and Greg Dash for years of much appreciated encouragement.

REFERENCES

1. M. Bretz, J. G. Dash, D. C. Hickernell, E. O. McLean, and O. E. Vilches, Phys. Rev. A8:1589 (1973); Erratum Phys. Rev. A9:2814 (1974).
2. S. V. Hering, S. W. Van Sciver, and O. E. Vilches, J. Low Temp. Phys. 25:793 (1976).
3. R. L. Elgin and D. L. Goodstein, Phys. Rev. A9:2657 (1974).
4. M. Nielson, J. P. McTague, and W. Ellenson, Journal de Physique 38:C4-10 (1977).
5. D. E. Hagen, A. D. Novaco, and F. J. Milford in Proceedings of the Symposium on Adsorption-Desorption Phenomena, F. Ricca, ed., Academic Press, New York, 1972.
6. Scattering results are reported by G. Boato, P. Cantini, and R. Tatarek, Phys. Rev. Lett. 40:887 (1978) and G. Boato,

P. Cantini, R. Tatarek and G. P. Fletcher, Surf. Sci. 80: 518 (1979). For reconstruction of the potential, see W. Carlos and M. W. Cole, Surf. Sci. 77:L173 (1978). I am indebted to Prof. Boato for providing Fig. 7.

7. R. L. Siddon and M. Schick, Phys. Rev. A9:907 (1974).

8. J. G. Dash and M. Schick, in Physics of Liquid and Solid Helium (part II), K. H. Benneman and J. B. Ketterson, ed., Wiley-Interscience, New York (1978).

9. J. R. Owers-Bradley, B. P. Cowan, M. G. Richards, and A. L. Thomson, Phys. Lett. 65A:424 (1978).

10. J. J. Rehr and M. Tejwani, Phys. Rev. B20:424 (1978).

11. A. D. Novaco, Phys. Rev. 7A:1653 (1973).

12. M. D. Miller and L. H. Nosanow, J. Low Temp. Phys. 32:145 (1978).

13. J. M. Kosterlitz and D. J. Thouless, J. Phys. C6:1181 (1973); J. M. Kosterlitz, J. Phys. C7:1046 (1974).

14. B. I. Halperin and D. R. Nelson, Phys. Rev. Lett. 41:121 (1978); E41:519 (1978); D. R. Nelson and B. I. Halperin Phys. Rev. B19:2457 (1979).

15. S. Alexander, Phys. Lett. A54:353 (1975).

16. M. Bretz, Phys. Rev. Lett. 38:501 (1977).

17. L. D. Landau and E. M. Lifshitz, Statistical Physics, Addison-Wesley, Reading, Mass. (1969); Chap. XIII, D. Mukamel and S. Krinsky, Phys. Rev. B13:5065 (1976); also see Rev. 19 below.

18. S. Goshen, D. Mukamel, and S. Shtrikman, Int. J. Magn. 6:221 (1974).

19. E. Domany, M. Schick, J. W. Walker, and R. B. Griffiths, Phys. Rev. B18:2209 (1978).

20. E. Domany, M. Schick (to be published in Phys. Rev. B, 1979).

21. J. V. José, L. P. Kadanoff, S. Kirkpatrick, and D. R. Nelson, Phys. Rev. B16:1217 (1977).

22. J. C. Buchholz and M. G. Lagally, Phys. Rev. Lett. 35:442 (1975).

23. M. P. M. den Nijs (to be published in J. Phys. A, 1979).

24. B. Nienhuis, A. N. Berker, E. K. Riedel, and M. Schick, Phys. Rev. Lett. 43:737 (1979).

25. P. M. Horn, R. J. Birgeneau, P. Heiney, and E. M. Hammonds, Phys. Rev. Lett. 41:961 (1978).

26. E. Domany, M. Schick, and J. S. Walker, Phys. Rev. Lett. 38: 1148 (1977).

27. M. Tejwani and O. E. Vilches (to be published).

28. E. Domany and E. K. Riedel, Phys. Rev. B19:5817 (1979).

29. M. Schick and R. B. Griffiths, J. Phys. A10:2123 (1977).

30. There is by now a voluminous literature on the Renormalization Group. Two articles which I like are K. G. Wilson, Phys. Rev. B4:3174 (1971), for the basic ideas and Th. Niemeijer and J. M. J. Van Leeuwen in Phase Transitions and Critical Phenomena, C. Domb and M. S. Green, ed., Academic, London 1976 Vol. 6 for an excellent discussion of position-space

      methods.
31. R. M. F. Houtappel, Physica (Utrecht) 16:425 (1950).
32. G. Wannier, Phys. Rev. 79:357 (1950). The analytic expression
     for the entropy is given correctly in this paper but the
     numerical value is not. The correct value is given in the
     text of these lectures.
33. J. M. J. van Leeuwen, Phys. Rev. Lett. 16:1056 (1975).
34. M. Schick, J. S. Walker and M. Wortis, Phys. Rev. B16:2205
     (1977).
35. D. S. Gaunt, J. Chem. Phys. 46:3237 (1967).
36. I am indebted to Dr. A. J. Berlinsky for the one shown in
     Fig. 32.
37. J. S. Walker and M. Schick (to be published in Phys. Rev. B,
     1979).
38. Contrast Ref. 41 below with the arguments of T. T. Chung and
     J. G. Dash, Surf. Sci. 66:559 (1977).
39. S. Ostlund and A. N. Berker, Phys. Rev. Lett. 42:843 (1979).
40. A. N. Berker, S. Ostlund, and F. A. Putnam, Phys. Rev. B17:
     3650 (1978).
41. D. M. Butler, J. A. Litzinger, G. A. Stewart, and R. B.
     Griffiths, Phys. Rev. Lett. 42:1289 (1979).
42. A. E. Ferdinand and M. E. Fisher, Phys. Rev. 185:832 (1969).
43. A. N. Berker and S. Ostlund (to be published in J. Phys.,
     1979).
44. D. J. E. Callaway and M. Schick (unpublished).
45. P. W. Stephens, P. Heiney, R. J. Birgeneau, and P. M. Horn,
     Phys. Rev. Lett. 43:47 (1979).

DOMAIN WALLS AND THE COMMENSURATE PHASE

A. Luther, J. Timonen, and V. Pokrovsky

NORDITA

Blegdamsvej 17, DK-2100 Copenhagen Ø, Denmark

This article is, in a sense, a review of work to be done. Strange as this may sound, it seems that many techniques have been developed recently which hold great promise for a further understanding of the commensurate-incommensurate transition of adsorbed layers. It is the intent here to review these techniques, discuss the applications to surface layers, and report on progress, such as it is, with a critical eye to future potential.

The particular techniques have had their heyday in the related discipline of quantum field theory, and are therefore not as well known in the field of two dimensional statistic mechanics as might be desired. The similarity of these two viewpoints seems to be quite deep, and is intuitively plausible if one reflects on the statistical mechanics of domain walls. Although it will be clearer later in this review, this domain wall problem is the same problem as tracing the space-time history of a fermion. In both cases, the problem deals with indestructible lines in a two dimensional manifold.

Of course the techniques to be discussed here would be unnecessary if the statistical mechanics problem could be solved using more traditional methods. Nearly all of the methods in use up to the present involve minimizing the free energy in certain configurations.[1] The major difference between this "mean field" procedure and the fermion method, resides in the inclusion of fluctuations in this latter approach. The famous quantum mechanical "Zitterbewegung", or quantum fluctuations, correspond to precisely the domain wall fluctuations omitted in the minimization procedure.

The domain wall configurations in the commensurate-incommen-

surate phase transition are most easily visualized in a one
dimensional system. The earliest work on this general problem
discussed precisely this problem.[2] Imagine a chain of mass points,
connected by ideal springs, in a periodic potential, with the
period near to the natural spacing between the points. As the
spacing is increased, there is a competition between two energies.
The periodic part would like to keep the points in the minima,
that is periodic. The springs would prefer for the positions to
be that of the natural spacing. Eventually, it becomes energetically favorable for a local distortion to occur, in which the mass
points sit in the bottoms of the periodic potential both far to
the left and right of the disturbance, but are displaced smoothly
by one period. This "disturbance" has many names, and the choice
here is to call it a domain wall.

A proper formulation of this problem, in two dimensions,
includes the two dimensional topology of these walls, separating
regions in which the mass points are locally "in the minima", it
includes the fluctuations about the minimum of the free energy,
and the proper statistical weight to the various possible domain
configurations. It must also include the small amplitude fluctuations within the domains. The fermion, or field theory, picture
does all of this, and can be solved in many circumstances.

Before proceeding to the technical details, it might be
interesting to consider the types of models which can be solved,
and the properties of the solutions. Typically, the models
consist of small amplitude fluctuations, treated in continuum
elasticity theory,[3] and a periodic piece resulting from the substrate interaction with these fluctuations. The simplest model,
for pedagogical purposes, consists of the two dimensional square
lattice in a "corrugated roof" potential along one of the symmetry
directions. The Fourier transform density operator is given by:

$$\rho(\vec{q}) = \sum_{i} e^{i\vec{q}\cdot\vec{R}_i^{op}} \longrightarrow \int d^2x \; e^{i\vec{q}\cdot\vec{x} + i\vec{q}\cdot\vec{\phi}(x)} \qquad (1)$$

where $\vec{R}_i^{op}$ is the position operator of the i-th particle in the
layer, the sum is over all sites and $\vec{q}$ is the wave vector of
a particular component. In the continuum theory, the sum is
replaced by an integral, as shown on the right hand side of Eq.
(1). In this version, $\vec{x}$ represents the mean positions of the
particles, and $\vec{\phi}(x)$ the deviation from mean.

The statistical weight to be given to any configuration is
the Boltzmann factor containing the harmonic deviations from the
mean positions, and the interaction with the substrate. The free
energy functional for this model is:

$$F\{\phi\} = \int d^2x \; \mu (\vec{\nabla}\phi)^2 + \int d^2q \; V(\vec{q})\rho(\vec{q}) \qquad (2)$$

where μ is an elastic constant. For the corrugated potential, $2V(\vec{q}) = V\delta(q_y)[\delta(q_x-q_o)+\delta(q_x+q_o)]$, leading to the result:

$$F\{\phi\} = \int d^2x \; \mu(\vec{\nabla}\phi)^2 + V\cos(q_o x + \phi(x)) \qquad (3)$$

where $\phi(x)$ represents the displacements along the axis of the potential, and terms representing displacements along the other direction will be discussed below, along with the $(\vec{\nabla}\cdot\vec{\phi})^2$ terms responsible for a finite compressibility. The model of Eq. (3) should be understood as an example of the domain-fermion problem in particularly simple form.

It is necessary to compute the partition function for this model, which means integrating over all configurations of the displacement fields:

$$Z = \int \delta\phi \; e^{-\frac{F\{\phi\}}{T}} \qquad (4)$$

This is performed by solving an equivalent problem, that of evaluating the ground state energy of the quantum mechanical problem associated with Eq. (3), summarized by:

$$Z = \lim_{\beta_o \to \infty} tr \; e^{-\beta_o H} \qquad (5)$$

The Hamiltonian in question turns out to be the quantum sine--Gordon problem in a uniform external field

$$H = \tfrac{1}{2}\int dx \; (\partial_x\varphi - q_o)^2 + \pi^2 + V\cos\beta\varphi \qquad (6)$$

Here $\varphi(x)$ and $\pi(x)$ are canonical variables, $[\varphi, \pi] = i\,\delta(x-x')$ and the parameter β, which plays the role of the coupling constant in the sine-Gordon problem,[4] is chosen to be an appropriate function of the temperature appearing in Eq. (4), to be discussed shortly. The integral in Eq. (6) is over the length of periodic direction, L, and the parameter β_o of Eq. (5) represents the other length.

The proofs that Eq. (6)[5] and Eq. (3) are equivalent will not be extensively reviewed here, for an ample literature exists on the subject when $q_o = 0$. It is primarily the extension to the "periodic" case which will be discussed and the rather technical

details surrounding these equivalences and the important further
equivalence to the fermion problem are not essential to this end.

The simplest relation between the two problems defined above
is derived via the two dimensional Coulomb gas.[5] In words of short
syllables, the formal expansion of Eq. (5) in powers of V is the
same as the expansion of Eq. (4) in V. In both cases, this
parameter takes the function of the fugacity, and the two expansions
are identical provided the expectation values of the phase-phase
correlation functions, each evaluated in its own ensemble, are
chosen to be equal. This determines the parameter β to be given
by:

$$\beta^2 \langle \varphi(x,t)\varphi - \varphi^2 \rangle = \langle \phi(x)\phi - \phi^2 \rangle \qquad (7)$$

Performing the indicated computations leads to the result:

$$\frac{\beta^2}{4\pi} \ln\left(\frac{\alpha^2}{x^2 - t^2}\right) = \frac{I}{2\pi\mu} \ln \frac{\alpha}{|x|} \qquad (8)$$

where these expectation values are computed in the harmonic part
of the appropriated problem. The parameter α is a short distance
cutoff, taken to be the same in both problems, representating the
fundamental lattice spacing back in the original problem of discrete particles. In the time-ordered products involved in the
expansion of Eq. (5), only imaginary times appear (perhaps the
words temperature ordered product are more correct), which makes
clear the meaning of the minus sign in Eq. (8). The time (or
temperature) evolution produces the other dimension in the problem,
and one dimensional quantum fluctuations correspond to the two
dimensional phase fluctuations in a direct way.

It is at this stage that the important role of the fermions
becomes clear. The fermions are used to diagonalize the cosine
term of Eq. (6). This leads, in turn, to the eigenvalue spectrum,
which can then be used to evaluate the partition function. The
physics is all fermionic - and through the equivalences, domain
walls. The eigenvalue spectrum, that of the quantum sine-Gordon
equation, is rather rich and complicated,[6] but known exactly.[7]
This enables the solution of the model.

In addition to the use of the exact eigenvalue spectrum,
it is necessary to understand the influence of the additional term
linear in q_0 in Eq. (6). It is shown below that it represents
a chemical potential, shifting the fermi energy for the fermions.
As the fermi level is shifted, there may or may not be states in
the interval. If there are, the population of these states as the

fermi level is shifted, corresponds to the appearance of domain walls
in the statistical mechanics problem. These two ingredients, eigenvalue spectrum and chemical potential, together lead to a satisfactory picture of the domain walls in this simple model.

The solution exhibits a continuous transition between a phase
which is commensurate with the underlying substrate potential, and
a phase which has a periodicity distinct from and incommensurate
with the substrate. The expectation value of the phase field $\varphi(x)$
increases linearly with distance in this latter phase, and the
coefficient, which is the inverse period, has typically the following form:

$$\langle \varphi(x) \rangle - \langle \varphi(\infty) \rangle = Ax$$
$$A^2 = q_0^2 - q_c^2(T) \tag{9}$$

where $q_c(T)$ is a complicated function of V, α , and T and is known
up to numerical factors.

The elastic constant corresponding to distortions along the
periodic axis is proportional to $A^{-\frac{1}{2}}$ near the transition line
$q_0 = q_c(T)$, as the lattice becomes more rigid to these displacements. For all values of $q_0 < q_c(T)$, in the commensurate phase,
this elastic constant is not definable since there is no elastic
mode for these displacements. Rather a gap has appeared in the
fluctuation spectrum, and elastic properties are determined by
the substrate.

How many of these features are preserved in the presence of
a real two dimensional periodic substrate potential, and the
$(\vec{\nabla} \cdot \vec{\phi})^2$ terms giving rise to a finite compressibility? There are
several situations for which the answers are known, and it is
possible to offer educated conjectures about the outcome. It
should be recognized that domain walls, and consequently fermions,
are an essential ingredient in all such periodic models. The
other terms can be expressed simply in the fermionic language –
that is, they provide a natural basis for the more complicated
problems of elasticity theory and substrate potentials. It is
somewhat natural to expect that the fermionic properties of the
transition, summarized by Eq. (9), will be preserved.

THE COMMENSURATE GAP

This section reviews the detailed connection between the
periodic substrate potential, fermion degrees of freedom, and the
commensurate phase. It is necessary to review many details of
these relations in order for the subsequent discussion to be
precise, for the central issues involve the sine-Gordon equation,

which displays the periodicity, and the interacting fermi problem,
which displays the solution. To the reader unfamiliar with this
lore, apologies are offered for the brevity. To the reader who
knows it all anyway, this section serves as a helpful summary of
the conventions to be used subsequently.

The starting point is the Hamiltonian of Eq. (6), which
contains the fermions, albeit safely hidden. It is first helpful
to define the operators:

$$\varphi(x) = i \pi^{1/2} L^{-1} \sum_{k} k^{-1} \left(\rho_1(k) + \rho_2(k) \right) e^{-ikx - \frac{\alpha}{2}|k|} \tag{10A}$$

$$\pi(x) = \dot{\varphi}(x) = \pi^{1/2} L^{-1} \sum_{k} \left(\rho_2(k) - \rho_1(k) \right) e^{-ikx - \frac{\alpha}{2}|k|}$$

where the auxiliary operators $\rho_1(k)$ and $\rho_2(k)$ satisfy the
commutation relations:

$$[\rho_1(-k), \rho_1(k')] = [\rho_2(k'), \rho_2(-k)] = kL(2\pi)^{-1} \delta_{k,k'} \tag{10B}$$

$$[\rho_1(k), \rho_2(k')] = 0$$

and k represents a wave number $k_n = 2\pi n\, L^{-1}$, for n integer.
These operators have direct physical significance, and provide
one of the links to fermions, for they will turn out to be the
Fourier transform of the fermion density operators.

A very useful canonical transformation of these operators is
generated by the relations:

$$\rho_1 \rightarrow \cosh\Theta \; \rho_1 + \sinh\Theta \; \rho_2$$

$$\rho_2 \rightarrow \cosh\Theta \; \rho_2 + \sinh\Theta \; \rho_1 \tag{11}$$

and the phase and canonical momentum, under this transformation,
are multiplicatively rescaled:

$$\varphi \rightarrow e^{\Theta} \varphi$$

$$\pi \rightarrow e^{-\Theta} \varphi \tag{12}$$

Direct substitution reveals that the harmonic part of the
Hamiltonian of Eq. (6), expressed in terms of the ρ operators,
is given by:

$$H = 2\pi L^{-1} \sum_{k>0} \rho_1(k)\rho_1(-k) + \rho_2(-k)\rho_2(k) \tag{13}$$

The transformation of Eq. (11) applied to this diagonal harmonic oscillator, Hamiltonian introduces cross terms of the form $\rho_1\rho_2$.

The equivalent fermi problem involves two types of particles on a line, moving in opposite directions. The fermi operators satisfy anticommutation relations:

$$[\psi_1(x), \psi_1^\dagger(x')]_+ = [\psi_2(x), \psi_2^\dagger(x')]_+ = \delta(x-x')$$

$$[\psi_1(x), \psi_2(x)]_+ = 0 \qquad (14)$$

and the density operators are defined by:

$$\rho_1(x) = \psi_1^\dagger(x)\psi_1(x)$$

$$\rho_2(x) = \psi_2^\dagger(x)\psi_2(x) \qquad (15)$$

The simplest local Hamiltonian involving all possible products of these operators is given by:

$$\mathcal{H} = \int \frac{dx}{c} \; v(\psi_1^\dagger \partial_x \psi_1 - \psi_2^\dagger \partial_x \psi_2) + g\,\rho_1\,\rho_2 + V\alpha(\psi_1^\dagger \psi_2 + h.c.) \qquad (16)$$

and the first two terms represent the free propagation of the particles, the "1" types to the right, with velocity v. The third term displays the fermion interaction, while the remaining generate a gap in the spectrum due to the mixing of the right moving particles with the left.

The surprise of one dimension is the identity of the sine--Gordon equation, Eq. (6) to the fermion problem Eq. (16) (ignoring the q_0 term). Substantial technical points underlying this identity will be omitted, but the essence is summarized by:[8]

$$\lambda \alpha^{-1} \cos 2\sqrt{\pi}\,\psi = \psi_1^\dagger \psi_2 + h.c. \qquad (17)$$

$$-i\int dx \; \psi_1^\dagger \partial_x \psi_1 - \psi_2^\dagger \partial_x \psi_2 = 2\pi L^{-1} \sum_{k>0} \rho_1(k)\rho_1(-k) + \rho_2(k)\rho_2(k)$$

$$(18)$$

These equations are to be understood in the appropriate representation - that is, the correlation functions of the cosine term in the sine-Gordon problem, are the same as the correlation functions of the $\psi_1^\dagger\psi_2$ + h.c. in the fermion problem. Solving the fermi problem thus solves the other as well, since any matrix element is given correctly using these prescriptions.

The parameters of Eq. (16) are related to the in Eq. (6) through the relations:

$$v = \cosh 2\theta$$
$$g = \sinh 2\theta$$
$$\beta^2 = 4\pi e^{2\theta} \qquad (19)$$

which implicitly determine the velocity and coupling constant in the fermion problem. The convention used here insists the true, or renormalized, velocity is chosen to equal to unity. That means the Lorentz invariant $x^2 - t^2$ will always remain, corresponding to the rotational invariant, $x^2 + y^2$, of the two dimensional surface layer.

These velocity renormalizations can be understood easily with the help of the canonical transform of Eq. (11) and Eq. (12). First apply it to the sine–Gordon Hamiltonian, Eq. (6), and express the result in terms of the ρ operators. Choosing the factor $e^{2\theta}$ to scale β^2 to 4π, and using the equivalence tables of Eq. (17) and Eq. (18), leads immediately to the fermi problem of Eq. (16).

Incommensurability is represented by the q_0 term of Eq. (6), and it is interesting to find the fermion equivalent to it. Using the definition of the phase, and integrating gives the simple result:

$$-2q_0 \int_0^L dx \, \partial_x \varphi = -2q_0 \sqrt{\pi} \, \rho_0 \qquad (20)$$

where ρ_0 is the total fermion density operator. A shift in the fermi level obviously results from changes in the intrinsic period of the surface layer, and the solution to the statistical mechanics problem requires filling the fermion states and adding up the energies.

Unfortunately, it is not quite that simple, for these particles are interacting, and the ground state can be expected to be rather complicated. It is instructive to consider a special case – the case with temperature determined such that $\beta^2 = 4\pi$. At that special temperature, the fermions are no longer interacting, the parameter g is zero, and the velocity is unity. The fermion problem is bilinear, and the eigenvalue spectrum is found by Fourier transforming the ψ operators and substituting into Eq. (16). The Hamiltonian is diagonalized by linear combinations of these operators, describing the mixing, or hybridization, of the "1" and "2" states. The eigenvalue spectrum is given by:

$$E_k^2 = k^2 + V^2 \qquad (21)$$

and, at this temperature, $T_1 = 2\pi\mu$, the effects of the periodic substrate can be calculated exactly.

As long as the shift in fermi energy per particle falls within the gap, $-V < 2q_0\pi^{+\frac{1}{2}} < V$, the surface layer is commensurate with the substrate. Commensurate is defined here by the ground state expectation value of the phase gradient:

$$\langle \partial_x \varphi \rangle = \sqrt{\pi} \langle \rho_0 \rangle \qquad (22)$$

and the object $\langle \rho_0 \rangle$ is found by filling the states up to the new fermi level. The density of states in energy of the one dimensional fermi gas has a square root divergence at the edge of the band, which integrates to a density expectation value vanishing as the square root. The exact expression is:

$$\sqrt{\pi} \langle \partial_x \varphi \rangle = \langle \rho_0 \rangle = (q_0^2 - V^2)^{\frac{1}{2}} \qquad (23)$$

as quoted in the introduction with $q_c(T_1) = V$.

In order to examine the behavior at other temperatures, it is necessary to have a more detailed picture of the ground state of the interacting fermion problem. Fortunately, the eigenvalue spectrum is known, as is the S-matrix, and certain correlation functions in the small β limit. The spectrum consists of massive fermions, with the energy eigenvalue as in Eq. (21), but with a renormalized gas (mass). The spectrum also contains bound fermion- -antifermion states, which lie within the band gap. The relation between the renormalized gap and the bare parameters is also known up to numerical constants. These facts can be assembled to produce a rather complete picture.

For temperatures greater than T_1 , corresponding to $\beta^2 > 4\pi$, there are no bound states, only the massive fermions. Although they interact, knowledge of the spectrum is sufficient to determine the nature of the ground state adequately for the present purposes. The system is commensurate until the fermi level reaches the renormalized gap,[7] given by:

$$V' = C \, V^{2\nu}\alpha^{-1} \qquad (24)$$

where C is a constant which generally depends weakly on β while ν is given by $\nu = 4\pi(8\pi - \beta^2)^{-1}$ which dramatically diverges as $\beta^2 \to 8\pi$, a singular point of the problem, corresponding to a temperature $T_2 = 2T_1$. Above this temperature, only the commensurate phase exists, even if $q_0 = 0$. (It is necessary to extend somewhat the commensurable criterion in this region to the elastic constant, for $\partial_x \varphi >$ is "accicently" zero. The region $T > T_2$ can be analyzed using the further equivalence of the sine-Gordon problem to the spin-$\frac{1}{2}$ chair. This review anticipates somewhat the future activity using this method to explore the high temperature region.)

It is clear, however, that the intermediate temperature region $T_1 < T < T_2$ is characterized by the square root dependence of Eq. (23), for that depends only on the density of states, which in turn depends only on the spectrum of Eq. (21). The characteristic gap is strikingly temperature dependent, and an accurate plot of q_0 at the transition versus temperature should provide a measure of the exponent ν, to be compared with the sine-Gordon prediction.

The low temperature region $T < T_1$ is complicated by the appearance of bound states in the quantum problem. These evidently correspond to the presence of "bound walls" or inclusions, which dominate the low temperature region. It is not clear what effects, if any, these have on the massive fermion picture result, found by the fermi level picture, predicting a universal square root behavior near the transition.

The results of this section have been summarized elsewhere,[9] although the full exposition of the fermion representation in the context of the surface layer problem has not. The discovery of the special temperature T_1 has also been reported in a recent treatment of the commensurate-incommensurate phase transition,[10] but without use of the fermions. In this work, the canonical ensemble has been used along with a variational principle, which leads to a result differing from the present. The difference resides in the choice of ensemble and identification of the ground state, and it is felt that the grand canonical ensemble, and consequent filling of the fermi sea, is the more appropriate one to use.

APPLICATION TO THE RECTANGULAR LATTICE AND DISCUSSION

An obvious application of the results in the previous section concerns the simple rectangular lattice, described by the free energy functional:

$$F\{\vec{\Phi}\} = \int d_{\vec{x}}^2 \quad \mu(\vec{\nabla}\varphi^x)^2 + \mu(\vec{\nabla}\varphi^y)^2 + \sum_G V_G e^{i\vec{G}\cdot\vec{x} + iG\Phi(x)} \qquad (25)$$

where $\varphi^x(\vec{x})$ and $\varphi^y(\vec{x})$ represent the different components of the displacement field, and V_G is the strength of the substrate potential with period $2\pi G^{-1}$. The sum is over all components. If only one period along each principal axis dominates, the substrate potential simplifies greatly, for the two displacement fields decouple. The resulting functional becomes:

$$F\{\phi\} = \sum_{j=x,y} \int d^2x \ \mu(\vec{\nabla}\phi^j)^2 + V_j \cos(\vec{G}^j x^j + \phi^j(\vec{x}))$$

and the sum over j represents the superposition of the x and y components of the separate free energies.

From this result, it is straightforward to apply the methods from the previous section to solve the problem exactly. For the general rectangular structure, there are two possible transitions. Each component may undergo a separate transition at the boundary line determined by the appropriate substrate potential. A more realistic model would include coupling between the two components, couplings which arise naturally if the additional invariants permitted in continuum elasticity theory are added to Eq. (25). However, when these are present, it still remains correct to describe the domain wall degrees of freedom in the fermion language. The additional couplings represent, by the use of Eq. (4) and Eq. (9), additional interactions between the two components of the fermion field.

Although the answers to the types of fermion problems, generated in this manner, are not known in most cases, several general festures seem to emerge. The existence of massive fermion states, as quantum mechanical equivalents of the domain wall configurations remains valid. The incommensurate transition is driven by a chemical potential of these fermi particles. Finally, the existence of the massive fermion states implies, through phase space considerations and dimension counting, the existence of square root behavior at the transition.

A situation of greater experimental interest involves the triangular or honeycomb lattices. The representations for domain wall excitations in the fermi language have not yet been constructed, and it is not clear that the reasoning based on counting the massive fermion states will continue to be correct. Complications due to bound states, or mixing between the different fermion representations, could become a complication. At the intuitive level, however, it is obviously correct to formulate these problems in the fermi language, and thereby include the domain wall statistics in the correct manner.

ACKNOWLEDGEMENTS

Much of this work was carried out while V.P. was a visitor at Nordita. A.L. wishes to acknowledge the support of the Alfred P. Sloan Foundation.

REFERENCES

1. Recent progress has been reviewed by J. Villain elsewhere in
        these proceedings.
2. F.C. Frank and J.H. van der Merwe, Proc. Roy. Soc. 198, 205
        (1949).
3. L.D. Landau and E.M. Lifshitz, Theory of Elasticity (Pergamon
        Press).
4. S. Coleman, Phys. Rev. D11, 2088 (1975). A. Luther and V.J.
        Emery, Phys. Rev. Lett. 33, 589 (1974).
5. P. Minnhagen, A. Rosengren, and G. Grinstein, Phys. Rev. B18,
        1356 (1978).
6. L.D. Faddeev and L.A. Takhtajan, Theor. Mat. Fiz. 21, 160
        (1974), R. Dashen, B. Hasslacher, and A. Neveu, Phys. Rev.
        D11, 3424 (1975).
7. A. Luther, Phys. Rev. B14, 2153 (1976)
8. D.C. Mattis and E.H. Lieb, J. Math. Phys. 6, 304 (1965),
        A. Luther and I. Peschel, Phys. Rev. B9, 2911 (1974).
9. V.L. Pokrovsky and A.L. Talapov, Phys. Rev. Lett. 42, 65
        (1979)
10. Hikaru Yamamoto, Kyoto University preprint.

NEUTRON SCATTERING STUDIES OF PHYSISORBED MONOLAYERS ON GRAPHITE

M. Nielsen

Risø National Laboratory
DK-4000 Roskilde
Denmark

J.P. McTague

University of California
Los Angeles
USA

L. Passell

Brookhaven National Laboratory
Upton, New York
USA

ABSTRACT

This article reviews neutron scattering work on films of N_2, D_2, H_2, He, Ar and Kr on graphite substrates. Elastic neutron diffraction measurements have determined the structure of the two dimensionally ordered layers and, in such favourable cases as ^{36}Ar and D_2 films, the dynamics of the films have been measured. Neutron scattering investigations of the melting transition, the commensurate - incommensurate transition and transitions between different incommensurate structures such as the α to β transition in O_2 films are reviewed.

I.   INTRODUCTION

The application of neutron scattering techniques to
the study of surfaces is severely restricted by the small-
ness of the neutron scattering cross section. Thus it is
never possible to do neutron scattering measurements with
samples consisting of only one monolayer, such as an adsorbed layer on the surface of a single crystal. In order
to get scattering intensity enough and at the same time
to be able to correct for all the scattering from the
substrate it is necessary to use substrate samples consisting of small particles so that the surface to volume
ratio is sufficiently large.

It was the discovery that exfoliated graphite products such as Grafoil[1] are rather ideal substrates in
experiments on physisorbed rare gas layers (by for instance vapour pressure measurements[2]) that showed that
neutron scattering might be used to study the structures
of adsorbed layers. Since then a series of such measurements has been performed contributing very significantly
to the knowledge of the phase diagrams of these systems[3].
When the neutron scattering measurements were started and
the applicability of the technique was proven the question arose whether it was possible to study the dynamics
of the adsorbed monolayers and the nature of the phase
transitions of the layers. With the techniques available
at present it is possible to study the phonon spectra of
ordered adsorbed monolayers in favourable cases, but for
the study of phase transitions better substrates are
needed to measure, for instance, critical exponents near
continuous phase transitions.

We will in this article describe the neutron scattering measurements done on simple gases physisorbed on
the exfoliated graphite substrate and concentrating on
the results most familiar to the authors. The technique
has been applied to quite different surface phenomena,
including hydrogen adsorption on a Ni-substrate, the
rotational state of H_2-molecules adsorbed on Al_2O_3 and
the orientation of butane molecules on graphite surfaces.
References may be found in Ref. 4.

II.   NEUTRON SCATTERING TECHNIQUE

A. Principles

Thermal neutron beams are produced in reactors or accelerators by slowing down high energy fission or spallation neutrons with a moderator containing H_2O at about room temperature. The resultant neutrons have an approximately Maxwell-Boltzmann velocity distribution with a characteristic temperature near the moderator temperature, typically 350 K. In many cases a considerably improved signal can be achieved by using a cold neutron beam which is obtained by replacing the H_2O-moderator by a liquid H_2 moderator. Thermal neutrons have typical energies E, wavevectors k, or wavelengths λ, given by

$$E = 3/2 \ k_B \cdot T = 1/2 \ M_N \ v^2 = \hbar^2 k^2 / (2M_N) = \hbar^2 / (2M_N \lambda^2)$$

$$E \sim 30 \ \text{meV} \sim 240 \ \text{cm}^{-1} \sim 350 \ \text{K} \qquad (1)$$

$$k = 3.8 \ \text{Å}^{-1}; \quad \lambda = 1.65 \ \text{Å}.$$

The fact that these energies and wavelengths are comparable to the typical vibrational energy per particle and to the repeat distance in crystals explains why neutrons are sensitive probes for measuring both the dynamics and the structure of solids.

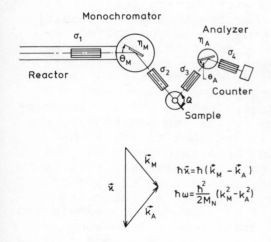

Fig. 1. Schematic drawing of a neutron spectrometer. σ_1 to σ_4 denote collimators, η_M and η_A monochromator and analyzer crystals, and θ_M and θ_A the two Bragg angles. The triangle shows how the momentum transfer ($\hbar\vec{\varkappa}$) and the energy transfer ($\hbar\omega$) are determined by the wave vectors of the incoming and outgoing neutrons, $\vec{k}_M$ and $\vec{k}_A$.

Fig. 1 shows schematically a typical neutron spectrometer. The monochromator separates out a nearly monochromatic beam from the broad distribution of neutrons coming from the moderator by Bragg scattering from a suitable crystal. For a 1% $\Delta\lambda/\lambda$ wavelength spread and $1/2^{\circ}$ collimation, typical fluxes on a sample from a high flux reactor are about 5×10^7 neutrons cm^{-2} sec $^{-1}$.

Neutrons interact with nonmagnetic atoms via short-range nuclear forces and the scattering length b is of the order of 10^{-4} Å. The scattering cross section $\sigma = 4\pi b^2$ is then of the order of 10^{-7} Å^2 (= 10 barn), which is about 10^{-8} that of the geometrical cross section of an atom. In other words neutrons interact weakly with nuclei. The potentials $V(\vec{r})$ are essentially δ functions

$$V(\vec{r}) = \frac{2\pi\hbar^2}{M_N} b\delta(\vec{r}-R(t)) \tag{2}$$

where $\vec{R}(t)$ is the time-dependent position of the nucleus and M_N the neutron mass. The scattering from a sample with nuclei i depends on

$$V = \sum_i V_i(\vec{r}) = \frac{2\pi\hbar^2}{M_N} \sum b_i \delta(\vec{r}-R_i(t)) \tag{3}$$

where b_i is the scattering length of the i'th nucleus. Even for a single element, b varies from one isotope to another; for a single isotope with nonzero spin, it depends on the spin state. Neither the isotopic composition nor the nuclear spin orientation are in general correlated with positions or the motion of the atoms in the sample, and as a consequence we can separate the scattering into coherent scattering, proportional to $|\bar{b}|^2$, and incoherent scattering, proportional to $(\bar{b^2}-\bar{b}^2)$.
The coherent scattering is determined by the total correlation function $G(\vec{r}_i,t)$ where $\vec{r}_i$ is the position of the i'th nucleus at time t relative to that of the j'th nucleus at t = 0; it measures the atomic positions and the dispersion relations of the collective excitations (i.e. the wavevector frequency relation of the phonons). The incoherent scattering is determined by the self-correlation functions $G_s(\vec{r}_i,t)$ of atoms in the sample and measures the single particle distribution functions and the dynamics of single particles (e.g. the phonon density of states).

Table 1

Neutron scattering cross sections (barn)

Element	σ_{coh}	σ_{inc}.
^{4}He	1.13	0
^{3}He	4.68	–
H	1.75	79.7
D	5.59	2.2
N	11.10	$\sim$0.4
O	4.23	$\sim$0
^{36}Ar	74.20	0

Values of typical cross sections $\sigma_{coh} = 4\pi|\bar{b}_i|^2$ and $\sigma_{inc} = 4\pi\overline{(b-\bar{b})^2}$, are given in table 1. Hydrogen has a large but almost purely incoherent cross section, and consequently the single particle dynamics can readily be studied in hydrogen-containing substances, but not the crystal structure or collective dynamics (deuterium can be used, however). ^{36}Ar and N_2, on the other hand, are strong coherent scatterers. In the case of ^{3}He the strong absorption makes measurements difficult.

The Bragg angles in the monochromator and in the analyzer θ_M and θ_A (see Fig. 1) can be varied to control respectively the energies of the incoming neutrons and of those scattered neutrons from the sample which are counted in the analyzer. Further, the variation of angle Q and sample rotation allow us to perform general scans in which we measure the intensity of scattering in the sample specified by a chosen momentum transfer $\hbar\vec{\kappa} = \hbar(\vec{k}_M - \vec{k}_A)$ and energy transfer

$$E = \hbar^2/2M_N \times (k_M^2 - k_A^2).$$

Thus, by coherent scattering in single crystalline samples we can measure the dispersion relations of the collective excitations point by point by for instance observing where in $\vec{\kappa}$, E space the neutrons undergo one-phonon absorption or one-phonon emission scattering. In incoherent scattering the intensity of the scattering as function of $\vec{\kappa}$ and E reflects the single particle state of the atoms in the sample and can be used, for instance, to measure particle diffusion. Formally, the double differential cross sections may be expressed:

$$(\frac{d^2\sigma}{d\Omega dE})_{coh} = \frac{\bar{b}^2 N}{2\pi\hbar} \frac{k_A}{k_M} \int d\vec{r}dt \ e^{i(\vec{\kappa}\cdot\vec{r}-\omega t)} G(\vec{r},t)$$

$$= \bar{b}^2 \cdot N \frac{k_A}{k_M} S_{coh}(\vec{\kappa},\omega) \qquad (4)$$

$$(\frac{d^2\sigma}{d\Omega dE})_{inc} = \frac{(\overline{b^2}-\bar{b}^2)N}{2\pi\hbar} \frac{k_A}{k_M} \int d\vec{r}dt \ e^{i(\vec{\kappa}\cdot\vec{r}-t)} G_s(\vec{r},t)$$

$$= (\overline{b^2}-\bar{b}^2)N \frac{k_A}{k_M} S_{inc}(\vec{\kappa},\omega),$$

where the two $S(\vec{\kappa},\omega)$ functions are referred to as the co-herent and the incoherent scattering laws. These two scattering contributions can only be separated by choosing samples where either the coherent cross section or the incoherent cross section is dominant.

Structural information is contained in the elastic scattering cross section

$$(\frac{d\sigma}{d\Omega})_{coh} = \bar{b}^2 |\sum_i e^{i\vec{\kappa}\cdot\vec{r}_i}|^2 \qquad (5)$$

and here no energy analyzer is needed, but often it is used to improve signal-to-background ratio. For single crystalline samples, the sum in equation (5) is non-zero only when $\vec{\kappa} = \vec{\tau}$, a reciprocal lattice vector. As in X-ray diffractometry, one has

$$(\frac{d\sigma}{d\Omega})_{coh} = N \frac{(2\pi)^2}{V_o} \sum_{\vec{\tau}} \delta(\vec{\kappa}-\vec{\tau}) |F_N(\vec{\tau})|^2 \qquad (6)$$

where N is the number of unit cells, of volume V_o, and $F_N(\vec{\tau})$ is the elastic unit cell form factor, which is simpler for neutrons than for X-rays:

$$F_N(\vec{\tau}) = \sum_j e^{i\vec{\tau}\cdot\vec{r}_j} \bar{b}_j \ e^{-2W(\vec{\kappa})}. \qquad (7)$$

The sum is over all atoms j in the unit cell with positions denoted by $\vec{r}_j$. The term $e^{-2W(\vec{r})}$ is the Debye-Waller factor, with $2W(\vec{\kappa}) = <(\vec{\kappa}\cdot\vec{u}_j)^2>$ where $<u^2_j>$ is the mean-square displacement of the j'th atom.

For simple crystalline solids with only one atom per unit cell the phonon scattering cross section is

$$(\frac{d^2\sigma}{d\Omega dE})_{coh} \propto \sigma_{coh}\ e^{-2W(\vec{\kappa})}\ \frac{\hbar^2\kappa^2}{2M_N\omega}\ (n(\omega)+1)\ \sum_j (\hat{\kappa}\cdot\hat{\xi}_j(\vec{q}))^2$$

$$\times\ \delta(\omega-\omega_j(\vec{q}))\delta(\vec{\kappa}-\vec{q}-\vec{\tau}). \qquad\qquad (8)$$

The Bose population factor is $n(\omega)$, $\hat{\xi}_j(\vec{q})$ is the phonon polarization vector of phonon mode j with wavevector $\vec{q}$. Thus, from a single crystalline sample we will get phonon scattering at the momentum transfer $\hbar\vec{\kappa}$ from phonons with wavevector $\vec{q} = \vec{\kappa}-\vec{\tau}$, $\vec{\tau}$ being the reciprocal lattice vector closest to $\vec{\kappa}$; and only the component of $\xi_j(\vec{q})$ along $\vec{\kappa}$ gives a contribution. The frequency of the selected phonon is found by performing a constant $\vec{\kappa}$ scan and observing an intensity maximum of $\omega = \omega(\vec{q})$. In case we have a multicrystalline sample the coherent phonon scattering will be given by averaging eq. 8 over a sphere with radius $|\vec{\kappa}|$ in reciprocal space.

B. Application to Adsorbed Monolayers

The exfoliated graphite products which are most commonly used as substrates in neutron scattering measurements are Grafoil, Papyex and UCAR-ZYX[1]. The first two have quite similar properties. As schematically shown in Fig. 2, they are built up of small flakes of graphite a few μm broad and 100-300 Å thick. The large flat surfaces of the flakes are hexagonal crystal faces with the graphite honeycomb structure, onto which the adsorption takes place. The surfaces have faults such as layer steps, grain boundaries, or dislocations which limit the size over which we can have two-dimensional positional order of the adsorbed monolayers. This size is characterized by the coherence length L, the value of which may be deduced from the line shape of the Bragg peaks from the adsorbed layers. It is about 110 Å for Grafoil and larger than

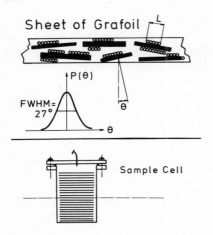

Fig. 2. Schematic drawing of
Grafoil sheet and the sample
cell. L indicates the typical
coherence length of the adsorbed layers. $P(\theta)$ describes
the broad distributions of
angles giving the non-paral-
lelism of the substrate surfaces. In the sample cell
the Grafoil sheets are oriented parallel to the neutron
scattering plane.

250 Å for UCAR-ZYX. The adsorbing surfaces are oriented
such that the angle θ defined by the normal to the surface and the exterior surface of the substrate falls in
a rather broad distribution with a full width at half
maximum of 27° for Grafoil (see Fig. 2) and 11° for UCAR-
ZYX. The former have in addition a certain
fraction of flakes which are randomly distributed. The
specific area for adsorption is about 30 m^2 per gram for
Grafoil and 3 m^2 per gram for UCAR-ZYX. Typically 50
grams of Grafoil are used in a sample cell for neutron
scattering, so the amount of gas adsorbed (referred to
as the filling or coverage) when a monolayer of e.g. H_2
is completed is 470 cc STP, the equivalent of about
0.5 cc of bulk solid. This gives a quite adequate sample
for structural measurements (2-D Bragg diffraction) even
considering that a severe background subtraction is un-
avoidable, but for inelastic measurements of the dynamics
of adsorbed layers the samples are only sufficient in
particularly favourable cases.

A typical sample cell for "in-plane-measurements"
is shown in Fig. 2. Circular discs of about 1 inch diameter are cut out of the 0.2 mm thick Grafoil sheets and
after outgassing in vacuum at 1000 $^{\circ}$C they are loaded
into an aluminum sample cell, without being exposed to
the atmosphere. The gas to be adsorbed is admitted through
a capillary tube from a gas handling cabinet to the sample
cell, mounted in a cryostat.

When doing diffraction measurements on a structurally ordered monolayer there is no restriction on the
momentum transfer perpendicular to the layer $h\vec{\kappa}_\perp$
Thus the Bragg condition is $\vec{\kappa} = \vec{\tau}(hk) + \vec{\kappa}_\perp$, where $\vec{\tau}(hk)$
are the reciprocal lattice vectors of the 2-D structure.
For monolayers tilted relative to the neutron scattering
plane the Bragg condition can be fulfilled when $\kappa =$
$\tau(hk)/\cos\gamma$, where γ is the angle between $\vec{\tau}$ and $\vec{\kappa}$ (see
Fig. 3). So when $\gamma = 0$ we then get Bragg scattering at
$|\vec{\kappa}| = |\vec{\tau}(hk)|$, but there is also scattering from the same
(hk) reflection at larger values of $|\vec{\kappa}|$ from the $\gamma \neq 0$
surfaces. Thus, the Bragg peak gets a sawtooth shape with
low κ onset at $\tau(hk)$. To calculate the line shape we
assume randomness of the in-plane orientation of monolayers and no correlation between the tilt angle γ and
neither the coherence length L nor the in-plane rotation. The scattering intensity from a Bragg reflection
at $\tau(hk)$ may then be expressed as

$$I_{hk}(\kappa) = \text{const.} \int_0^{\pi/2} d\psi \int_0^\kappa d\tau P(\gamma)\, e^{-(\tau - \tau_{hk})^2/\Delta^2} \frac{1}{\kappa\sqrt{\kappa^2 - \tau^2}}$$

$$(9)$$

where $P(\gamma)$ gives the probability of the tilt angle γ,
and $\Delta \equiv 2\sqrt{\pi}/L$. To calculate the integral, the value of γ
given by $\cos\gamma = \tau/(\kappa\cos\psi)$ should be inserted in $P(\gamma)$.
The resultant line shape is a rounded sawtooth with a
steep edge on the low κ side, determined predominantly
by L, and a high κ tail which is essentially controlled
by $P(\gamma)$. If the two-dimensional unit cell of the adsorbed
layer contains more than one atom the line shape is
modified by a form factor.

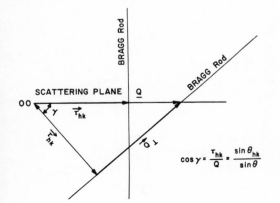

Fig. 3. Scattering diagram
in reciprocal space for a
pair of 2-D polycrystals;
one with its plane parallel
to the neutron scattering
plane and one with $\vec{\tau}_{hk}$
tilted from the plane by an
angle γ.

e^{-2W} Finally we have to include the Debye-Waller factor due to the thermal vibration of the adsorbed parti-cles. For infinite 2-D systems e^{-2W} diverges at all finite temperatures[5] but in practice this does not happen because the divergence is due to acoustic modes with frequencies going to zero at long wavelengths and even for rather large 2-D systems the boundaries can effectively surpress the divergence at all T below melting. Quantitatively the Debye-Waller factor may be expressed by $2W \simeq 2T/T_Q \times \log(L/a)$, where a is the interparticle distance in the layer and a characteristic temperature is given by $T_Q = 4\pi MC^2/(V\kappa^2)$. Here C is the in-plane acoustic sound velocity, V the unit cell area and M the particle mass. T_Q is of order of 10^3 K to 10^4 K, making $2W \sim 10^{-2}$-$10^{-3} \times T$ for $L \sim 100$ Å. Even for very large layers this effect is numerically unimportant, at least at low temperatures when the sound velocity is well defined, and as we shall see well behaved Bragg peaks are observed for many ad-sorbed layers. In cases where the adsorbed monolayer has a two dimensional lattice which periodically "locks in" in perfect phase with the substrate surface potential this registry will prevent free floating of the adsorbed layer and an energy gap removes the acoustic phonons near zero wavevector, $q = 0$, and thereby the divergence problem of e^{2W} at low temperature, even for infinitely large systems.

In all elastic neutron diffraction measurements on adsorbed layers the scattering from the substrate is a severe problem. Carbon atoms have a purely coherent neu-tron scattering cross section, meaning that elastic scat-tering from the graphite products only occurs where the condition for Bragg reflection from graphite is ful-filled. For basal plane reflections, the smallest reci-procal lattice vector $\tau(h,k,0)$ is $\tau(100) = 2.95$ Å^{-1}. The only reflection at smaller scattering vector κ is the c-axis reflection $\tau(002) = 1.873$ Å^{-1}. If the parallelism of the graphite flakes was sufficient that practically none had the c-axis in the scattering plane there should be from such substrates no scattering for $\kappa < \tau(100)$. Unfortunately, this is not so. For Grafoil and Papyex the (002) scattering from graphite is very strong around $\kappa = \tau(002)$, and makes structural measurement on the adsorbed layers almost impossible in this range. For UCAR-ZYX the parallelism of graphite c-axes is much better, so the background scattering is

smaller and narrower. In addition to the (002) scattering
there is a rather unstructured scattering with intensity
strongest at small κ-values which comes from defects on
the surfaces and in the bulk of the graphite flakes.
Another significant contribution to the observed intensity
can arise from multiple scattering where the incoming or
outgoing neutron has been scattered more than once. This
scattering can be greatly reduced by using neutron energies below cut-off of in-plane reflections in graphite,
i.e. neutron wavevectors smaller than half of $\tau(100)$
($k_M < 1.47$ Å^{-1}).

       In most cases the substrate scattering can be ac-
counted for by doing diffraction measurements both with
and without the adsorbent. For favourable cases like Ar,
N_2, O_2 or D_2 on Grafoil the peak intensity from the first
Bragg reflection of the adsorbed monolayers can be 1/4 -
1/2 of the scattering from the substrate, but in limiting
cases, peaks being only a few per cent of the substrate
scattering intensity can be measured.

III.   EXPERIMENTAL RESULTS

A. Phase Diagrams

The structure of several simple gases adsorbed on graphite
has been determined by neutron diffraction, and table 2
gives a list of these and the structures found. We will
in the following give a brief description of the results
and present them where possible in density-temperature-
phase diagrams. In these measurements the density is
determined by the total surface area of the graphite
substrate and the amount of gas admitted to the sample
cell. Below the melting point of the 2-D structures the
vapour pressure of the 3-D gas in equilibrium with the
adsorbed films will in most cases be small enough that
the amount of gas in the 3-D gas phase (the dead volume
gas) is insignificant for the density determination. The
total surface area is found through the diffraction
results themselves or by doing vapour pressure isotherm
measurements[2].

       Crystallography in two dimensions is very simple
because not very many structures are possible; most of
the adsorbed monolayers we shall discuss have the equi-

Table 2

Structures of monolayers adsorbed on graphite at low T.

Adsorbent	Structures	→	ρ increasing	Reference
^{3}He, ^{4}He	Fluid	$\sqrt{3}$	HIΔ	6, 15
H$_2$, D$_2$	$\sqrt{3}$		HIΔ	6, 14
Ne			(HIΔ)	
Ar		HIΔ		7
Kr	$\sqrt{3}$		HIΔ	8, 16
Xe	HIΔ	$\sqrt{3}$		9
N$_2$	$\sqrt{3}$		HIΔ	10
O$_2$	δ		α(β)	11
NO	γ		β	13
CH$_4$	$\sqrt{3}$		HI	12

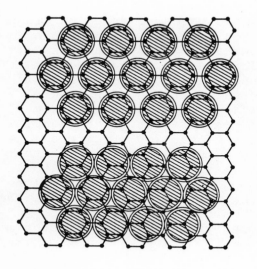

Fig. 4. Schematic representation of the $\sqrt{3} \times \sqrt{3}, 30°$ registered phase (top) and of an incommensurate dense monolayer structure (bottom).

lateral triangular (or hexagonal) structure and in fact
frequently only the τ(10) Bragg-reflection then needs
to be measured. Fig. 4 shows schematically two such structures. The hexagons show the honeycomb lattice of the
graphite surface and the dashed circles illustrate adsorbed particles. In the upper part of the figure there
is one particle adsorbed "above" every third hexagon.
This represents the registered $\sqrt{3} \times \sqrt{3}$ phase, also called
the commensurate $\sqrt{3}$ phase (HC), and as seen in table 2 this
is found in several of the simple gas films on graphite.
In the lower part of Fig. 4 an equilateral triangular
structure which is not in registry with the honeycomb
lattice is shown and this is called hexagonal incommensurate (HI) structure.

$\underline{D_2,\ H_2\ films\ on\ graphite}$. A diffraction group
measured from a monolayer of D_2 on graphite is shown in
Fig. 5. The full line is a least-squares-fit to the
measured points using formula (9) and varying τ(10), L,
and the peak intensity. The scattering from the bare
substrate has been subtracted. The position of the peak
is $\tau = 1.703$ $Å^{-1}$, which is the value of the commensurate
$\sqrt{3}$ structure (Fig. 4). Similar groups were observed for
different values of filling, and the resultant phase diagram is given in Fig. 6. The diagram shows the regions
in which particular structures were found, the points
indicating where diffraction groups were observed, or
where no Bragg peak could be seen. The filling (ρ) is
measured in units of that amount of gas which completes
a monolayer of the $\sqrt{3}$ structure.

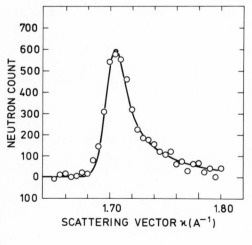

Fig. 5. Diffraction group
from HC monolayer of D_2 adsorbed on ZYX. The full curve
is the fitted line shape
using formula (9), and τ(10)
= 1.703 $Å^{-1}$, L = 300 Å.

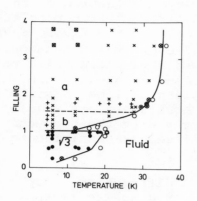

Fig. 6. Phase diagram for D_2
and H_2 layers adsorbed on
Grafoil. Filling is the
amount of gas adsorbed,
measured in units of that gas
amount which is needed to
complete the commensurate
$\sqrt{3}$ structure. The broken line
indicates where promotion to
the second layer starts. The
plotted points show where
diffraction peaks have been
observed or looked for and
the symbols correspond to:
● D_2 and ▲ H_2 layers in the
$\sqrt{3}$ phase, x D_2 and + H_2
layers in the b- and a-phase,
o D_2 and Δ H_2 in the fluid
phase, ⊠ D_2 layers giving
deformed groups with high
background (from Ref. 6).

Both the H_2 and D_2 molecules are to a very good
approximation quantum-mechanical free rotors in the adsorbed phase. This has the consequence that the H_2 molecules, being in their rotational ground state, have zero
nuclear spin and the strong spin incoherent scattering
by neutrons is quenched. The H_2-gas has been converted
to pure para H_2 prior to adsorption. Thus in the adsorbed
phase the molecules behave like spherically symmetric
particles which have a small purely coherent scattering
cross section. (The strong incoherent scattering re-
appears in inelastic scattering when the transferred
energy is larger than 14.6 meV, the energy of the first
rotational energy level of the H_2 molecule)[14].

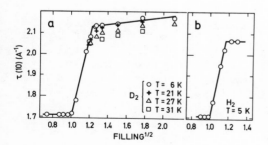

Fig. 7. The position of the
diffraction peaks from monolayers of D_2 and H_2 as function of the square root of
the amount of gas adsorbed
(ρ) (from Ref. 6).

As seen in Fig. 6 the monolayers of H_2 and D_2 behave quite similarly. At submonolayer coverages the commensurate $\sqrt{3}$ structure (HC) is found, but when ρ is increased beyond 1 the $\tau(10)$ Bragg peak is continuously shifted towards higher κ values, signifying that as the $\sqrt{3}$ structure is "overfilled" a new incommensurate structure is formed with a lattice parameter a_{nn} smaller than 4.26 Å (Fig. 7). To a first approximation the value of $\tau(10)$ is proportional to $\sqrt{\rho}$ in region b of Fig. 6, indicating that the first adsorbed layer accommodates all molecules within the homogeneous triangular structure. There are no data points shown in Fig. 7 in the region of $\tau(10)$ values around 1.9 $Å^{-1}$. This is because here the scattering from the substrate is very large due to the (002) reflection from graphite grains which are $90°$ out of alignment with the standard orientation. At the upper knee of the lines in Fig. 7 the promotion to the second adsorbed layer starts as shown by the dashed line between the a and b regions in Fig. 6. In this region the density of the first adsorbed layer corresponds to that of the hexagonal planes of the bulk solid D_2 and H_2 at a pressure of 700 bar, and both the monolayer and the bulk melt at about 28 K. In region (a) of Fig. 6 the effect of the molecules adsorbed in the second layer is seen as a small further compression of the first layer and an accompanying increase of melting temperature, but little is known about the second layer.

The nearest neighbour distances in different phases are shown in table 3.

Table 3

Nearest neighbour distance a_{nn} in Angstrøm, at low temperature.

Element	HC($\sqrt{3}$)	HI structure*	Bulk solid (hcp)
H_2	4.26	3.51	3.745
D_2	4.26	3.40	3.569
^{3}He	4.26	3.32	3.50**
^{4}He	4.26	3.19	3.42

* The values of a_{nn} are taken at the upper knee point of $\tau(10)$ vs. $\sqrt{\rho}$ in Figs. 7 and 9 where second layer promotion starts.
** It is the values at P = 160 bar and T = 3.5 K.

   Measurements of the dynamics and the commensurate-
incommensurate phase transition of the D_2, H_2 layers are
described in sections B and C.

   He-films on graphite. Helium adsorbed on graphite
has been studied extensively with techniques other than
spectroscopy, and the phase diagram is well known[2]. The
neutron scattering technique has been applied in studies
of the superfluidity in adsorbed films of ^{4}He[15] which
we shall not describe here and to a few structural
measurements[6,15]. Figs. 8 and 9 show the results for ^{3}He.

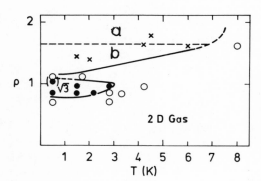

Fig. 8. Phase diagram for
adsorbed ^{3}He layers. The
dotted line indicates where
second layer promotion
starts, (from Ref. 6).

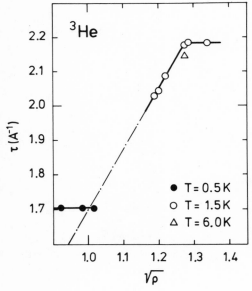

Fig. 9. The position of the
diffraction peak from ad-
sorbed layers of ^{3}He as function of the square root of
the filling, (from Ref. 6).

The phase diagram is drawn in the same way as above for
D_2 and H_2 and so is the graph of $\tau(10)$ vs. square root
of the filling. The very large neutron absorption cross
section of ^{3}He makes these measurements difficult. However, in measurements on thin adsorbed films only a small
fraction of the scattering comes from the film, and scattering from the substrate dominates. The exponential
length for the decay of the neutron intensity in Grafoil
is about 3 cm (at 5 meV). Filled with a monolayer of ^{3}He
this has decreased to about 1 cm. The coherent scattering
cross section of ^{3}He is reasonably large, 4.9 barn,
making the measurements feasible.

A special sample cell was used, 8 mm thick, 62 mm
wide and 41 mm high, containing 11 grams of Grafoil. The
substrate scattering was accounted for by subtracting
the scattering from Grafoil + ^{3}He at elevated temperatures.

For ^{3}He the same phases as for adsorbed D_2, H_2 are
identified. In phase a we have a completed first layer
with the HI structure and, in addition, some atoms in
the second layer. Table 3 gives the neighbour distances
near the phase boundary between the a and b regions of
Fig. 8. The value for ^{4}He was measured at Brookhaven[15]
for a filling 14 per cent higher than that of the phase
line, but $\tau(10)$ is varying slowly in the a region.

In the $\sqrt{3}$ region of Fig. 8 it was confirmed that the
HC structure exists but it is only stable in a limited
region of fillings. At $\rho < 0.75$ no structural reflections
were found down to temperatures of 0.5 K.

N_2 films on graphite. The first neutron scattering
experiment in which Grafoil was used as substrate was
performed by Kjems et al. to study the structures of N_2
films[10]. In the monolayer regime they found both the HC
and the HI structure. The phase diagram is shown in Fig.
10. The nearest neighbour distance in the two structures
is 4.26 for HC and 4.04 A for HI. For N_2 layers the
energetically favourable monolayer density, ignoring the
in-plane substrate potential, ($\sim$ the HI-density) is much
closer to the density of the HC structure than what is
the case for He and H_2 and thus the region (b) of Fig. 6
and Fig. 8 for N_2 layers corresponds to the filling between the two arrows of Fig. 10.

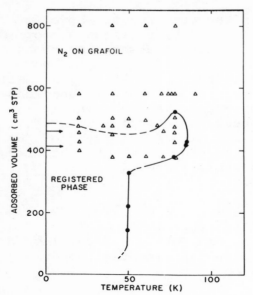

Fig. 10. The "phase diagram" for N_2 adsorbed on Grafoil (from Ref. 10). Plot indicates where diffraction scans were made (Δ) and where transitions involving the registered $\sqrt{3}$ phase were observed (●). The arrows show the fillings for which the $\sqrt{3}$-phase and dense monolayer phase are completed.

Ar films on graphite. [36]Ar is the isotope with the largest coherent neutron scattering cross section per atom of all the common elements and thus has been studied in detail in the monolayer phase on Grafoil[7]. In contrast to all other studied adsorbed monolayers on graphite, the crystalline phase of Ar was observed always to have the HI structure. The measurements of 2-D phonons and the 2-D melting transition are described in the next sections.

Kr, CD_4 and Xe films on graphite. These three species have increasing size as given by the Lennard-Jones radii $\sigma \times 2^{1/6} = 4.04$, 4.28 and 4.60 Å, meaning that Kr atoms are a little too small for just fitting into the HC-structure which has the neighbour distance 4.26 Å. CD_4 molecules almost fit and Xe atoms are too large to fit into this lattice. Monolayers of Kr have been studied by both neutron scattering[8], LEEDS[16] and X-ray scattering[17]. They have both the HC and HI structures, the HI structure being the densest. CD_4 and Xe layers have been studied by neutron scattering[12,9], and CD_4 has the HC structure at low temperatures but close to the melting temperature it expands into a more open HI structure. Xe-layers have an expanded HI structure at monolayer densities and it is still uncertain whether it is compressed into the HC structure above the $\rho = 1$ filling.

$\underline{O_2\ layers\ on\ graphite}$. The phase diagram of O_2 layers adsorbed on Grafoil has been measured by neutron scattering[11] and is shown in Fig. 11. The first layer of the adsorbed O_2 molecules has three distinct phases α, β and δ, none of which is commensurate with the substrate lattice. The filling is still measured in units of that amount of gas which would complete a monolayer of the commensurate $\sqrt{3}$ phase (HC) although this structure is not seen for O_2 layers.

In Fig. 11 the phase "δ" is observed for $\rho < 1.61$ and for temperatures lower than indicated by the melting line. It has a HI-structure much denser than the $\sqrt{3}$ structure. The length of the $\tau(10)$-vector as function of filling is shown in Fig. 12; it corresponds to nearest neighbour distances a_{nn} between 3.37 and 3.47 Å. This means that the molecules do not have sufficient space to be free rotors parallel with the substrate, and if they were forming domains with parallel molecular axes within each domain then we would have observed more than one Bragg reflection. We infer rather that the molecules form an equilateral triangular structure where the average tilt

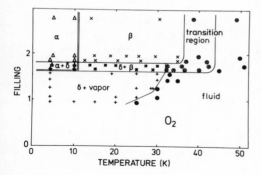

Fig. 11. The phase diagram of adsorbed O_2 layers. The filling is measured in units of the gas amount needed to complete a commensurate $\sqrt{3}$ structure. The points show where diffraction groups have been observed, (from Ref. 11).

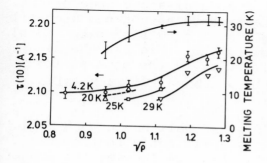

Fig. 12. O_2 layers in the δ-phase. The length of the (10) reciprocal lattice vector plotted vs. the square root of the coverage ρ. The lines are guides to the eye, (from Ref. 11).

angle between the molecular axis and the normal to the
substrate probably varies with the value of a_{nn}. Packing
arguments indicate that the tilt angle could be as large
as 40°, in the "δ" phase, but there has been no direct
measurement of this.

When the filling increases beyond $\rho = 1.68$ then the
(10) Bragg peak splits at low temperatures into a double
peak, signalling a distortion of the triangular lattice.
This sets in at $\rho = 1.61$ and coincides with the completion
of a monolayer of the "δ"-phase. The region of ρ and T
in which the splitting is observed is indicated in Fig.
11 as the α-phase, which has a structure almost identical to the basal plane structure of bulk solid O_2 at
low temperature (also called the α-phase). The molecular
axes are oriented perpendicular to the planes and the
centres of the molecules are located on an (isosceles)
skew triangular lattice, as shown in Fig. 13.

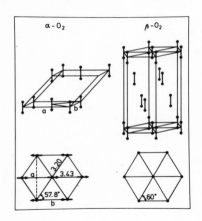

Fig. 13. a. The structure of
bulk α- and β-O_2. b. The
structures of the densest
packed planes of bulk α-
and β-O_2.

At 11.9 K there is a phase transition from the α to
the β-phase which is observed as a disappearance of the
Bragg peak splitting. In the β-phase at not too high fillings a single Bragg peak giving a good fit to the ideal
line shape is observed, showing that this has the HI
structure. The density is found to be the same as for
the basal plane of bulk solid O_2 above T = 23 K (also
called the β-phase, see Fig. 13).

When the filling is increased well beyond $\rho = 1.61$,
the second layer contributes significantly to the scattering groups. This is seen in Fig. 14 showing the Bragg

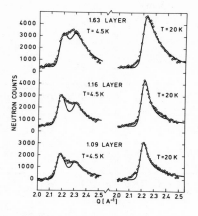

Fig. 14. Neutron diffraction groups from O_2 layers of three different fillings with partial second layer occupancy as shown in the figures. The density of the first layer is determined by the diffraction peaks. The substrate scattering is subtracted. The full lines are fitted curves. (From Ref. 11).

peaks observed for different values of ρ. The molecules of the second layer must be adsorbed in positions which are in registry with the first layer since a rather strong variation of the scattering profile is seen; the full curves are calculated Bragg scattering profiles assuming this, second layer registry.

When the temperature is increased in the β-phase the Bragg scattering intensity decreases and the groups get broader. This happens in a 7-10 K broad temperature range and is denoted the "transition region" in Fig. 11, signifying that the "melting" occurs over a broad temperature range.

As described here both the δ-phase and the β-phase have HI structures which differ only by having different densities as can be seen in table 4. Nevertheless the neutron measurements show that a very distinct coexistence phase exists between the δ and β-phase. In this region the scattering profile has two peaks, one coming from δ-phase layers and the other from the β-phase layers, their relative intensities given by the expected ratio of molecules going into the two phases. The phase transition is discussed in chapter 5.

The coexistence region δ-β extends into the corresponding $\delta+\alpha$ coexistence region in Fig. 11. This is experimentally more difficult to observe because the low κ component of the α-phase double Bragg peak is only slightly higher in κ-position than the δ-phase peak. Only measurements in which the ZYX UCAR substrate was used allowed the $\alpha+\delta$-phase to be observed.

Table 4

Phase diagram data, O_2 on Grafoil

	$\tau(10)\text{Å}^{-1}$	ρ	a_{nn} (Å)
At completion of δ-phase	2.15±0.01	1.59±0.01	3.37
At phase boundary between δ and δ+β		1.61±0.02	
At completion of α-phase monolayer	2.170±0.005 2.30±0.02	1.69±0.01	3.41 3.22
At completion of β-phase monolayer (15 K)	2.210±0.005	1.69±0.01	3.28
At phase boundary between δ+β and β		1.76±0.02	

The close resemblance of the basal plane structures of bulk solid O_2 to the α and β-phase monolayer structures suggests that the monolayer α-phase has an antiferromagnetic ordering like that of the bulk α-phase and shown in Fig. 13. The simplest model for explaining the skew triangular structure of the α-phase is to ascribe it to a magnetostriction which shortens the a_{nn} distance between molecules having antiparallel spins and increases the a_{nn} values of parallel spin molecules[18], (see Fig. 13). The existence of the antiferromagnetic ordering in the α-phase was observed by measuring the magnetic Bragg peak in neutron scattering at the position of half the momentum transfer of the high κ-component of the split nuclear Bragg peak of the α-phase. Fig. 15 shows the measured magnetic groups.

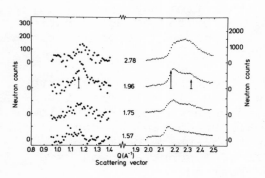

Fig. 15. Neutron diffraction groups from O_2 layers at T = 4.2 K. The fillings ρ are given in the figure. Displayed on the right-hand side is the difference spectrum I($\vec{\kappa}$)(Grafoil + O_2) = I(κ)(Grafoil), while the left-hand side is I(κ,T = 4.2) - I(κ,T = 20 K)(Grafoil + O_2), (from Ref. 11).

Only the $\rho = 1.96$ group is well resolved. The arrows in the figure indicating positions and relative intensities of the groups are calculated under the assumption that the structure is identical to the basal plane structure of the bulk O_2, with the intensity normalized to the low κ-component of the split nuclear group. Even at fillings below $\rho = 1.61$ where we are within the δ-phase some scattering is observed near the position of the magnetic Bragg peak. The origin of this could be short range magnetic fluctuations.

The antiferromagnetism of the O_2-layers is caused by the two unpaired electrons ($\pi^* (2p_x)_g \ \pi^* (2p_y)_g$) which are coupled in an $S = 1$ triplet. The molecular ground state has the magnetic moment perpendicular to the molecular axis. In condensed phases these antibonding π^* electrons can interact via a direct overlap coupling. According to the Pauli principle, two such molecules can have extra attractive interactions through virtual low lying ionic $O_2^- - O_2^+$ states only if their spins S are not parallel. To lowest order, then, there is an interaction between O_2 molecules with an effective Heisenberg Hamiltonian with J strongly dependent on overlap. This direct exchange mechanism causes bulk O_2 to order antiferromagnetically below 23 K and the adsorbed monolayers to order below 11.9 K.

Recent X-ray measurements on O_2 layers adsorbed on ZYX indicate that the β-phase may not have the pure HI structure[25].

B. The Dynamics of Ordered Adsorbed Layers

In three dimensional crystals the neutron scattering technique is the standard tool for measuring phonon dispersion curves. Using single crystals and 3-axes spectrometers the dispersion curves are determined point by point, typically in constant $\vec{\kappa}$-scans. We cannot do this with 2-D crystals like the ordered monolayers because of randomness especially of the in-plane orientation. Nevertheless we can perform inelastic neutron scattering scans, keeping the momentum transfer constant. After subtracting the background scattering from the substrate alone the scattering intensity reflects the phonon scattering from 2-D crystals with all in-plane orientations. We are thus

observing the phonon scattering averaged along a circle
of radius $|\vec{\kappa}|$ in the reciprocal lattice of the 2-D crys-
tal. If the nonparallelism of the adsorbing surfaces is
included the averaging will have to include a part of
the sphere with radius $|\vec{\kappa}|$. Although no "pure" scans
can be done as in 3-D single crystals it is possible by
choosing different values of $|\vec{\kappa}|$ to measure phonon spectra
which are predominantly zone-boundary transverse phonons
or which are dominated by longitudinal phonons. This
refers to cases where the adsorbed particles have coherent cross sections. For incoherent scatterers we will
observe the 2-D phonon density of states and the $|\vec{\kappa}|$ de-
pendence will be a weighing factor $\kappa^2 \times e^{-2W}$. The out of
plane motion of the adsorbed particles will contribute
to the neutron scattering when the tilt angle is non-
zero but only by a fraction $\sin^2\gamma$ (where γ is the tilt
angle) of their maximum contribution.

   Such phonon measurements have been performed on ^{36}Ar
layers[7] and H_2, HD and D_2[6] layers on Grafoil. ^{36}Ar is
representative for HI-layers and the hydrogen layers are
representative for the HC layers with the $\sqrt{3}$ structure,
although in the latter case large differences must be
expected for particles (adatoms) of different size.

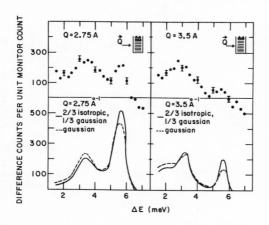

Fig. 16. Inelastic neutron
scattering spectra from
^{36}Ar monolayers adsorbed
on Grafoil at T = 5 K, observed with the scattering
vector $\vec{\kappa}$ parallel to the
plane of the Grafoil discs.
At the top the circles are
the experimental data and
below are the computed
spectra, (from Ref. 7).

   Fig. 16 shows some of the results for the Ar layers
together with a model calculation of the expected scattering profiles. From these results it was concluded
that the in-plane Ar-substrate forces play a very minor
role and that the response is that of an ideal 2-D harmonic crystal. The out of plane motion of the Ar mono-

layer was measured using a geometry in which the sheets were perpendicular to the scattering plane of the neutrons. A localized oscillator response was observed here with an energy of 5.75 meV, but with a tail of intensity extending down to low energies. The latter is interpreted as an effect of the combined motion of the substrate and the Ar layer.

The inelastic neutron scattering response of the adsorbed D_2 and H_2 layers in the registered $\sqrt{3}$ phase, HC, is very different from the Ar result. Only the in-plane motion has been studied, for which some results are shown in Fig. 17. For both H_2 and D_2 it is the coherent scattering function which is measured, and the profiles show no distinct change of shape with $|\vec{\kappa}|$. These results show that the molecules in the $\sqrt{3}$ phase behave like Einstein oscillators with the energies $\hbar\omega_{D_2} = 4.0$ meV and $\hbar\omega_{H_2} = 4.9$ meV, and the full dispersion of the phonon energies is contained within the width of the groups shown in Fig. 17 for $\rho \lesssim 1$. The very simplest picture of the adsorbed molecules, the isolated single particle in a harmonic in-plane potential from the substrate, cannot be correct, however, because the total amplitude of this potential for hydrogen molecules is about 3 meV. Therefore, the neighbour forces between the H_2 (or D_2) molecules in the layer must be included in the model, but no calculation has been performed so far.

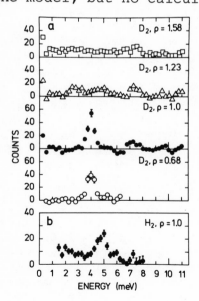

Fig. 17. Inelastic neutron scattering spectra from D_2 and H_2 monolayers adsorbed on Grafoil at T = 5 K. The scattering vector was κ = 2.25 Å^{-1} for the D_2 scans and κ = 1.6 Å^{-1} for the H_2 scan (from Ref. 6).

It is worth noting that the isotope shift is much less than $\sqrt{2}$ and the width of the phonon group is about 1 meV for H_2 but small, equal to the experimental resolution for the D_2 layers. Both facts must be explained by the larger zero point motion of the H_2 molecules. The root mean square displacement of the H_2 molecules, assuming a harmonic oscillator, is 30 per cent of the distance to the boundary of the hexagons of the graphite potential.

Near completion of a D_2 monolayer of the densest HI structure (ρ = 1.58 in Fig. 17) the phonon response resembles that of the Ar-layers, but the intensity is not sufficient to adequately determine the spectra.

C. Phase Transitions of the Adsorbed Layers on Graphite

The nature of the phase transitions of the adsorbed films is more difficult to study experimentally than is the study of structures. When approaching continuous transitions the systems have fluctuations extending over large distances and thus very large coherent areas are needed with no faults on the substrate. Existing neutron scattering data gives only an approximate rather uncertain picture of the transitions.

Ar and N_2 layers: As described above Ar monolayers on Grafoil have the HI structure at all fillings and temperatures up to the melting temperature. The neutron scattering Bragg peak reflects the effective crystallite linear dimension, or correlation length L. For submonolayers at low temperatures L was found to be independent of filling and of the order L $\sim$ 110 Å for all monolayers studied on Grafoil. This means that the adsorbate clusters into relatively large crystallites (L > 110 Å), with L controlled by substrate surface imperfections. This is confirmed by measurement in which other graphite substrates like UCAR ZYX are used. In such measurements values of L at low temperatures are independent of T, and larger than 300 Å.

The temperature dependences of L (Ar) and L (N_2) are shown in Fig. 18. Registered N_2 maintains L $\sim$ 110 Å up to a sharp, possibly first order "melting" region, where L decreases to L $\sim$ 20 Å. In contrast, for the incommensurate Ar films a continuous decrease of L was observed over the range 40 < T < 80 K. This behaviour indicates

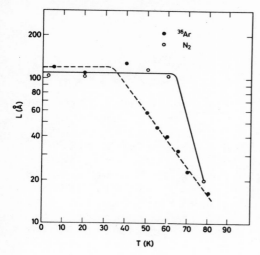

Fig. 18. Temperature dependence of the correlation range (or cluster size) L in adsorbed ^{36}Ar and N_2 layers (from Ref. 7).

that there is no qualitative difference between the "crystalline" and the "fluid" states for submonolayer Ar. 2-D computer molecular dynamics (MD) simulations[19] are in excellent agreement with both the neutron scattering and heat capacity measurements on Ar, indicating that such continuous behaviour is a property of 2-D atomic systems. Neither the heat capacity nor the MD calculations showed any evidence for a gas-liquid coexistence.

Inelastic coherent phonon measurements from the Ar[7] films offer further insight into this melting transition. At low temperatures the in-plane measurements show a transverse phonon dominated peak at 3 meV and a longitudinal phonon dominated peak at 5.5 meV (Fig. 19). For a 2-D harmonic solid the profiles shown on the right hand side of Fig. 19 would be temperature independent. The figure shows that around 40 K the transverse phonon peak moves to smaller energies, suggesting that the melting process involves a continuous decrease in the resistance to shear.

The Ar-monolayer density is strongly temperature dependent with a total linear expansion of 9 per cent up to 60 K. In contrast, the bulk solid has only a 2 per cent expansion in the same range. Approximately 2/3 of the increase of a_{nn} occurs in the range 40 < T < 60 K above which the Bragg peak is too broad to determine a_{nn}[7].

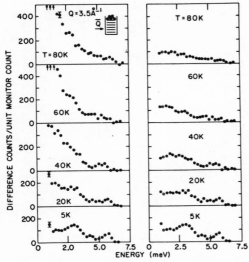

Fig. 19. On the left: the
temperature dependence of
in-plane spectra from Ar
monolayers at Q = 3.5 Å^{-1}.
On the right: the tempe-
rature dependence of the
nominal "density of phonon
states" obtained by dividing
the observed spectra by the
phonon population factor
$n(\omega)$ + 1, (from Ref. 7).

Kr-layers: The melting transition of the HC-phase
is expected to be continuous close to ρ = 1, but changing
to be of first order at a value of ρ smaller than 1[20].
This behaviour is in agreement with recent X-ray measurements on Kr-monolayers on UCAR-ZYX[17]. These measurements
showed close to ρ = 1 a sharp melting transition which
could be parametrized with $\beta \sim 0.08$.

O_2-layers: In O_2-films on graphite three different
melting transitions can be studied. In the submonolayer
δ-phase we have a system equivalent to the Ar-layers,
but the melting transition of the O_2 δ-phase is much
sharper, the Bragg peak intensity decreasing from about
its full low temperature value to zero within 2-3 °K.
No accompanying variation of correlation length L was
observed.

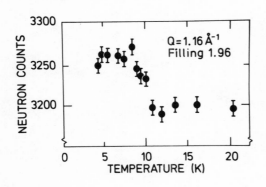

Fig. 20. The neutron scattering intensity at the
position of the maximum of
the magnetic superlattice
reflection from adsorbed O_2
layers in the α-phase. The
substrate scattering is not
subtracted, (from Ref. 11).

The melting of the β-phase of O_2 layer is a somewhat different problem because here we always have some 2nd layer population. In the phase diagram, Fig. 11, the "transition region" indicates that the β-phase Bragg intensity disappears over a 5-10 °K broad region. No significant change of L was observed.

The transition between the α-phase and the β-phase in O_2 films on graphite has been investigated both by measuring the nuclear and the magnetic Bragg peaks. In Fig. 20 the scattering intensity at the position of the low temperature magnetic Bragg peak maximum is shown as a function of temperature. The substrate scattering was not subtracted. The variation of the magnetic Bragg intensity indicates that the α-β transition is continuous, but the magnetic scattering is too small to allow this to be studied in detail.

The structural aspects of the α-β transition were studied by measuring the splitting of the nuclear peak from O_2 films on UCAR ZYX. The splitting $\tau_2-\tau_1$ is a

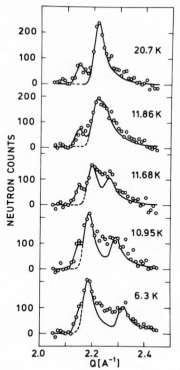

Fig. 21. Neutron diffraction groups observed from O_2 monolayers at constant filling but at varying temperatures through the δ+α and δ+β regions of the phase diagram (fig. 3). The ZYX UCAR graphite was used as substrate. The peak at Q = 2.16 Å^{-1} is a coexisting δ-phase peak. The curves are calculated lineshapes and the splitting of the two α-phase peaks is varied to obtain the best fit to the observed points, (from Ref. 11).

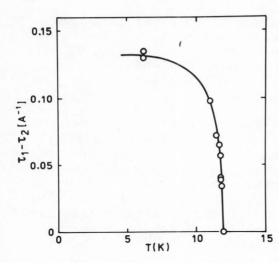

Fig. 22. The splitting of the (10) Bragg peak observed from monolayers of O_2 in the α-phase. The values of the splitting $\tau_1 - \tau_2$ are determined by fitting the curves to the observed points in fig. 12, (from Ref. 11).

measure of the <u>magnetic</u> order parameter. The temperature dependence of this splitting is shown in Figs. 21 and 22. The sharpness of the transition corresponds to a continuous one with small β ($\beta \approx 0.1$ to 0.2), but since the data only goes to $\Delta\tau/\Delta\tau (T = 0) \approx 0.25$ it is not possible to derive a reliable value. The major difficulty in obtaining values of the critical exponents from measurements on adsorbed films stems from the finite coherence length L of the 2-D layers, which gives rise to a spread of the transition temperature. This effect dominates the measurements in the critical regime even when the UCAR ZYX substrate is used.

The continuous nature of the α-β transition is also demonstrated by the absence of a coexistence region between the two phases. In a melting transition such a coexistence is often difficult to observe, if present, because the fluid phase has no distinct diffraction result. In constrast to this a transition between two ordered 2-D phases can be followed by observing the intensities of the two sets of Bragg peaks. This was very clearly seen in the δ-β phase change of the O_2 layers (Fig. 11) as described above.

The α-β transition has also been studied by magnetic susceptibility[21] and specific heat measurements[22] and it has been analyzed by Riedel and Domany[23].

$\underline{H_2, D_2 \text{ and HD layers}}$: At submonolayer densities, $\rho < 1$, the hydrogens have the HC structure. The melting of this is seen as a disappearance of the Bragg intensity

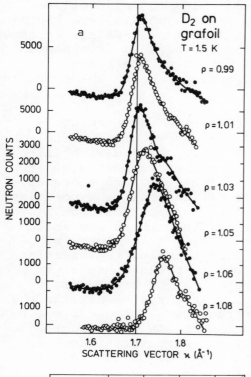

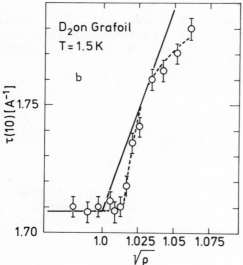

Fig. 23. (a) Neutron scattering groups from D_2 monolayers on Grafoil near $\rho = 1$. (b) The position of the Bragg peaks of fig. (a) as given by $\tau(10)$ vs. the square root of ρ.

in a $\sim$ 2 ^{O}K broad region around the melting line in
Figure 6. In principle it is possible by analyzing the
Bragg intensity as function of filling to distinguish a
region of coexistence of a dense ordered and a dilute
fluid phase from a continuous phase with disolved vacancies, the two situations having the same ρ-value. Attempts
to do this have not been successful because it appeared
difficult to get reproducible values of the Bragg intensity for small ρ-values, probably because the annealing
process is very slow.

Near ρ = 1 the layers undergo the commensurate -
incommensurate transition. Fig. 23 shows how the (10)
Bragg peak from D_2 layers on Grafoil changes with fillings near and just above ρ = 1 at low temperatures. The
main features are that the peak intensity drops dramatically by increasing ρ beyond ρ = 1 and the peak position
gradually moves out from the commensurate position,
$\tau(10) = 1.703$ A^{-1}. Fig. 23b shows that there are small
but significant deviations from the $\tau(10)$-values, as
derived from the Bragg peaks, and the line representing
the ideal homogeneous structure. It needs about 3 per
cent overfilling to get the monolayer to break off registry. Then around ρ = 1.06-1.08 the system attains the
τ-values of the homogeneous structure but above ρ = 1.08
again the τ-values get too small. At low temperature the
Bragg peaks did not show satellite structures. However,
at higher temperatures a drastic change of the melting
behaviour was observed as shown in Fig. 24 and at the
same time very distorted groups were seen, Fig. 25. These
distortions should be interpreted as originating from coexisting phases or from either hexagonal superstructures
or a unidirectional distortion of the layers[24]. So far no
fitting of the distorted groups to the models has been
done.

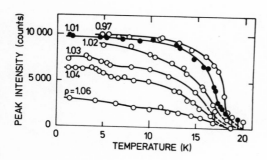

Fig. 24. The peak intensity
of the (10) Bragg peaks
from monolayer of D_2 on
Grafoil plotted as function
of temperature for different fillings near the commensurate-incommensurate
transition at ρ = 1.

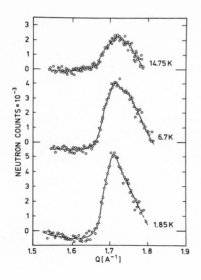

Fig. 25. Neutron scattering groups at the (10) Bragg reflection from monolayer D_2 on Grafoil measured at the different temperatures given in the figure. The filling is $\rho = 1.04$.

The picture which emerges from the diffraction results for the hydrogen layers is: At $\rho \leq 1$ we have the
HC structure, the melting transition is rather narrow
and T_m is filling dependent. At $\rho \gtrsim 1$ and $T < 5 \, ^\circ K$ the
HC phase goes into an HI phase after $\rho > 1.03$, but little
or no distortion of the Bragg peaks is seen (the intensity however decreases abruptly); At $\rho \gtrsim 1$, but $T > 5$ K
distorted groups are observed and a change of intensity
with temperature (Fig. 24) indicates that a phase change
may occur. At values of ρ between 1.13 and 1.3 $\AA^{-1}$
another phase change may occur between a dilute HI phase,
still strongly influenced by the in-plane substrate potential and an HI much denser phase, which is dominated
by the admolecule neighbour forces and have densities
close to that of the dense packed planes of bulk solid
D_2.

The dynamic response of the $\rho \gtrsim 1$ layers has been
measured by inelastic scattering experiments on HD layers
adsorbed on Grafoil. By using the HD molecules we get the
full incoherent scattering from the H-atom per molecule
and thus the phonon density of states can be observed
with good intensity. The results can be compared with
the coherent response shown in Fig. 17, and they confirm that for $\rho < 1$ the hydrogen layers behave like local-

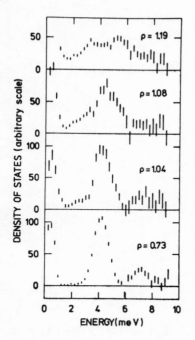

Fig. 26. Phonon density of states (T = 4.2 K) for HD adsorbed on Grafoil as function of filling ρ.

ized oscillators with frequencies $h\omega_{H_2}$ = 4.9 meV, $h\omega_{HD}$ = 4.2 meV, and $h\omega_{D_2}$ = 4.0 meV. The widths of the frequency bands are $w_{H_2} \sim w_{HD} \sim$ 1 meV whereas w_{D_2} < 0.5 meV. The higher intensity of the HD measurements shows a second peak around 7 meV with 1/5 of the intensity of the 4.2 meV peak.

At $\rho \gtrsim$ 1 the HD-phonon density of states broadens in the region of fillings up to $\rho \sim$ 1.08, and for higher ρ-values a splitting of the frequency spectrum is seen. The setting in of the splitting coincides roughly with breaking away from the ideal τ-line in Fig. 23b at τ > 1.76 Å^{-1}, which supports the assumption of a phase transition between ρ > 1.13 and ρ > 1.3 in addition to the commensurate - incommensurate transition at near ρ = 1.

The same commensurate - incommensurate phase transition has been studied by X-ray diffraction on Kr adsorbed on UCAR-ZYX[17]. These measurements show that at 80 K and at 89 K the transition is at least nearly continuous and they give a value of the mean exponent β = 0.30±0.06. The X-ray results are in good agreement with LEED measurements[25] on Kr adsorbed on single crystals of graphite at T = 54-59 K and both conclude that there appears to be a single hexagonal-commensurate to hexagonal-incommensurate transition which is of second order.

Thus at present the commensurate - incommensurate phase transition seems to be different in the adsorbed layers of the light gases H_2, D_2 (and probably He) and in the layers of the heavier gases like Kr.

IV.   CONCLUSION

The neutron scattering experiments have given a considerable amount of data on structures and the dynamics of monolayers adsorbed on graphite, and the nature of several phase transitions has been studied. Many of these results probably should be considered as showing the gross features only and many details can have been hidden in poor resolution. The main reason for this is the limited coherence size of the 2-D structures caused by the faults on the adsorbing surfaces. Therefore, in many cases a next generation of measurements are wanted in which better quality substrate than Grafoil is used and UCAR ZYX can sometimes be a useful substrate but often intensities get too small. However, X-ray scattering has already proven to be a very fruitful technique for adsorbants heavier than helium and the hydrogens and much improved intensities are obtained. Thus for the future the X-ray technique is preferable in many diffraction measurements whereas the neutron scattering technique still will be superior in studying light elements and in any case of the dynamics of the physisorbed layers.

References

1. Grafoil, Graphon and UCAR-ZYX are products made by
       Union Carbide Corp., Carbon Product Div.,
       270 Park Ave., N.Y., USA. Papyex is a product
       similar to Grafoil, made by Le Carbon Lorraine,
       45 rue des Acacois, 75821, Paris Cedex 17.
2. J. G. Dash, "Films on Solid Surfaces", Academic
       Press (1975).
3. J. P. McTague, M. Nielsen, and L. Passell, CRC Crit.
       Rev. Solid State Mat. Sci. 8: 135 (1978).
4. M. Nielsen, W. D. Ellenson, and J. P. McTague, Neutron
       Inelastic Scattering, JAEA, Vienna 1977 p. 433.
5. Y. Imry, Phys. Rev. B3, 3939 (1971).
6. M. Nielsen, J. P. McTague, and W. D. Ellenson, Jour.
       de Phys. Suppl. Col 38, C4-10 (1977).
7. H. Taub, K. Carneiro, J. K. Kjems, L. Passell, and
       J. P. McTague, Phys. Rev. B16, 4551 (1977).
8. P. Thorel, B. Croset, C. Marti, and J. P. Coulomb,
       Proc. Conf. Neutron Scattering, R. M. Moon, Ed.,
       Nat. Tech. Inf. Serv., Springfield, Va, USA
       (1976) p. 85.
9. L. Passell, W. D. Ellenson, J. P. McTague, and
       M. Nielsen, unpublished.
10. J. K. Kjems, L. Passell, H. Taub, and J. G. Dash,
       Phys. Rev. Lett. 32, 724 (1974) and J. K. Kjems,
       L. Passell, H. Taub, J. G. Dash, and A. D. Novaco,
       Phys. Rev. B13, 1446 (1976).
11. M. Nielsen and J. P. McTague, Phys. Rev. B19, 3096
       (1979).
12. S. Sinha, to be published, and J. P. Coulomb, M.
       Bienfait, and P. Thorel, Phys. Rev. Lett. 42,
       733 (1979).
13. J. Suzanne, J. P. Coulomb, M. Bienfait, M. Matecki,
       A. Thorny, B. Croset, and C. Marti, Phys. Rev.
       Lett. 41, 760 (1978).
14. M. Nielsen and W. D. Ellenson in Proceedings of the
       14th International Conference on Low Temperature
       Physics, Otaniemi, Finland 1975, edited by M.
       Krusins and M. Vuorio (North Holland, Amsterdam,
       1975), Vol. 4, p. 437.
15. K. Carneiro, W. D. Ellenson, L. Passell, J. P.
       McTague, and H. Taub, Phys. Rev. Lett. 37, 1695
       (1976).
16. M. D. Chinn and S. C. Fain, Phys. Rev. Lett. 39, 146
       (1977).

17. P. W. Stephens, P. Heiney, R. J. Birgeneau, and
     P. M. Horn, Phys. Rev. Lett. 43, 47 (1979), and
     P. M. Horn, R. J. Birgeneau, P. Heiney, and E. M.
     Hammand, Phys. Rev. Lett. 41, 961 (1978).
18. C. A. English, J. A. Venables, and D. R. Salahub,
     Proc. R. Soc. Lond. A. 340, 81 (1974).
19. F. E. Hanson, M. J. Mandell, and J. P. McTague,
     J. Phys. (Paris) 38: C4-76 (1977).
20. S. Ostlund and A. N. Berker, Phys. Rev. Lett. 42,
     843 (1979).
21. S. Gregory, Phys. Rev. Lett. 40, 723 (1978).
22. O. Vilches, to be published.
23. E. Domany and E. K. Riedel, J. Appl. Phys. 49, 1315
     (1978).
24. P. Bak, D. Mukamel, J. Villain, and K. Wentewska,
     Phys. Rev., to appear.
25. Private communication R. J. Birgeneau and P. W.
     Stephens.

NUCLEAR MAGNETIC RESONANCE IN ADSORBED HELIUM

MICHAEL RICHARDS

School of Mathematical and Physical Sciences, University
of Sussex, Falmer, Brighton, BN1 9QH, England

1. INTRODUCTION

The study of surface films - particularly molecular monolayers - has advanced very significantly in the last few years.
This advance has come about through the use of an increasingly
wide range of techniques for the study of films and also through
the development of better characterized substrates. Nuclear
magnetic resonance (NMR) is a relatively new technique to be applied
in this field though NMR measurements on adsorbed ^{3}He were first
reported[1] in 1963. Whereas in the 1960's the field was characterized by isolated studies of different systems by different techniques,
it is now possible to build comprehensive models of surface films
through the use of complementary techniques applied to well-
characterized systems.

The field can be sub-divided in several ways but with a view
to discussing the contribution of NMR it is useful to divide the
techniques up into two sets of two categories. The first category
is determined by whether the technique can be applied to a single
macroscopic surface of area 10^{-6} to 10^{-4} m^2 or whether an area 10^0
to 10^2 m^2 is needed which calls for some finely divided surface such
as a powder or porous adsorbate. Electron and molecular spectroscopy studies fall into the former category, whereas specific heat,
vapour pressure, nuclear spectroscopy, Mössbauer spectroscopy as
well as NMR need the larger areas in order to get useful signals.
The second category concerns the type of information obtained. Most
techniques provide either <u>spatial</u> information about the surface
layer molecules or <u>temporal</u> information. Some techniques, at least
in principle, yield both types of information and a few techniques
such as the molecular spectroscopy experiments of Boato et al[2]

165

don't readily fall into either set.

The reason why NMR is potentially an important tool for the study of surface films is because it can yield fairly directly temporal information about the molecules in the surface films. Quasi-elastic neutron scattering is also able to yield temporal information and the time scale to which it is sensitive (10^{-8} to 10^{-14} sec) complements that of NMR (10^0 to 10^{-10} sec).

Most spectroscopic techniques involve an incident beam or an applied field and the response of a plane of atoms is anisotropic with respect to the angle between the plane normal and the field or beam direction. This raises some problems for those spectroscopic techniques like NMR which require porous substrates since some geometrical information about the disposition of substrate surfaces may be needed to interpret fully the spectroscopic data.

2. NMR IN TWO DIMENSIONS

(a) General

A large (10^{-2} to 10^1 Tesla) field B_O along the z direction is applied to the system and the response to a small (10^{-7} to 10^{-4} Tesla) oscillating field $2B_1 \cos \omega t$ applied in the x-y plane is studied. Close to $\omega = \gamma B_O$ where γ is the magnetogyric ratio for the nucleus studied, the rf field is found to have a large effect on the magnetic moment $\underline{M}$ of the system. In particular energy will be absorbed from the field over a range of frequencies $\Delta\omega$ and on removing the rf field, M_z, the magnetisation parallel to B_O, will return to equi-librium with a time constant T_1, while the magnetization perpendicular to z relaxes to zero with a time constant T_2.

NMR signals can either be observed in the time domain (pulsed NMR) when pulses of B_1 field are used to tip $\underline{M}$ away from its equi-librium value $\chi_O B_O$ (χ_O is the static magnetic susceptibility). The creation of a component of $\underline{M}$ rotating in the x-y plane produces in a signal coil a voltage whose time development is studied. Alter-natively, the steady state response of $\underline{M}$ to a small continuous B_1 field can be studied as a function of the frequency of this field. In principle both experiments yield the same information but in practice the time domain technique is the more powerful because weak broadening of the resonance due to spin dependent interactions can be studied in the presence of much larger broadening due for instance to magnet (i.e. B_O) inhomogeneity.

(b) Information from NMR Measurements
(i) Susceptibility. The strength of the signal is proportional

to the equilibrium magnetisation and hence to χ_o. In many situations the resonance line has structure indicating that the nuclei are not interchanging frequently enough between different positions to yield a single resonance line. In adsorbed films this applies to the signals from first and second layer atoms whose dynamics may be very different but the interlayer exchange rate may not be high enough to create a single set of NMR properties.

For all films other than those composed of ^{3}He atoms, χ_o will obey Curie's law, $\chi_o \propto T^{-1}$ and this allows a check to be made on whether a significant number of molecules have "disappeared", i.e. have moved to positions where the NMR properties are different enough for them not to contribute significantly to the main NMR line. Below 1K, ^{3}He films show quantum degeneracy effects and the temperature dependence of χ_o can be used to determine whether the atoms are localised or not. A two dimensional (2D) gas of non-interacting ^{3}He atoms at a temperature T well below the 2D Fermi temperature T_F has a susceptibility $n\mu^2/k_B T_F$ rather than the Curie value $n\mu^2/k_B T$ where n is the areal density of fermions of magnetic moment μ. Switching on the spin-independent interactions causes the effects of quantum degeneracy to be reduced and the susceptibility is increased towards the Curie value.

(ii) Relaxation Times. T_1 and T_2 are determined by the fluctuations in magnetic field experienced by a single spin. Starting with the classic work of Bloembergen, Purcell and Pound[3], there have been many approaches to the theory connecting parameters describing the motion of particles such as correlation times and diffusion coefficients with the observed parameters, that is the relaxation times T_1 and T_2. We follow here the approach of Kubo and Tomita.[4]

The decay of longitudinal and transverse magnetism is governed by time correlations existing between components of the spin vectors, i.e.

(a) $M_{long}(t)/M_o \equiv L(t) = \mathrm{Tr}\ \{I_z(t)\ I_z(0)\}/\mathrm{Tr}\ \{I_z^2\}$

(1)

(b) $M_{trans}(t)/M_o \equiv F(t) = \mathrm{Tr}\ \{I_+(t)\ I_-(0)\}/\mathrm{Tr}\ \{I_+ I_-\}$

where the time development of the I operators is generated by the total spin Hamiltonian, H,

$$\frac{d}{dt}\ \hbar I_\alpha(t) = i\ \left[H,\ I_\alpha(t)\right]$$

The Hamiltonian is made up of Zeeman and dipole parts, $H = H_z + H_{dip}$ where

$$H_z = \hbar I_z\ \omega_o$$

$\hbar I_z$ being the total nuclear spin angular momentum in the z direction.

$$H_{dip} = \sum_{i<j} (h^2\gamma^2/r_{ij}^3) \{ \underline{I}^i \cdot \underline{I}^j - 3(\underline{I}^j \cdot \underline{r}_{ij}) (\underline{I}^j \cdot \underline{r}_{ij})/r_{ij}^2 \}$$

where $\underline{r}_{ij}$ is the vector joining the ith and j^{th} spins whose angular momentum are, respectively, $h\underline{I}^i$ and $h\underline{I}^j$. We separate H_{dip} into its various spin-flip components,

$$H_{dip} = h^2\gamma^2\sqrt{4\pi/5} \sum_{m=-2}^{2} \sum_{i<j} Y_2^{-m} (\theta_{ij}, \phi_{ij}) T_{ij}^m (-1)^m/r_{ij}^3 \qquad (2)$$

where Y_2^{-m} are spherical harmonics and T_{ij}^m are the corresponding tensor operators in spin space.

The essence of the Kubo method is to use an approximate relation (involving the second cumulant approximation) between the relaxation functions and the $G_m(\tau)$ as follows,

$$F(t) = \exp - \int_0^t (t - \tau) \sum_{m=-2}^{2} (3 - m^2/2) \ G_m(\tau) \exp im\omega\tau d\tau \qquad (3a)$$

$$L(t) = \exp - \int (t - \tau) \sum_{m=-2}^{2} m^2 G_m(\tau) \exp im\omega\tau d\tau \qquad (3b)$$

where $G_m(\tau)$ are the local field autocorrelation functions defined by

$$G_m(\tau) = \frac{3\pi h^2\gamma^4}{5} \sum_{i<j} \{Y_2^{-m}(0)/r_{ij}^3(0)\} \ \{Y_2^m(t)/r_{ij}^3(t)\} \qquad (4)$$

The time dependence of $Y_2^{\pm m}(t)$ arises from the time dependence of θ_{ij} and ϕ_{ij}.

F(t) and L(t) will decay exponentially for times of interest provided the expressions inside the summation signs decay fast enough. For $m \neq 0$ this is ensured by the oscillating factor $\exp im\omega\tau$. For the $m = 0$ (adiabatic) term it is necessary for $G_0(\tau)$ to decay sufficiently fast for $(t - \tau)$ to be replaced by t leading to

$$(1/T_2)_{ad} = (3/2)J_0(0) \qquad (5)$$

where the spectral density functions $J_m(\omega)$ are defined as the Fourier Transforms of the correlation functions $G_m(\tau)$, i.e.

$$J_m(\omega) = \int_{-\infty}^{\infty} G_m(\tau) \exp im\omega\tau \ d\tau \qquad (6)$$

If we include the non-adiabatic terms we now obtain

$$1/T_2 = (3/2) \ J_0(0) + (5/2)J_1(\omega) + J_2(2\omega) \qquad (7a)$$

$$1/T_1 = J_1(\omega) + 4J_2(2\omega) \qquad (7b)$$

It should be stressed that these expressions apply to a system subject only to Zeeman and dipole fields. The substrate may create magnetic fields either through its instrinsic diamagnetism or due to impurities. Mobile adsorbed atoms will then experience time dependent local fields which will have to be treated in a way similar to the dipole field analysed above.

In most isotropic spin systems all three spectral density functions have the same analytic form and it is usually found that a single time constant, usually called the correlation time τ_c governs the decay of $G_m(\tau)$ and hence of $J_m(\omega)$. Exponential and Gaussian functions are often used for $G_m(\tau)$ and lead, respectively, to Lorentzian and Gaussian forms for $J_m(\omega)$. In the former case this leads to the well known expressions for T_1 and T_2.

$$\frac{1}{T_2} = M_2 \left[\tau_c + \frac{5\tau_c/3}{1 + \omega^2\tau_c^2} + \frac{2\tau_c/3}{1 + \omega^2\tau_c^2} \right] \tag{8a}$$

$$\frac{1}{T_1} = \frac{2M_2}{3} \left[\frac{\tau_c}{1 + \omega^2\tau_c^2} + \frac{4\tau_c}{1 + 4\omega^2\tau_c^2} \right] \tag{8b}$$

where M_2 is the second moment of the NMR absorption line and is equal to $3G_0(0)$. The expression for T_2 is only correct if $G_0(\tau)$ decays fast enough so that $\sqrt{M_2}\tau_c < 1$. Otherwise the line width $1/T_2$ is given by its static value $\sim \sqrt{M_2}$. We are thus lead to the classic BPP picture of how relaxation times vary with τ_c, as shown in Figure 1.

M_2 is calculable if the positions of the spins are known (e.g. on a solid lattice) and τ_c can be easily determined from T_1 or T_2. By measuring T_1 at different frequencies one directly measures $J(\omega)$ and this can be used to check the assumed form for $G(\tau)$. If the T_1 minimum is observed it gives direct information about M_2 and τ_c.

The correlation time is the time for which two spin dipolar field correlations last and is thus closely related to the mobility in fluids and the hop time in solids, i.e. to the diffusion coefficient in both cases.

(iii) Diffusion. The diffusion coefficient D can be measured via the correlation time τ_c using the relation $D \approx r^2/6\tau_c$ for spins randomly hopping a distance r every τ_c. However this assumes some microscopic model for motion, e.g. vacancy - assisted hopping in solids or transitional diffusion in liquids. A more direct measurement of translational diffusion in liquids can be obtained by applying a linear field gradient G and observing the height of the spin echo at τ as a function of τ or G.

$$h(\tau) = h(0) \exp(-\tau/T_2) \exp(-\gamma^2 G^2 D\tau^3/12).$$

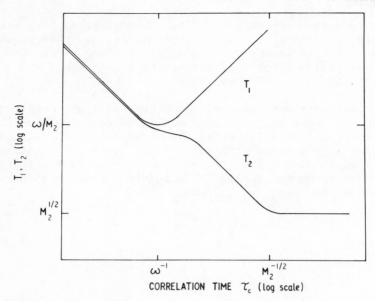

Figure 1. T_1 and T_2 as functions of the correlation time for an isotropic NMR system according to BPP theory(Ref.3). $\omega/2\pi$ is the Larmor
frequency and M_2 the second moment of the resonance line.

(c) Special Problems in Two Dimensions

      (i) Inhomogeneity and substrate effects. Dipolar coupling is
the only source of relaxation in a spin 1/2 atomic system like ^{3}He
and yet it is extremely weak, particularly in fluids. In liquid
helium for instance, $M_2 \sim \hbar^2\gamma^4/a^6 \sim 10^8$ sec^{-2} and $\tau_c \sim 10^{-11}$ sec
giving $T_1 = T_2 \sim 10^3$ sec from Eqs. (8) since $\omega\tau_c \ll 1$. For this
reason dipolar coupling rarely predominates in bulk fluid systems:
wall relaxation, paramagnetic impurities or in molecules, some inter-
molecular process usually provides a stronger relaxation path. In
2D it is even less likely that dipolar relaxation will predominate
since a substrate is required to confine the motion to a plane and
this may provide magnetic fields either through its intrinsic
magnetic properties or through impurities. In addition, adsorbed
spins are not likely to travel macroscopic distances/meeting a "wall"
i.e. an interruption in the perfect plane, and here both the local
fields and/or the correlation time are likely to change radically.
It was approximately five years after NMR studies were begun on
liquid ^{3}He before intrinsic (i.e. dipolar) T_1 values were obtained.[5]
For gaseous ^{3}He (where M_2 and τ_c are even smaller) it was about 15
years before ways were found to reduce wall relaxation sufficiently
to allow dipolar fields to predominate.[6]

To show how much worse the problem is for adsorbed ^{3}He we can use an expression for relaxation of a system of spins a small fraction α of which are found in a region where their relaxation time is T_2'. Provided that the lifetime in this region is $\ll T_2'$ then the observed relaxation time for the whole system due to visits to these boundary regions will be T_2'/α. In a macroscopic 3D system the fraction of spins adsorbed on the wall is usually less than 10^{-6} but to have a similar factor for an adsorbed system it would be necessary to have domains of perfect surface of size $L \sim 10^6 a$ where a is the atomic size, i.e. $L \sim 300\mu m$, whereas the greatest lengths of faultless crystalline surface reported in high specific area substrates is about $0.1 \mu m$. These considerations make it unlikely that that intrinsic relaxation will be observed in the fluid phases of adsorbed ^{3}He until there is a further dramatic improvement in substrate preparation.

The problem is markedly less severe in the solid phases because τ_c is here 10^{-5} to 10^{-8} sec making relaxation by the dipolar mechanism correspondingly stronger and boundary relaxation correspondingly weaker, and often a spin will not get to a boundary before it has relaxed by the dipolar mechanism.

Similar considerations apply to the effects of paramagnetic impurities which may be adsorbed on the surface or embedded in the substrate. Para or diamagnetism of the substrate is another problem whose effect increases as the mobility of the spin increases. Discussion of this is deferred until Section 4(d).

(ii) Anisotropy. The effect of dipolar coupling between two spins on the NMR relaxation times is changed if the spins are confined to move in a plane whose normal is inclined at an angle β to the DC field B_0. For instance if $\beta = 0$ then θ_{ij} in Eq.(2) is always $\pi/2$ and $Y_2^{\pm 1}(\theta_{ij} \, \phi_{ij}) = J_1(\omega) = 0$.

The effect of this is to make all the NMR parameters M_2, τ_c, T_1 and T_2 functions of β.

Several discussions of these effects have been published[7] but any relation to experiment has to consider an entirely separate effect.

(iii) Effects of Diffusion. In the decaying correlation functions of Eq. 4 there are two separate regions: The short time part reflects the microscopic nature of the motion (e.g. vacancy-assisted hopping or quantum mechanical tunnelling in solids, Brownian motion in liquids). For times longer than the correlation time τ_c the motion can be described by the diffusion equation and this leads to the correlation functions having a long time tail varying as $t^{-d/2}$ where d is the dimensionality of the system[8]. The case of $d = 2$ is one of critical dimensionality since integrals like

$\int_{t_c}^{\infty} G_o(\tau) d\tau$ will diverge logarithmically. For $d = 3$ there is no
divergence and for $d = 1$ the divergence is more severe. The
divergence in 2D is sufficient to prevent one using the formalism
of section 2b) (ii) and an expression like Eq.5 is incorrect because
the system is not fully motionally narrowed. The effect of this is
not to give a singularity in $J_o(\omega)$ at $\omega = 0$ as suggested by Sato
and Sugawara[9] but to cause $J_o(\omega)$ to continue to rise in an easily
calculable manner as ω falls below $1/\tau_c$. The main effect of this
is to prevent T_2 from becoming equal to T_1 as τ_c is shortened below
ω^{-1} because of the low frequency components always present in $J_o(\omega)$
due to diffusion, however fast the microscope motion may be. Figure
2 illustrates schematically the effects of diffusion and how it leads
to different effects in 2D and 3D.

To obtain an expression for T_2 it is necessary to return to
Eq.(3a) and evaluate

$$3 \int_o^t (t - \tau) \; G_o(\tau) d\tau$$

including the τ^{-1} tail which has to be joined onto the short   part
of the decay in some arbitrary manner. The effect[10] is to cause the

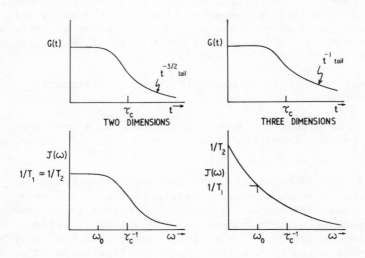

Figure 2. Correlation functions G(t) and spectral densities J( )
for two and three dimensional systems.

transverse magnetism to decay non-exponentially because of a term in $\ln t/2\tau_c$ in the exponent. Since measurements are made with $t \sim T_2$ we can say that the effect is to increase $1/T_2$ by a factor of about $\ln T_2/2\tau_c$. T_1 and the non-adiabatic contribution to T_2 will not be affected since the long time tail in the $G_m(\tau)$ does not contribute significantly to $J_1(\omega)$ or $J_2(\omega)$ unless $\omega < 1/\tau_c$.

3. REVIEW OF NMR MEASUREMENTS ON ADSORBED HELIUM

The first published work[1] on NMR of adsorbed ^{3}He appeared in 1963 and since that time there has been a development towards obtaining data which relate to 2D phases of ^{3}He. Table I lists the main groups who have worked in the field together with details of substrates, frequencies, etc. and type of data collected.

The early work of Careri and coworkers[11], Monod and Cowan[12], and Weaver[16] make use of the properties of the zeolite molecular sieves which provide a large specific area for adsorption because they consist of a complex matrix of interconnected small pores (typically about 13A in diameter). Later work by Brewer, Thomson and coworkers[13 - 15] used Vycor, a porous glass with pores of about 70A diameter.

These early studies showed that good signals can be obtained from adsorbed ^{3}He, that it is not difficult to find substrates where paramagnetic impurities do not play an important role and that there is no difficulty in cooling the adsorbed spins down to at least 0.4K. In some cases measurements of T_1 and T_2 taken at varying temperatures and/or areal density showed qualitative resemblance to Fig. 1 and some approximate τ_c values could be obtained. However, the data do not give clear indication of surface phases of ^{3}He because features of the substrate dominate the motion of the ^{3}He atoms. Thus for coverages corresponding to less than a complete monolayer the system behaves like a solid but this is because of the localizing effect of the inhomogeneous Van de Waals potential binding the ^{3}He atoms to the wall. As more atoms are put in the pores, motion increases but it is not usually character-ized by a single correlation time and the adatoms do not behave as a 2D system moving primarily under the influence of the adatom-adatom forces.

These early studies illustrate the point that if it is the search for 2D behaviour which motivates the study of adsorbed systems (there are, of course, many other reasons for studying them) then amorphous substrates which offer a multiply connected system of small pores (less than 100A in diameter) are not an attractive approach except in yielding large signals. Nearly all the most recent work has been on graphite in various forms that offer usefully

TABLE I   Summary of NMR Studies on Adsorbed ^{3}He

Group	Date[a]	References	Substrate	Frequency MHz	cw or pulse	parameters measured[b]	Temperature range, K	coverage range	Other variables
Careri(Rome) Santini	1963	11	13 × Zeolite	5	cw	χ	0.42 - 4.2	0.42 - 4.2	
Monod (Paris) Cowan	1966	12	13 × Zeolite	0.4 - 15	pulse	T_1, T_2	1.25 - 22	0.1 - 1	
Brewer (Sussex) Thomson, Reed, Creswell, Goto	1966	13,14,15	Vycor glass	0.6 - 7	cw, pulse	χ, T_1, T_2	0.3 - 4.2	0.3 - 3	^{3}He/^{4}He mixtures
Weaver(Albuquerque)	1972	16	3A 13 × Zeolites	12	pulse	χ, T_1, T_2	1.25 - 4.2	-	pressure in vapour
Rollefson (Seattle)	1972	17	Graphitized C black	20.5	cw	ΔB	1.3 - 4.2	0.3 - 0.9	
Grimmer (St. Louis) Luczszynksi	1972	18, 19	Grafoil	10,20	pulse	χ, ΔB, T_1	0.35 - 4.2	0.3 - 6	^{3}He/^{4}He mixtures
Daunt (New York) Husa, Hegde, Hickernell	1973	20-24	Grafoil and Argon-coated grafoil	30	pulse	χ, ΔB	0.07 - 4.2	0.2 - 1	
				5.5	cw	χ. ΔB	0.35 - 4.2	0.2 - 1	
Richards (Sussex) Cowan, Thomson, Owers-Bradley	1977	25-28	Grafoil, ZYX	0.3 - 2	cw, pulse	χ, T_1, T_2	0.4 - 10	0.2 - 2	$\pi/2 > \beta > 0$ ^{3}He/^{4}He mixtures
Sato (Tokyo) Sugawara	1978	9	Grafoil	10	pulse	χ, T_1, T_2	1.2 - 4.2	0.2 - 2	$\pi/2 > \beta > 0$

[a] Refers to year of first publication.

[b] ΔB refers to cw line width or to $1/\gamma T_2^*$ in pulse methods where T_2^* is the length of free induction decay.

[c] Fractions of a monolayer.

large areas for adsorption. The reason for the widespread use of
graphite is that adsorption isotherms have shown[29] that basal planes
of graphite offer an exceptionally uniform surface to rare gas
adatoms resulting in as many as seven steps in the adsorption
isotherms, corresponding to the clear build-up of distinguishable
layers. An additional point is that the close spacing of carbon
atoms in a basal plane leads to very little "corrugation" of the
Van der Waals potential between the graphite surface and a ^{3}He
atom as it moves across the surface. The corrugation depth[30] is
about 10K but this does not dominate lateral motions of a ^{3}He atom
at 1K because the thermal de Broglie wavelength at that temperature
(about 10A) is much greater than the wavelengths present in the
corrugation (mainly 2.8A).

The graphite is needed in some form that presents an area of
more than 1m per cm and powders or compressed exfoliated sheets
are used. The latter are obtainable as commercial products (Grafoil[31]
and Papyrex[32] and have the advantage of offering some alignment of
the exposed crystallite surfaces. A more recently available form of
exfoliated graphite, called ZYX UCAR oriented graphite has been found[33]
to have larger areas of perfectly crystalline surfaces and to have
greater uniformity of plane direction than grafoil. However these
forms of graphite must be regarded as intermediate between the small
pore amorphous substrates like zeolite and Vycor and an ideal
substrate which would offer crystallite surfaces free of steps for
macroscopic distances and aligned within about 1º. Unfortunately it
is clear that the progressive improvement in substrates which is
particularly important for NMR (because it is primarily mobility
that is studied) is accompanied necessarily by a progressive reduction
in specific area for adsorption and hence in signal strength. Table
II summarises some important features of substrates so far used in
NMR studies of adsorbate ^{3}He.

Five groups have carried out NMR studies on ^{3}He adsorbed on
graphite. Rollefson[17], working with graphitized carbon black powder,
observed very broad resonance lines whose widths varied with coverage
and temperature in ways suggesting that high density monolayers are
solid between 1K and 4K with mobilities caused by thermally activated
defects. Grimmer and Luszczynski[18, 19] showed that the anisotropic
diamagnetism of the graphite creates field gradients which broaden
the resonance and reduce the spin echo lifetime when the spins are
sufficiently mobile. Daunt and coworkers[20] measuring susceptibilities
of ^{3}He films in grafoil found a distinction between high coverage
monolayers which obeyed Curie's law down to below 0.1K (a conclusion
in conflict with Grimmer and Luszczynksi) and low coverage films
which showed the marked departures from Curie's law expected for a
degenerate Fermi system. More recent work from this group[22,23] has
included line width measurements which conflict with those of
Rollefson.

TABLE II

Characteristics of Different Substrates

Substrate	Coherence length[a]	Spins/cm^3 in Monolayer[b]	Angular Distribution of Planes
Zeolite 13x	Amorphous (13A pores)	1.5×10^{22} (full pores)	Isotropic
Vycor glass	Amorphous (70A pores)	2×10^{21}	Isotropic
Grafoil/Papyex	100A	2×10^{20}	30° for 50% Isotropic for 50%
ZYX UCAR	300A	1×10^{19}	3° - 10°
Foam[c]	500A	3×10^{19}	Isotropic
Stack of 50 Silicon Wafers[d]	> 10μ	4×10^{16}	< 1°

[a] Defined as length of perfect crystalline surface.

[b] This is a measure of NMR sensitivity.

[c] This is another product of Union Carbide.

[d] Wafers are 0.025 cm thick with 0.025 cm spaces.

The most recent work has been carried out in the author's laboratory at the University of Sussex[25-27] and by Sato and Sugawara[29] at the University of Tokyo. In both cases grafoil has been used as the substrate. The Sussex group have used both pulse and continuous wave techniques to obtain T_1, T_2 and χ at frequencies between 0.3 and 2.0 MHz. The temperature range was 10K to 0.4K, the coverage ranged from 0.05 to 2.0 completed monolayers (this parameter is called x in this article)* and the angle β between the normal to the grafoil sheets and the applied field B_0 direction was 0, $\pi/4$ or $\pi/2$. The Tokyo group using pulse methods have measured T_1 and T_2, at 10MHz varying T between 4.2K and 1.2K, β between 0 and $\pi/2$ and x between 0.2 and 2 monolayers.

Over most of the range of parameters where there is overlap, the two groups obtain very similar data and draw similar conclusions. Figure 3 for instance shows T_2 data at 1.2K as the coverage is varied from 0 to 1.7 monolayers. Both sets of data have the same

*A completed monolayer is a somewhat arbitrary quantity because it varies with temperature. In this paper it refers to a 4.2K isotherm using the point B criterion (see Ref.34).

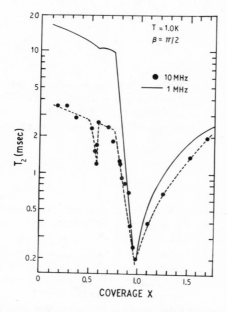

Figure 3. T_2 as a function of coverage at 1K. The data are from two groups. Sussex (Ref.25) at 1MHz (solid line) and Tokyo (Ref.9) at 10 MHz (dashed line and circles).

shape showing the features at x ≈ 0.6 and x ≈ 1.0. The difference
in the values of T_2 for x < 0.7 is due to the different frequencies
used. From such figures it is clear that we have now high quality
reproducible data from which to build a temporal map for 2D helium
describing how the motion of atoms in the system varies with areal
density and temperature. Much of the data is believed to reflect
only the properties of helium atoms but, particularly at low
coverages and high temperatures, it is clear that the substrate
used is still affecting both the motion of the adatoms and NMR data
obtained. In this respect again grafoil is intermediate between
the old amorphous substrates and an ideal substrate.

No mention has been made of a number of studies involving bulk
^{3}He in contact with a surface. In many situations the relaxation
of the bulk helium is controlled by the atoms near the surface. In
this way one can probe surface phases by studying bulk helium.
However in many cases too much interpretation is needed to be sure
of the surface properties deduced and one is in any case restricted
to those surface phases which exist in equilibrium with a high
pressure bulk phase. For this reason studies of ^{3}He gas in contact
with pyrex glass,[35] liquid ^{3}He in a cell filled with carbon
particles,[36] superfluid ^{3}He[37] and solid ^{3}He[38] in contact with various
powders will be omitted in the review, significant though they are
in other contexts. By the same token, no mention is made of work
carried out at x > 1 when there are one or more atomic layers
present in addition to the first layer.

4. SELECTED NMR DATA

Limitations of space preclude the discussion of all published
NMR on adsorbed helium. Instead a selection of topics is made
where NMR provides useful new information about surface phases.
The regions to be discussed are indicated on the phase diagram of
figure 4.

(a) The Low Density Phase

A single helium atom is attracted to an exposed basal plane of
graphite by a potential which varies periodically across the surface.
A band theory calculation shows[30] that the adatom is effectively free
and specific heat studies indicate[34] a heat capacity quite close to
k_B per atom as expected for a perfect classical 2D gas. However,
any classical model giving two degrees of freedom per atom will
yield this value (e.g. a 1D simple harmonic oscillator). NMR
measurements can be expected to show gas-like behaviour more
directly from the temperature dependence of the magnetic susceptibility χ and the relaxation times T_1 and T_2.

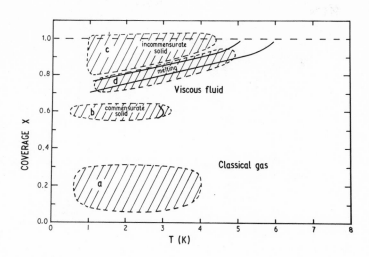

<u>Figure 4</u>. The regions of coverage and temperature discussed in detail in this article are indicated on the phase diagram by symbols a, b, c, d.

Figure 5 shows X as a function of temperature for a film corresponding to about a quarter of a completed monolayer. The effects of quantum degeneracy can clearly be seen at temperatures below about 1.5K. As is clear in the figure, neither a perfect Fermi gas model (degeneracy temperature T_F = 1.6K) nor a model in which interactions are supposed to lead merely to a change in T_F (to about 0.5K) fit the data well. However, Siddon and Schick[39] have carried out a virial coefficient calculation using the known helium – helium potential and found that the main features of the specific heat data are obtained from the second virial coefficient. Applied to the susceptibility they show

$$X = X_c \{1 - n(B_B(T) - B_F(T))\} \qquad (9)$$

where X_c is the Curie value, n is the adatom areal number density and $B_B(T)$, $B_F(T)$ are the second virial coefficients for gases of respectively spinless bosons and fermions of the ^{3}He mass.

The predicted curve for X is shown in Figure 5 and the theory gives a good indication of where XT begins to depart from the Curie value (it is not expected to work when the departures are large). A better indication that we are dealing with a 2D imperfect Fermi gas unaffected by substrate effects comes from studying the onset of departures from Curie's law at different areal densities. Eq.(9)

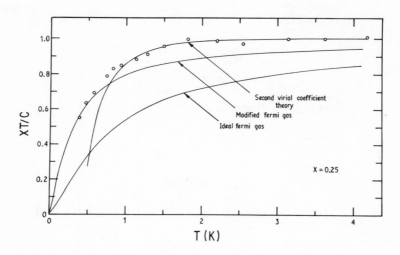

<u>Figure 5</u>. The magnetic susceptibility χ of ^{3}He on grafoil as a function of temperature. The data are plotted so that if Curie's law were obeyed the points would lie along $\chi T/C = 1$.

shows that $(\chi_c - \chi)/\chi_c$ should be proportional to areal density at any one temperature and Figure 6 shows data at two temperatures for a range of coverages up to 0.8 of a completed monolayer. For values of x less than 0.3 the fit to this theory which has no adjustable parameters is excellent. Further support for the conclusion that the motion of the adsorbed atoms is not affected by the substrate comes from the relaxation time data shown in Figure 7. As discussed in section 2(c) (i) above, substrate fields and imperfections are likely to dominate relaxation for highly mobile films such as we are dealing with here. This makes it impossible to obtain expressions for T_1 and T_2 unless the imperfections are fully characterized. However it is clear that the correlation time τ_c for relaxation will be the time to travel some distance along the substrate such as the distance to some step on the surface. For a classical gas $\tau_c^{-1} \propto v \propto \sqrt{T}$. For $\omega_0 \tau_c \ll 1$, Eqs. (8) show we should expect $T_1 = T_2 \propto \tau_c^{-1} \propto \sqrt{T}$ and this temperature variation is seen between about 4K and 0.7K, though $T_1 \approx 40\ T_2$. This could be due to 2D effects as shown in Fig. 2 or to different sites being effective for T_1 and for T_2.

The quantum degeneracy effects apparent in Fig. 5 below 1K should manifest themselves in T_1 and T_2 since below T_F atoms will move at the Fermi velocity rather than at thermal velocities. The apparent levelling off of T_1 and T_2 below 0.7K may be due to this

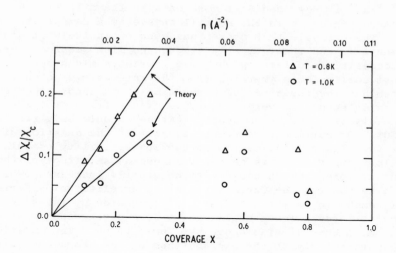

Figure 6. The fractional departure from Curie's law as a function of areal density n. n and x are related by n = 0.110 x.

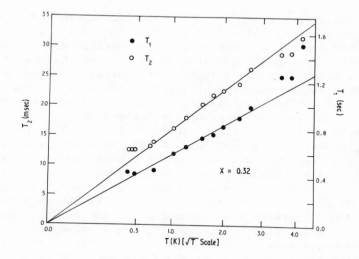

Figure 7. T_1 and T_2 as a function of temperature for a film of density 0.32 of a completed monolayer.

effect. What is now needed is more data in the millikelvin
temperature region where 2D ^{3}He will behave as a degenerate Fermi
liquid. The equivalent 3D system has been extensively studied both
experimentally and theoretically though no microscopic theory has
been successful in obtaining observed thermodynamic properties or
transport coefficients starting from the interatomic potential. The
2D system has the advantage of being simpler to handle theoretically
and of existing over a very wide range of densities from zero up to
values equivalent to the 3D density which can only be varied by
about 30%. Of particular current interest is the question of
whether the high constant value of the susceptibility of bulk liquid
^{3}He at low temperatures is better regarded as an effect of the spin-
independent forces which cause χ to be greatly increased over the
Pauli value <u>towards</u> the Curie value or as an effect of a ferromagnetic exchange term (itself of course created by the spin-
independent forces) which could, if it were strong enough, cause
the system to have a value <u>above</u> the Curie value. The question
relates particularly to the explanations of the anomalous values
of χ found[37] for liquid ^{3}He at very low temperatures in confined
geometries.

Some susceptibility data have been reported[20] at temperatures
down to 0.07K but it is not clear whether this was low enough to
have reached the constant values expected.

(b) The registered phase on graphite

At a value of x close to 0.6, called x_c, there are just enough
^{3}He atoms to be localised on one in three of the hexagons provided
by a graphite basal plane surface. Below 3K this ordered solid
phase appears from specific heat studies[34] to have become stabilised,
and Bragg peaks are found in neutron scattering experiments[40]
corresponding to the expected structure. NMR data observed in
this region of the phase diagram are shown in Figures 8 and 9.
Working at 1 MHz the Sussex group saw sharp dips in T_1 at x = x_c
(Fig. 8) with very little effect in T_2 while the Tokyo group saw
dips in T_2 measured at 10 MHz (Fig. 9) but very little effect in
T_1. From the frequency and temperature dependence it seems likely
that the T_2 are determined by substrate effects while T_1 in this
region is determined by dipolar fields. The cause of the dips is
a slowing up of the motion caused by the system changing from a
highly viscous liquid to a solid with rapid tunnelling. By
comparing the value of T_1 at the minimum with values found in the
2D incommensurate solid (see Section 4(c) below) we can deduce
that the tunnelling frequency at x = x_c is about 10^7 sec^{-1}.

By working near to x = x_c we have a system with a controlled
fixed number of defects which are probably vacancies for x < x_c
and interstitials for x > x_c. Such defects will have high quantum

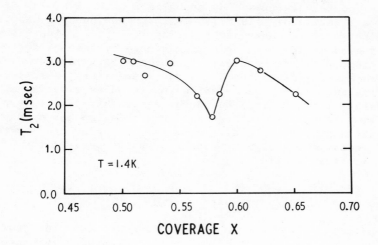

Figure 8. T_2 as a function of coverage close to registry at 1.4K. Data taken by Tokyo group at 10 MHz (Ref.9).

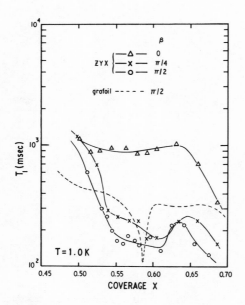

Figure 9. T_1 as a function of coverage close to registry at 1K. Data taken at 1.5 MHz on ZYX (Ref. 27) and 1 MHz on grafoil (Ref. 26).

mobility at low temperatures. By comparison with 3D solid ^{3}He we may expect a lifetime of $\sim 10^{-10}/|x - x_c|$ sec and hence a cusp in the relaxation rates as is seen. At temperatures approaching the melting of these ordered solids, the defects may move by thermal activation and more defects will be created. This leads to a further shortening of τ_c and increasing of T_2 and T_1 as the temperature is raised as shown for instance in Figure 10 which was taken as close to $x = x_c$ as was possible. The activation energy W for this effect is plotted as x is varied through x_c in Figure 11 and this too has a cusp at $x = x_c$ suggesting it is more difficult to create defects, or that existing defects have a higher barrier to pass over, in the perfectly registered solid.

(c) The incommensurate solid phase

Neutron scattering[40] has confirmed the deduction made from specific heat data that the low temperature high density phase of 2D helium is solid with a triangular lattice. Measurements of NMR relaxation times are able to deduce information about the tunnelling processes occurring in this solid. Figure 12 shows measurements

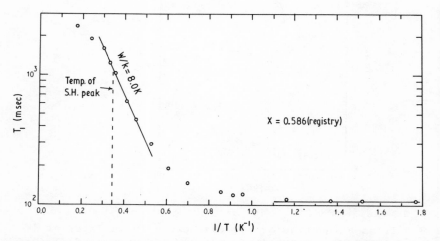

Figure 10. T_1 at 1 MHz as a function of temperature for a coverage corresponding to registry.

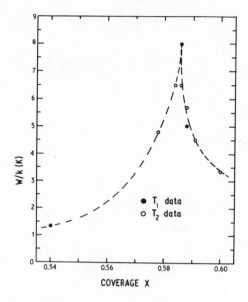

<u>Figure 11</u>. Activation energy for thermal excitations in the commensurate solid at different coverages close to registry.

taken at 1K for three different frequencies as the areal density is changed. The data show strong resemblances to Fig. 1 suggesting that τ_c increases as x increases. This is what one would expect from our understanding[41] of 3D solid ^{3}He in which below 1K quantum tunnelling dominates the motion of the spins and the tunnelling frequency decreases rapidly from MHz to kHz as the density increases above the value on the melting line.

The main differences between Figs. 1 and 12 are that in the experiments τ_c never quite gets long enough to give the rigid lattice value for T_2 (a mobile second layer begins to form first) and T_1 and T_2 do not merge on the low τ_c side of the T_1 minimum. However we recall that the result for $\omega_o \tau_c \ll 1$, i.e. $T_1 = T_2 = 10M_2\tau_c/3$ (Eqs. 8), only applies to an isotropic system and as discussed in section 1(c)(iii), diffusion in 2D creates a low frequency part to $J(\omega)$ which shortens T_2 below T_1 and gives some frequency dependence to T_1 for $\omega < \tau_c^{-1}$.

At present the theory is not complete enough to extract tunnelling frequencies unambiguously from T_1 or T_2 data even if the samples had a definite orientation or a well verified distribution of orientations. However we can obtain values of τ_c accurate to within a factor of about three by using the high x values of T_2 where the factor $\ln(T_2/2\tau_c)$ by which $1/T_2$ is increased

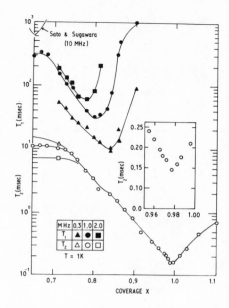

<u>Figure 12.</u> T_1 and T_2 at 1K and three different frequencies as a function of coverage.

due to 2D diffusion effects is close to unity and the low x values of T_1 for which we know that $\omega\tau_c \approx 1$ at the T_1 minimum. The values of τ_c which vary from about 2×10^{-8} sec at x = 0.7 to 2×10^{-5} at x = 1 are plotted in Figure 13 where τ_c^{-1} is compared to the exchange frequency in 3D solid ^{3}He and to theoretical values at comparable interatomic distances.

     As the temperature rises above 1K, thermally activated defects shorten τ_c below its quantum value and the effect of this is seen in Figure 14 for one particular areal density. From a comparison with 3D solid ^{3}He one deduces that the defects are almost certainly vacancies. The activation energy for their creation (plus any barrier height over which they have to pass) is plotted as a function of lattice spacing in Figure 15 and compared to the values[42] in the 3D hcp crystal. Again we may note the wide range of density available in the 2D case: to obtain a lattice spacing of 3.3A in 3D solid ^{3}He a pressure of about 400 bar is required.

(d) The nature of the melting transition

     There are further features of Fig. 14 that merit attention. In contrast to Fig. 12 we find that T_2 comes closer to T_1 in the low τ_c side of the T_1 minimum. Since the difference at 1K was

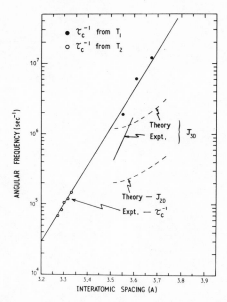

Figure 13. Tunnelling frequencies in two and three dimensional solid ^{3}He. Values of τ_c^{-1} are deduced from T_2 at high x, using $1/T_2 = M_2\tau_c$ with $M_2 = 5 \times 10^8$ sec^{-2}, and from T_1 at low x, using $\omega\tau_c = 1$ at the T_1 minimum. The other three curves are taken from Ref. 25.

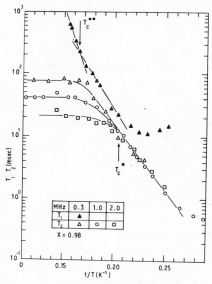

Figure 14. T_1 and T_2 at different frequencies as a function of temperature for a completed monolayer.

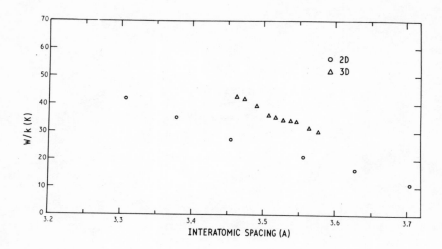

<u>Figure 15</u>. Activation energies for thermal excitation in the incommensurate 2D solid compared with the 3D solid (Ref.42) at different lattice spacings.

attributed to 2D effects we now need to see why the isotropic results of Eqs.(8) are obtained. At 5K the vapour pressure in equilibrium with a completed monolayer is several torr whereas at 1K it is less than 10^{-4} torr.[43] This change will drastically shorten the lifetime of an adsorbate atom against evaporation to the 3D vapour and when it falls below 10^{-6} sec this process will begin to modify the long time part of the correlation functions and make them more like the 3D case for which T_1 and T_2 are equal for $\omega \tau_c \ll 1$. Since at a pressure of 1 torr the rate at which an adatom is struck by an atom in the vapour at 5K is $\sim 10^7$ sec^{-1} we can see that for pressures much smaller than this, interchange with the 3D vapour will not affect the correlation functions significantly while for pressure much greater than this, evaporation may be important depending on the sticking coefficient, i.e. what fraction of collisions result in an exchange of 3D vapour and 2D solid atoms.

Other features of significance in Fig. 14 are the sharp onset of a frequency dependence in T_2 at T_c^* and a change in slope in T_1 at T_c^{**}. These features occur also in films of lower areal density and we can plot the loci of T_c^* and T_c^{**} as a function of x as is done in Figure 16. The two lines drawn in that figure correspond to the line of specific heat peaks[44] ascribed to melting of the 2D solid films and a theoretical melting line based on a theory of

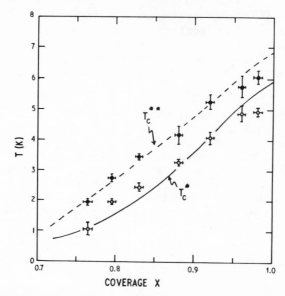

Figure 16. The loci of the two temperatures T_C^*, T_C^{**} from Fig.14 for films of different coverages. The dashed line corresponds with the position of specific heat peaks (Ref.44) and the full line is a theoretical curve based on a dislocation theory of melting (Ref.45).

melting due to Kosterlitz and Thouless[45] in which the unbinding of dislocations occurs at a temperature T_{K-T} and causes a loss of shear rigidity in the 2D solid.

$$T_{K-T} = k_B(m/32\pi^2\hbar^2n)\,\theta_D^2$$

where n is the 2D number density and θ_D is the Debye temperature.

The close fit of T_C^* to T_{K-T} suggest this theory may apply to the melting of ^{3}He films adsorbed on graphite. The reason why T_2 acquires a frequency dependence above T_C^* is that the phase here is a liquid crystal like state which retains some orientational order but has lost the ability to resist shear over macroscopic distances due to the presence of unbound dislocations. Hence patches of solid can flow over the substrate leading to a new contribution to $1/T_2$ from the substrate diamagnetism. If the substrate induced local field is characterized by an RMS magnitude B_{graf} and a correlation length l_c, then when the spin moves a distance l_c in a time less than $(\gamma B_{graf})^{-1}$, the substrate fields will begin to reduce the spin echo lifetime. Under these circumstances, $1/T_2 \sim (\gamma B_{graf})^2\tau(l_c)$ where $\tau(l_c)$ is the time to travel l_c. Since $B_{graf} \propto B_0$, this leads to a frequency dependent T_2. $B_{graf} \sim 10^{-4} B_0 \sim 3 \times 10^{-6}$ Tesla and l_c is probably ~ 100A (the length over which translational order is found[46] in the 2D solid) so we find that a velocity greater than about 10^{-3} cm sec^{-1} (or a diffusion coefficient greater than about 10^{-9} cm^2 sec^{-1}) would be sufficient to explain the data. If patches of solid are floating over the substrate at this sort of velocity no effect on T_1 would be seen since there would be no significant Fourier component in the field fluctuations at frequencies near to 1 MHz.

At T_c^{**}, the data suggest that the system begins to acquire the mobility of a normal liquid and loses the short range order associated with the liquid crystal phase since these changes would lead respectively to a change of slope in T_1 and a specific heat anomaly. It is unlikely that this second phase change is related to the unbinding of disinclinations discussed by Halperin and Nelson[47] since in their model the liquid crystal phase with its orientational order would be stabilised indefinitely by the periodic lattice.

5. FUTURE WORK

(a) Low temperature work (T < 0.4K) to study the 2D Fermi fluid over the wide range of density available. Also to seek clarification of the phase diagram including the question of whether 2D liquefies.

(b) Study of the isotropic phase transition in ^{3}He - ^{3}He mixtures.

(c) Behaviour of the second and subsequent layers. NMR and specific heat data suggest that the first atoms added above a completed mono-layer behave like a 2D gas. Mullin and Landesman[48] have discussed the NMR properties of the complex system of a rather immobile 2D solid with a highly mobile 2D fluid moving over it. Specific heat data suggest[44] that below 0.98K the second layer solidifies at high densities.

(d) There is a pressing need to find improved substrates for a number of reasons:

  (i) To avoid the diamagnetism present in graphite.
  (ii) To achieve larger areas of perfect exposed
       crystalline surface with a view to reducing
       wall effects and the destruction of long range
       order in the adsorbed system.
  (iii) To achieve improved alignment of substrate
        surfaces for studying the anisotropy in NMR
        properties as the direction of the applied B_O
        field is varied.

6. ACKNOWLEDGEMENTS

The writing of this review has been greatly assisted by stimulating conversations with many colleagues including Brian Cowan, John Owers-Bradley, Allan Widom, Bill Mullin, André Landesman and Peter Richards. The experimental work reported has been carried out in collaboration with Low Thomson, Brian Cowan and John Owers-Bradley and with financial support from the Science Research Council.

REFERENCES

1. M. Santini, M. Girera and G. Careri, Phys. Lett. $\underline{5}$, 102 (1963).
2. G. Boato, P. Cantini and R. Tatarek, Phys. Rev. Lett. $\underline{40}$, 887 (1978).
3. N. Bloembergen, E.M. Purcell and R.V. Pound, Phys. Rev. $\underline{73}$, 679 (1948).
4. R. Kubo and K. Tomita, J. Phys. Soc. Japan, $\underline{9}$, 888 (1954).
5. R.H. Romer, Phys. Rev. $\underline{115}$, 1415 (1959).
6. R. Chapman and M.G. Richards, Phys. Rev. Lett. $\underline{33}$, 18 (1974).
7. W.J. Mullin, D.J. Creswell and B.P. Cowan, J. Low Temp. Phys. $\underline{25}$, 247 (1976).
8. P.M. Richards has given a very full discussion of this problem in the 1973 Fermi Summer School at Varenna on Local Properties at Phase Transitions published by Editrice Compositori, Bologna 1975.
9. K. Satoh and T. Sugawara, 15th Inst. Conf. on Low Temp. Phys., J. de Phys. $\underline{39}$, C6, Supp.8, 281 (1978), and J. Low Temp. Phys., to be published.
10. B.P. Cowan, to be published.
11. G. Careri, M. Santini and G. Signorelli in "Low Temperature Physics" LT9, Vol.A. P.364, Plenum Press, New York, (1965).
12. P. Monod and J.A. Cowan, Internal Report No.000571 from the Service de Physique du Solide et de Résonance Magnétique, Centre D'Etudes Nucléaires de Saclay, Paris 1966.
13. D.F. Brewer, D.J. Creswell and A.L. Thomson, Proc. 12th Int. Conf. on Low Temp. Phys. p 157, Acad. Press of Japan, Tokyo (1970).
14. D.J. Creswell, D.F. Brewer and A.L. Thomson, Phys. Rev. Lett. $\underline{29}$, 1144 (1972).
15. A.L. Thomson, D.F. Brewer and Y. Goto, Proc. 14th Int. Conf. on Low Temp. Phys. $\underline{1}$, 463, North Holland, Amsterdam (1975).
16. H.T. Weaver, J. Phys. Chem. Solids $\underline{34}$, 421 (1972).
17. R.J. Rollefson, Phys. Rev. Lett. $\underline{29}$, 410 (1972).
18. D.P. Grimmer and K. Luczszynski, J. Low Temp. Phys. $\underline{26}$, 19 (1977).
19. D.P. Grimmer and K. Luczszynski, J. Low Temp. Phys. $\underline{30}$, 153 (1978).

20. D.C. Hickernell, D.L. Husa and J.G. Daunt, Phys. Lett. 49A, 435 (1974).

21. S.G. Hegde, E. Lerner and J.G. Daunt, Phys. Lett. 49A, 437 (1974).

22. S.G. Hegde, J.G. Daunt, J. Low Temp. Phys. 32, 765 (1978).

23. S.G. Hegde and J.G. Daunt, J. Low Temp. Phys. 34, 233 (1978).

24. D.L. Husa, D.C. Hickernell and J.G. Daunt, J. Low Temp. Phys. 34, 157 (1979).

25. B.P. Cowan, M.G. Richards, A.L. Thomson and W.J. Mullin, Phys. Rev. Lett. 38, 165 (1977).

26. M.G. Richards, 15th Int. Conf. on Low Temp. Phys., J. de Phys. 39, C6, Supp.8, 1342 (1978).

27. J.R. Owers-Bradley, A.L. Thomson, M.G. Richards, 15th Int. Conf. on Low Temp. Phys., J. de Phys. 39, C6, Supp.8, 298 (1978).

28. J.R. Owers-Bradley, B.P. Cowan, M.G. Richards and A.L. Thomson, Phys. Lett. 65A, 424 (1978).

29. A. Thomy and X. Duval, J. Chim. Phys. Physicochem. Biol. 66, 1966 (1969); 67, 286 and 1101 (1970).

30. D.E. Hogan, A.D.Novaco and F.J. Milford in "Adsorption – Desorption phenomena", Ed. F. Ricca, Acad. Press, London (1972).

31. Grafoil is the trademark of a gasket product manufactured by Union Carbide, New York.

32. Papyex is produced by Le Carbone Lorraine, Paris.

33. M. Bretz, Phys. Rev. Lett. 38, 501 (1977).

34. M. Bretz, J.G. Dash, D.C. Hickernell, E.P, McLean and O.E. Vilches, Phys. Rev. A8, 1589 (1973).

35. R. Chapman, Phys. Rev. A12, 2333 (1975).

36. J.F. Kelly and R.C. Richardson, in "Low Temperature Physics LT 13", I, 167, Plenum Press, New York, (1974).

37. A.I. Ahonen, T.A. Alvesalo, T. Haavasoja and M.C. Veuro, 15th Int. Conf. on Low Temp. Phys., J. de Phys. 39, C6, Supp.8, 285 (1978).

38. N.S. Sullivan, J. Low Temp. Phys. 313 (1976).

39. R.L. Siddon and M. Schick, Phys. Rev. A9, 907 (1974).

40. M. Nielson, J.P. McTague and W. Ellenson, J. de Phys. 38, C4, Supp.10, 10 (1977).

41. M.G. Richards, Adv. Mag. Res. 5, 305 (1971).

42. N.S. Sullivan, G. Deville and A. Landesman, Phys. Rev. B11, 1858, (1975).

43. G.J. Goellner, J.G. Daunt and E. Lerner, J. Low Temp. Phys. 21, 347 (1975).

44. S.W. Van Sciver and O.E. Vilches, Phys. Rev. B18, 285 (1978).

45. J.M. Kosterlitz and D.J. Thouless, J. Phys. C6, 1181 (1973).

46. H. Taub, K. Carneiro, J.K. Kjems, L. Passell and J.P. McTague, Phys. Rev. B16, 4551 (1977).

47. B.I. Halperin and D.R. Nelson, Phys. Rev. Lett. 41, 121 (1978).

48. W.J. Mullin and A. Landesman, J. Low Temp. Phys. to be published.

ORDERING IN TWO DIMENSIONS

J.M. Kosterlitz

Department of Mathematical Physics
Birmingham University
Birmingham B15 2TT, England

INTRODUCTION

During the last few years, the study of phase transitions in two dimensional systems has absorbed a great deal of effort by both theorists and experimentalists. Although such an activity is rather esoteric in the sense that these systems are rather special and do not occur in every day life, they are a theorist's paradise because they form a very special class of systems for which theory is capable of yielding quantitative predictions. With the increase of the sensitivity of experiments, these predictions can now be checked by the experimentalists.

There have been various important results for two dimensional systems obtained over the years such as Langmuir's theory of gas adsorption on surfaces (Langmuir 1938), Peierls' argument for the absence of long range order in two dimensional solids and isotropic ferromagnets (Peierls 1934, 1935) and Onsager's solution of the two dimensional Ising model (Onsager 1944).   In recent years, a wide variety of two dimensional systems have become available for detailed experimental study so that the relevance of various theoretical ideas could be assessed.   At the same time there has grown up a great interest in the influence of dimensionality on the behaviour of physical systems, particularly in relation to critical phenomena near phase transitions.

Probably the main reason for the interest in two dimensional systems is that, while they are broadly similar in many respects to three dimensional systems, the theoretical analysis is somewhat simpler.   The geometry of a plane is simpler, easier to visualise, and more familiar than the geometry of a volume,

integrals are easier to evaluate and fewer particles are needed
in a molecular dynamics or similar numerical computation.   All
this is more true of one dimensional systems but they are known
to have peculiar properties not shared by three dimensional
systems.   There are particular theories that can be solved in
two but not higher dimensions such as the Ising model, eight
vertex model etc.   These topics are reviewed by Lieb and Wu (1972).
Problems of classical statistical mechanics in two dimensions can
be related to quantum ground state problems in one dimension.

     A major part of the theoretical interest in two-dimensional
systems has been concerned with the possibility of phase transitions
and the nature of the long-range order that may occur in these
systems and a large portion of these lectures will be devoted to
this aspect.   It has been known for over forty years that the
Ising model in two dimensions has a phase transition to a state
with finite magnetisation at low temperature and it is generally
believed that in analogous cases, where the order is described by
a single real scalar quantity, that there will be a phase transition
from an ordered state at low temperatures to a disordered one at
high temperatures.   In many cases, however, the order is described
by a quantity with more than one degree of freedom the number of
which will be denoted by n.   For example, in superfluids and
superconductors the ordered state is described by a complex
variable so that n = 2 while for the isotropic Heisenberg model
of a ferromagnet or antiferromagnet the order is described by a
vector with n = 3.   Less obviously, n is also 2 for the two
dimensional solid since there are two degrees of freedom determining the absolute position of the lattice in space.   Peierls (1934,
35) argued, and later Mermin and Wagner (1966), Hohenberg (1967)
and Mermin (1967,1968) proved that there is no long range order of
the usual type in such systems except at zero temperature,
essentially because thermal motion destroys long range order at
arbitrarily low temperatures.   The interpretation of this formal
result is not so clear.   It may be that at low temperatures the
system is almost ordered with, in the case of the crystal, a Debye-
like specific heat with a continuous transition to a high temperature
state with no abrupt change of phase.

     When the number of degrees of freedom n is two or more, the
two dimensional systems are of particular interest because they are
on the borderline of having a sharp phase transition.   We say that
two is the lower critical dimension.   This is particularly clear
in the case of the exactly soluble spherical model which is known
to be like an n-component system in the limit of infinite n.
This model has no transition for dimensionality two or less owing
to the failure of an integral to converge.   Above two dimensions
this integral converges and there is a normal transition to an
ordered phase at low temperatures.   In two dimensions the

susceptibility diverges exponentially which seems to be characteristic of a lower critical dimension.   For systems of finite n in
two dimensions, although at the lower critical dimension, further
study is required to determine whether or not there is a finite
temperature transition.   In order to resolve this problem
attention has been directed at the behaviour of defects in the
ordered state.   It is argued on these grounds that the two
dimensional solid and superfluid should have a sharp transition
whereas systems with n greater than two hould not.

2   Examples of two-dimensional systems

It is one thing to construct theories of systems in two
dimensions but it is quite another thing to find an experimental
realisation of your theory.   It is not possible to make a system
a single atomic layer or so thick without a substrate to put it on.
Then the system becomes different from the theoretical one and in
order to compare with experiment the effect of the substrate must
be assessed.   Unfortunately, since the systems we shall be
considering are on the edge of having a transition it is not
surprising that they are very sensitive to disturbances.   In
other words, the substrate is likely to have a large effect on
the system.   This would not matter too much if one had a "perfect"
substrate whose effect would in principle be accessible to
theoretical analysis.   Unfortunately the word perfect here may
mean atomically perfect over thousands of atomic spacings, a
situation which is not easy to realise.   Fortunately, there are
two important systems in which the substrate is unimportant or
easily taken into account and which are excellent realisations of
the ideal theoretical models - superfluidity in thin helium films
(Bishop and Reppy 1978, Rudnick 1978, Scholtz et.al 1974, Chester
and Yang 1973, Webster et al 1979) and the melting of a two
dimensional crystal of electrons on the surface of helium (Grimes
and Adams 1979).   In the superfluid case, the substrate appears
to play no role whatever because the superfluid order parameter
does not couple and in the second the helium forms a clean substrate
whose effect can be analysed (Fisher et al 1979).

2.1   Classical Fluids and Solids

The most obvious examples of a two dimensional system are
provided by very thin, monolayer or less, films of atoms or molecules on the surface of a solid.   There is an excellent review
of this topic by Dash (1975) so we shall only mention some features
of such systems and the theoretical problems which arise when they
are regarded as two dimensional systems.

The usual way of forming and studying a thin film is to have
a substrate with a large surface to volume ratio while the material

to be studied is in the form of a dilute gas which condenses on
the substrate because of the lower potential energy of the surface.
Obviously, for a thin film to form, the substrate must bind the
adsorbate atoms or molecules more tightly than they are bound in
the bulk liquid.   The thermodynamic parameters of such a system
are readily controllable.   The temperature and chemical potential
of the molecules are controlled by the temperature and pressure
of the gas and the number of condensed atoms deduced from the
decrease in volume of the gas as a result of the presence of the
substrate.   The area occupied by the surface layer is the area of
the substrate which can be directly measured in certain cases and
in others deduced if the properties of a standard surface layer
are known.

It is possible to perform various types of experiment on such
systems.   Perhaps the easiest one is a specific heat measurement
which can be deduced by subtracting from the specific heat of the
combined system of substrate and gas the contributions from the
bulk substances.   Over the past few years there has been a great
deal of experimental effort by a number of different workers.
Dash (1975) describes a number of measurements in his book on thin
films.   Unfortunately, specific heat measurements are rather
uninformative about one of the most interesting phenomena in
monolayer films - the melting of an unregistered solid (Kosterlitz
and Thouless 1973, 1978 Nelson 1978, Halperin and Nelson 1978,1979).
In principle, neutron scattering is a very direct probe of the way
in which the positional ordering of the adsorbate atoms varies
with temperature and coverage.   If the atoms are ordered as a
two-dimensional crystal the intensity spectrum will have well
defined peaks at reciprocal lattice vectors while in a liquid
there will be no peaks (Jancovici 1967, Mikeska and Schmidt 1970,
Imry and Gunther 1971) so that neutron scattering appears to be
an excellent method of studying a phase transition to a crystalline
solid.

These measurements have shown that even the simpler materials
such as the inert gases have a variety of phases corresponding to
gaseous, liquid and solid phases in the bulk.   Various types of
solid phase have been observed some of which have lattice constants
commensurate with the substrate lattice while others are incommensurate.   Which solid phase is formed depends on various factors
such as the coverage and the relative sizes of the adsorbate atom
and the substrate lattice spacing.   One can make the system still
more complicated by adsorbing molecules which interact magnetically
so that one has competing crystalline and magnetic ordering.   For
example oxygen forms a triangular lattice which is distorted from
the close packed form below $10^{\circ}K$ and is closed packed above $10^{\circ}K$
(McTague and Nielsen 1976).

A number of interesting theoretical questions are raised by

these experimental results.   Does the existence of something like
a gas-liquid critical point indicate that there are phases which
behave like two dimensional gases and liquids?   Does Peierl's
argument for the non-existence of long-range order in two dimensions
mean that the substrate plays an essential role in the formation of
a crystal lattice of adsorbed gases?   Is there a sharp distinction
between phases commensurate and incommensurate with the lattice
and how are incommensurate phases modified by the substrate?

    A partial answer to the first question is that the phase
diagram tells us rather little about the dynamics of the molecules
since in classical statistical mechanics the kinetic and potential
energies make separate contributions to the free energy.   Thus,
the thermodynamic properties of the lattice gas, which would be
a realistic model if molecules were bound to particular lattice
sites of the substrate, and a model in which the molecules move
freely on the surface are quite similar to one another.   This
fact has of course been used in the Ising lattice gas model of
the critical properties of fluids.   To settle this question
experimentally, it is necessary to study the diffusion of
molecules on the surface, which will be roughly independent of
concentration if the molecules are tightly bound to particular
sites and will be more rapid the diluter the system if the molecules are free to move.   A strong substrate interaction will be
shown up in a frictional force between a flowing monolayer and
the substrate.

    Over the past few years there has been an enormous amount of
experimental and theoretical work done on this type of system and
the situation is becoming clearer, but is still not entirely
resolved.   It is obvious that even on a naive level there are a
large number of possible phases for classical atoms regarded as
small spherical balls interacting only via an isotropic Lennard-
Jones type of potential adsorbed on a perfect substrate with a
periodic binding potential.   The most common substrate is
graphite in one form or another which, in a theorist's idealisation
may be regarded as a set of binding sites arranged on a triangular
lattice with a nearest neighbour separation of 2.46 $\overset{\circ}{A}$.

    The size of an atom of an inert gas is rather larger than
this which causes nearest neighbour exclusion and prevents every
site of the substrate being occupied.   For example, a Krypton
atom is about 4$\overset{\circ}{A}$ in diameter.   Thus, if one insists that each
adsorbate atom sits on a binding site, the maximum coverage one
can obtain in such a situation corresponds to every third site
being occupied - the $\sqrt{3}$ x $\sqrt{3}$ epitaxial structure.   This is an
excellent experimental realisation of a lattice gas which corresponds to a three state Potts (1952) model since there are three

equivalent ordered states (Alexander 1975). Since these lattice
gas transitions have a pronounced specific heat anomaly associated
with them, measurements of this will provide information about this
part of the phase diagram. At lower coverages, one expects that
the competition between the attractive interatomic potential and
the binding to the substrate will lead to a two phase region of
regions of registered overlayer interspersed with more or less
unoccupied (two-dimensional gas) regions at low temperatures. This
leads one to expect a phase diagram of the type shown in figure 2.1a.

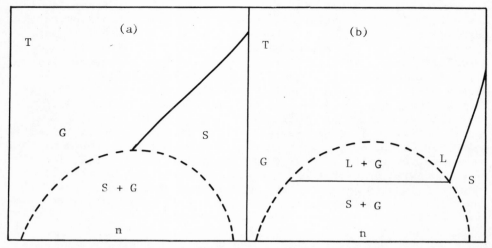

Fig.2.1 Phase diagram of registered monolayer.

It is not entirely obvious what the difference is between the
solid and gas at the same coverage but it can be understood in terms
of the three types of ordering of the solid. If one calls the
substrate lattice sites types A,B and C, where each type forms a
triangular lattice of spacing a, then the substrate is the triangular
lattice of spacing $a/\sqrt{3}$ formed by the three interpenetrating sublattices. In the solid phase, only one of the three sublattices
is occupied while the gas phase has all three equally occupied on
average. The type of phase diagram shown in figure 2.1a has been
deduced from experimental adsorption vapour pressure studies
pioneered by Thomy and Duval (1969,70) and specific heat measurements (e.g. Huff and Dash 1976). There is an enormous experimental
literature on these topics for different adsorbates, but I am afraid
I am rather ignorant of it. However, other phase diagrams are
possible, such as that depicted in fig. 2.1b. This looks more
like the phase diagram for bulk solid, liquid and gas, but to my
knowledge no real adsorbed layer system displays such a behaviour.

In this phase diagram, the only difference between a two

dimensional gas and liquid is in the density of atoms.   It displays
a conventional (Ising-like) gas-liquid critical point which
terminates a first order line.   The melting line is second order,
unlike bulk systems, and terminates in a critical endpoint.   There
is an excellent theoretical paper on the krypton-graphite system
using real space renormalisation group methods (Berker et al 1978)
which can be improved to yield a remarkable agreement with experiment when finite size effects are taken into account with no
adjustable parameters (Ostlund and Berker 1979).   Many further
references on this subject can be found there.

      So far, our discussion has been limited to relatively low
coverages.   At higher coverages, the finite size of the more or
less hard core of the atoms causes each atom to be pushed out of
registry with its substrate binding potential and a hexagonal close
packed structure with its own periodic lattice incommensurate with
the substrate results.   At sufficiently high coverages, the
arrangement of the adsorbate atoms have many equivalent positions
relative to the substrate.   Because such configurations have the
same energy, the phonons have an energy proportional to wavenumber at low frequencies, despite the effect of the substrate and
so the specific heat is Debye-like, proportional to T^2 at low
temperatures.   Although there are structure measurements which
show incommensurate crystal lattices for adsorbed molecules, e.g.
argon on graphite (Taub et al 1977), there do not seem to be clean
cases of T^2 specific heats at very low temperatures for classical
systems.

      Peierl's argument (1934,35) for the non-existence of conventional
long range order in the positions of the atoms in a two dimensional
solid depends on the effect of long wavelength phonons and the
argument of Mermin (1968) depends on a related property of the
Hamiltonian.   If phonons of wavenumber q have an energy hcq,
then thermal motion gives rise to a mean square displacement of
the atoms proportional to kT/c^2q^2 and, if this is valid for values
of q right down to the order of the inverse of the linear
dimension of the system, the sum over q gives a mean square
displacement which increases logarithmically with the area of the
system.   This argument seems to remain valid in the presence of a
substrate if the lattice is incommensurate and the specific heat
is proportional to T^2 at low temperatures.   If the molecules are
in registry with the substrate, there are no low temperature phonons
and the argument does not apply.

      It has been argued that other properties of the solid state
may exist even without the long range order in the positions of
the atoms which gives rise to a sharp Bragg peak in scattering of
X-rays and neutrons (Berezinskii 1970, Kosterlitz and Thouless
1973,78, Jancovici 1967, Imry and Gunther 1971).   It is argued
that long range fluctuations that give rise to the smearing of the

Bragg peak will still leave the structure of the solid distinct
from the fluid phase so that there is a quadratic specific heat
at low temperatures, and that the solid can still resist shearing
stresses.

The mechanism which causes the transition from the low
temperature incommensurate solid phase to the fluid phase is an
unusual one, quite different from the types of transition
discussed earlier (Berezinskii 1970, Kosterlitz and Thouless 1973,
Nelson 1978, Halperin and Nelson 1978,79).   It is argued that
in two dimensions, dislocations in the crystal are present in
thermal equilibrium.   This is not possible in a bulk crystal
because of the large energy required and so this is a very special
property of such two dimensional systems.   At sufficiently low
temperatures, these dislocations are tightly bound together in
pairs whose effect cancels each other out, so that to all intents
and purposes their presence does not have any effect.   As the
temperature is increased, these pairs become more widely separated
until eventually they become free.   While these dislocations are
tightly bound, the system has the characteristics described
earlier - Debye-like specific heat, resistance to shear etc.
Once they are widely separated, however, the resistance to shear
disappears because the dislocations can move freely to the edges
and the system behaves like a viscous liquid.   This unbinding of
the dislocations is interpreted as the melting of the solid to a
fluid phase, which does have some very peculiar properties (Halperin
and Nelson 1978,79).   The detailed theory of such a transition
shows that the density of dislocations is very low even at the
transition so that there is very little entropy associated with it,
and the specific heat will be smooth.   Moreover, in such a model,
these dislocations are the only thing which destroy the positional
ordering of the atoms over and above the effect of long wavelength
phonons.   Since all the long wavelength phonons do is to smear
out the Bragg peak without destroying the local crystalline
structure, the presence of a few dislocations will not change the
local structure very much. These very crude arguments suggest
that, if this mechanism is the correct one, the melting transition
will be rather difficult to see because the positional order will
disappear very gradually. In fact this is consistent with neutron
scattering measurements on argon adsorbed on graphite which, owing
to the size of the argon atom and the much stronger interatomic
potential, forms a lattice incommensurate with the substrate. In
this system, the Bragg peaks disappear rather gradually and there
seems to be no sharp signature of the melting (Taub et al. 1977).
These topics will be discussed in more detail later in these
lectures and by other speakers.

We have discussed some of the transitions from a registered
phase to a fluid and the transition from an unregistered phase to

a fluid.   What remains is the transition between the commensurate
and incommensurate solids.   This is a subject currently receiving
much experimental and theoretical attention and is not yet
completely resolved, and will not be discussed any further in these
lectures.

     A lot of space has been devoted to classical solids and fluids
in two dimensions for two reasons.   First of all, it is relatively
easy to get a physical picture of what is happening by regarding
the atoms of rare gas as little hard balls, an N_2 molecule as a
rigid dumbell or an ammonia molecule as a rigid equilateral triangle.
Secondly, it is a good system to illustrate the richness of
possibilities for different types of ordered phases and transitions
between them and the associated complexity of the problem.   It is
also useful to appreciate the importance of the substrate and how
the periodic structure and imperfections can affect the overlayer.

2.2   Quantum Fluids and Solids

     When atoms of helium or molecules of hydrogen are adsorbed on
to a surface their small mass is likely to lead to a large zero-
point motion so that such films are likely to behave rather
differently from films of heavier molecules.   It may be less easy
for the molecules to be attached to a particular site of the
substrate so that they may be regarded, to a first approximation,
as moving freely on the surface.   Also, in the case of helium,
which in the bulk remains liquid down to zero temperature, various
phenomena characteristic of quantum fluids can be seen.

     At temperatures above $2^{O}K$, dilute monolayer helium films have
a specific heat of k per atom, characteristic of a two dimensional
perfect gas.   At lower temperatures the specific heat of ^4He
rises to a peak and there appears to be a condensation to a liquid
phase with a critical temperature of around $1.4^{O}K$.   The specific
heat of ^3He however drops and there is no sign of condensation.
These results have been explained in detail by Siddon and Schick
(1974) using the quantum virial expansion.

     At higher coverages, still in the monolayer region, solid
phases of helium can be formed.   These can be either commensurate
with the substrate at intermediate coverages and these can be
discussed in terms of a Potts lattice gas exactly as for the
classical gases such as krypton.   At still higher coverages near
monolayer completion, it appears that a solid is formed which is
really incommensurate with the substrate since the specific heat
is quadratic in temperature down to quite low temperatures.   When
the temperature is raised the specific heat rises to a peak and
then drops again to its value in the fluid phase.   One may expect

that this peak is associated with a melting transition but now
there is a good deal of evidence, both experimental from NMR
measurements on He^3 on grafoil (Richards 1979) and theoretical
(Berker and Nelson 1979), that this peak occurs well above the
melting temperature.

Of course the most characteristic manifestation of the
quantum nature of He^4 is superfluidity and the question is can
a thin film be superfluid. Although there is some evidence from
the specific heat measurements of Stewart and Dash (1970) that
films with a very low coverage of helium atoms behave like quantum
fluids with specific heats proportional to T^2 for He^4 and to
T for He^3, there is no evidence for superfluidity in monolayer
films. Superfluidity can be observed in thin films a few atoms
thick, and in such films there appears to be a monolayer or so of
helium atoms forming a solid on the substrate, the exact coverage
depending on the composition of the substrate. Superfluidity
occurs in the helium above this layer whose superfluid mass
increases linearly with coverage.

Experiments have been done on various substrates such as vycor
(Brewer and Mendelssohn 1961, Henkel et al 1969), which is a porous
glass, on grafoil (Herb and Dash 1972) on the surface of glass
(Scholtz et al 1974; Telschow and Hallock 1976), on the surface
of quartz (Chester and Yang 1973) and on the surface of Mylar
(Bishop and Reppy 1978). Superfluid flow is detected in various
ways which may involve the observation of mass flow under more or
less steady conditions or the observation of third sound in the
film, which involves an alternating supercurrent coupled to changes
of thickness of the film. For example, the quartz microbalance
experiment of Chester and Yang (1973) involves a measurement of
the shift in the resonant frequency of a polished quartz crystal
with a thin film of He^4 adsorbed on the crystal. At low
temperatures, when the film is superfluid, the crystal oscillates
but the superfluid remains at rest. As the temperature is raised,
superfluidity disappears and the film is locked to the crystal
thereby increasing its mass and lowering its resonant frequency.
The shift in the resonant frequency is a direct measure of the
total mass of the superfluid component. The experiment of Bishop
and Reppy (1978) is a similar idea except that it is performed
at frequencies in the kHz range while that of Chester and Yang is
in the MHz range. Also, the former experiment has a major advantage
in that the power dissipation in the film can be simultaneously
measured giving a more stringent test of theory.

When the film is very thin, the superfluid can be regarded as
two dimensional since the coherence length is much greater than
the film thickness so that the phase of the superfluid wave function
cannot vary appreciably between the substrate and the free surface

of the film.   The connectivity of the film has an important effect
on what is observed.   In vycor there are multiple connections
between the pores and one may expect that helium adsorbed in vycor
should behave more like the bulk, which is in fact observed
(Berthold et al 1977).   The surface of the quartz crystal or a
roll of Mylar are simply connected and so may be regarded as truly
two dimensional and helium on such a substrate shows quite different
and characteristic behaviour.   The onset temperature at which
superfluidity is observed is lower in helium films than in the
bulk, and decreases as the thickness is decreased.   It is not
clear from the experiments alone whether the ideal transition is a
sharp transition or a more gradual quantitative change since actual
measurements involve averages over an inhomogeneous substrate with
varying film thickness.   One interesting feature of the results is
that the superfluid density obtained from third sound measurements
appear to jump discontinuously to zero, which behaviour is quite
consistent with the Josephson relation $\rho_s(T) \alpha \xi^{2-d}$ (Josephson 1966).
However, there is a large attenuation of third sound near the onset
temperature (Scholtz et al 1974, Telschow and Hallock 1976) and in
the quartz microbalance experiment the superfluid density goes
continuously to zero.

All the methods of measuring the superfluid density described
above are actually measuring a finite frequency response function
(Hohenberg and Martin 1965) which, as a bit of thought will tell
you, is not possible at zero frequency.   The ideal theorist's
measurement is at zero frequency and it is possible that the non-
zero frequency has some dramatic effect on what is observed.   The
resolution to this finally arrived when the experimental data of
Bishop and Reppy (1978) was fitted to a proper dynamical theory
(Ambegaokar et al 1978).

Specific heat measurements have also been made in the neigh-
bourhood of the transition.   These show that the high peak of the
bulk transition becomes a low lump for films of five atomic layers
or so, the height of the lump decreasing with decreasing coverage.
(Bretz 1973).   In precise analogy with the dislocation mediated
melting of incommensurate monolayers it is now believed that the
specific heat should be completely smooth at the onset of superfluidity.

One of the most important experimental features of superfluidity in thin helium films is that the superfluid order
parameter does not couple to the substrate so that almost anything
is a suitable candidate - glass, plastic etc.   This is in contrast
with the systems described in the previous section where there is
a strong coupling to the substrate and presumably an inhomogeneous
substrate will just wash out everything of interest.

2.3 Magnetic Compounds

At first sight, a magnet in two dimensions is an ideal system for studying phase transitions and ordering. There are magnetic materials available with different numbers of degrees of freedom n while crystals and superfluids are restricted to n = 2.   One has both ferromagnets and antiferromagnets which opens up the possibility of studying the effect of non ordering fields which can induce anisotropies in the coupling between the degrees of freedom.   These effects have been extensively studied by various techniques, both theoretical and experimental, in three dimensions. Experimental probes are immediately available which couple directly to the order parameter, again in contrast to superfluids which is the ideal system from an experimental viewpoint.   All this makes two dimensional magnetic systems a theorist's paradise and a lot of theoretical effort has been put into them by various workers.

Unfortunately, it is extremely difficult to make a two dimensional magnetic system.   Even if one could, the substrate will play a very important role inducing all sorts of deviations from the theorist's idealisation.   The nearest one can get to this is a magnetic layer compound which is a material in which the magnetic ions are arranged in planes such that there is a strong coupling between the spins within a plane but only weak coupling between spins in different planes.   A survey of earlier experimental and theoretical work on magnetic layer compounds is contained in the review by de Jongh and Miedema (1974).

One can find layered magnets in which the ratio of the inter-planar to intraplanar coupling is as low as one part in a million and the (Ising-like) deviation from ideal isotropic Heisenberg-like coupling between spins is 0.1%.   At first sight, this is an experimental realisation of the theorists ideal two dimensional Heisenberg magnet.   Remember that this is the system we are interested in because of the Mermin-Wagner theorem which tells us there can be no conventional long range order or spontaneous magnetisation in such a system, but, since we are at the lower critical dimensionality of two, the thermal fluctuations only just destroy the order.   We want to know if there can be a transition at finite temperature despite the absence of a spontaneous magnetisation, and to find a system on which any theoretical predictions can be checked.

A few moments reflection will immediately tell us that the presence of an interplanar coupling or Ising-like anisotropy invalidates the Mermin-Wagner theorem.   Firstly, the coupling between planes makes the system three dimensional and there is bound to be a spontaneous magnetisation at low temperatures even without the anisotropy and secondly, even in the absence of interplanar coupling the anisotropic coupling between spin components will make the thermal fluctuations anisotropic and the system will behave like an Ising model at sufficiently low temperatures.   We know that the two dimensional Ising system orders (Onsager 1944), and so this system should also.   Since the two dimensional isotropic system is on the verge of ordering anyway, one then expects that even a small perturbation will have a large effect.   Thus, before using these systems to investigate experimentally the ideal two dimensional model one must have some idea of how important are these effects.

It has been shown recently (Kosterlitz and Santos 1978) that the experimental data on the magnetic susceptibility of a class of quasi two dimensional nearly Heisenberg magnets (de Jongh 1976, Navarro and de Jongh 1976) is consistent with the theoretical prediction that the critical temperature is zero and that the susceptibility diverges exponentially (Polyakov 1975) as the temperature vanishes.   However, this is not a very stringent test of the theory because of large deviations from ideality in the experimental systems but it is the best available at present.

2.4   Other Systems

There are a number of other systems which are more or less good realisations of two dimensional model systems.   From our previous discussion, the most obvious is a charged superfluid or super-conducting film.   Theoretically it turns out that this does not have a true superconducting state (Kosterlitz and Thouless 1973) but in a film of realistic dimension there should be very little difference (Beasley et al 1979, Doniach and Huberman 1979, Halperin and Nelson 1979).

This appears to be a very promising system for future experi-mental work because, in principle at least, the same sample can display a melting ransition (in a magnetic field) and the onset of superconductivity.   It is a more flexible system than the helium film because there are more experimental probes available: an electric field which drives a superflow and is analogous to moving the substrate in the case of helium and a magnetic field.

Then there is the system of electrons on the surface of helium which is important because it gives the possibility of studying the melting transition in two dimensions on a substrate whose effects are easily analysed.   Because of the long ranged electrostatic repulsion between the electrons, this behaves rather like an incompressible crystal to a first approximation.   The electrons can be regarded as classical particles because of the low density of electrons of approximately 10^8 cm^{-2} so quantum effects are unimportant.   It is believed that the melting of the lattice has been observed experimentally (Grimes and Adams 1979).

Low energy diffraction studies have shown that the surfaces
of several clean metals have their atoms arranged differently from
the bulk i.e. are "reconstructed", (Felter et al 1977, Yonehara
and Schmidt 1971, Blakeley and Somorjai 1977, Chan et al 1978,
Debe and King 1977, Barker et al 1979). Bak (1979) has recently
shown how models of such systems can be mapped into various two
dimensional theoretical models of current interest such as the
Ising model, the planar rotor model, and the planar rotor model
with cubic anisotropy. It should be possible to study the
transitions experimentally by surface scattering techniques such
as LEED and X-rays.

The last system I want to mention is at first sight very
different to the others namely the roughening transition in crystal
growth. This is important when one asks about the structure of
the surface of a crystal in contact with vapour. If the surface
is smooth, the rate of growth will be controlled by nucleation
processes and will be slow but if rough growth does not require
activation and will be more rapid. Surprisingly, fairly realistic
models of the interface can be mapped into the same model describing
superfluidity in helium films (Chui and Weeks 1976, Villain 1975,
Jose et al 1977) which in turn is the prototype planar rotor model.

3   Theory of Ordering in Two Dimensions

As we have already discussed, two is a lower critical dimension
for many systems and the existence or otherwise of an ordered low
temperature phase is a delicate matter. In quasi two dimensional
systems such as layered magnets with a very weak coupling between
spins in different planes or with a weak Ising anisotropy in a
plane or a thin film interacting with the substrate it is quite
possible that these couplings modify in a drastic way the behaviour
near a transition. Before trying to assess these effects we must
first understand the ideal system.

There are a number of two dimensional models which have a
phase transition at finite temperatures and for which exact
results exist. The best known of these is the Ising model which
Onsager (1944) showed to have a continuous transition of what one
could call the common or garden variety. This is a special case
of the p-state Potts models described by a nearest neighbour
interaction on a lattice of the form

$$H = - J\sum \delta(s_i - s_j) \qquad\qquad s_i = 1...p \qquad\qquad (3.1)$$

The Ising model has $p = 2$. These are known to have transitions
to an ordered state at low temperatures where the majority of the

spins s_i have a particular value. In the Ising system, this
corresponds to a state of finite magnetisation. This type of
transition is familiar because it is analogous to a transition in
a three dimensional system to a low temperature phase with complete
long range order. Such models are important as theoretical
idealisations of gases adsorbed on graphite when each molecule of
adsorbate sits on a binding site of the substrate. Such an
idealisation is a reasonable description of the actual physical
system because it is not too sensitive to deviations from ideality
such as the finite size of substrate crystallites etc.

Of more theoretical interest are models at their lower critical
dimension since the ordering is a much more delicate matter. A
class of two dimensional systems displaying many of the different
behaviours possible and of experimental relevance, is the set of
classical ferromagnets with nearest neighbour interactions on a
square lattice described by the Hamiltonian

$$H = - J\sum \underset{\sim}{s}_i \cdot \underset{\sim}{s}_j - \underset{\sim}{h} \cdot \sum \underset{\sim}{s}_i \qquad (3.2)$$

where $\underset{\sim}{s}_i$ is an n-component spin of unit length and $\underset{\sim}{h}$ the applied
magnetic field. The restriction to nearest neighbour interactions
and a square lattice is not essential provided the interaction is
ferromagnetic $J > 0$.

For $n = 1$, $s_i = \pm 1$ and we have the Ising model with true
long range order below the critical temperature T_c. The
magnetisation in zero field vanishes as T approaches T_c from
below as $(T_c-T)^\beta$, and the susceptibility diverges as $|T-T_c|^{-\gamma}$.
The exponents β and γ are dependent on the spatial dimension
but the point I want to emphasise is that the qualitative nature
of the transition (power law behaviour of magnetisation, true long
range order etc.) is not very different to that in three.

The situation for $n = 2$ (superfluids, crystal) and higher
(Heisenberg magnet) is very different. In more than two dimensions,
we know that there is a state with true long range order at low
temperature but in exactly two there is not. The Mermin-Wagner-
Hohenberg theorem tells us that the magnetisation (order parameter)
is zero for finite T. Note that this theorem does not exclude
the possibility of distinct high and low temperature phases but
says that there is no long range order in either phase. In the
case of an interacting Bose gas, it says that there is no Bose
Einstein condensation in two dimensions but says nothing about
the superfluidity of the system. For a crystal it says that there
are no sharp Bragg peaks in the neutron scattering spectrum but
nothing about the resistance to shear. In other words, the
theorem does not exclude the possibility of a two dimensional
superfluid or crystal but does say that if they do exist they are
rather unusual.

The only other exactly soluble model described by eq.(3.2) is the n = ∞ limit which is equivalent to the spherical model (Berlin and Kac 1952, Stanley 1968, Kac 1971). In this case one can easily show that not only is there no long range order at finite T but also that the critical temperature is at T = 0. One finds that the susceptibility diverges as $\exp(1/T)$ and the correlation length, which is a measure of the distance over which spins are correlated with each other, also diverges exponentially.

The theory of models of more physical significance is not in such good shape because very few exact results exist for inter-mediate values of n. At present, our knowledge is based on some approximate calculations based on renormalisation group ideas (Polyakov 1975, Migdal 1975, Pelcovitz and Nelson, 1976, 77, Brezin and Zinn-Justin 1976, Kosterlitz and Thouless 1973, Kosterlitz 1974, Jose et al 1976). These lead to the conclusion for this class of systems at their lower critical dimension that Heisenberg models with n ≳ 3 behave in a very similar manner to the spherical model. There is a single phase at all temperatures and $T_c = 0$ with an exponentially diverging correlation length and susceptibility as T → 0. These models are very sensitive to deviations from the theoretical ideal implying that it is almost impossible to find an experimental realisation of such models. Any real system is bound to have anisotropies induced by a substrate or by crystal field effects etc.

The planar rotor model with n = 2 is rather different. Theory tells us that there is a phase transition at a finite temperature but that the nature of the high and low temperature phases differ in a rather subtle way. There is no true long range order in either phase but the rate of decay of order changes from a power law decay below T_c to an exponential decay above

$$\langle \underset{\sim}{s}(r) . \underset{\sim}{s}(o) \rangle \sim r^{-\eta(T)} \qquad T < T_c$$

$$\exp(-r/\xi(T)) \qquad T > T_c \qquad\qquad (3.3)$$

Since this model is a theoretical idealisation of certain systems of current experimental interest, namely thin helium films, and a generalisation of the model is relevant to the properties of adsorbed monolayers, we must look at which quantities show a characteristic behaviour as we go through the transition. Fortun-ately, it turns out that these are the superfluid density and the shear modulus which are simultaneously accessible to theory and experiment.

3.1 Planar Rotor and Superfluid

It is generally believed that a superfluid can be described by a single complex order parameter $\psi(r)$ with a free energy functional of a Ginzburg–Landau form

$$F[\psi] = (1/2T) \int d^2\underset{\sim}{r} \{|\nabla\psi|^2 + \alpha|\psi|^2 + \beta|\psi|^4\} \qquad (3.4)$$

where $\psi(r) = |\psi|\exp(i\phi)$ and the superfluid velocity $\underset{\sim}{v} = (\hbar/m)\nabla\phi$. This Ginzburg–Landau form can only be valid for fluctuations taking place on a length scale large compared to the de Broglie wavelength so that quantum fluctuations have already been taken into account in eq. (3.4). This sort of phenomenological free energy has been successful in treating the critical behaviour of a superfluid in higher dimensions (Hohenberg and Halperin 1977). In higher dimensions, the magnitude $|\psi|$ is taken to be small near the transition and fluctuations in both phase and magnitude are responsible for the various critical anomalies near T_c.

In two dimensions the situation is rather diffent. Let us suppose that an expansion of the form 3.4 is correct for all values of ψ and consider what happens for very low values of temperature. The coefficient α is negative for $T < T_o$ where we can take $T_0 \simeq T_\lambda$, the critical temperature of bulk helium. When $T \ll T_o$, to a first approximation $|\psi|$ is spatially uniform function of temperature and so in this region only phase fluctuations are important. This crude argument can be put on a firmer footing by renormalisation group calculations (Pelcovitz 1978). Thus we will consider the system at low temperatures and study the stability of the ordered phase against fluctuations in the phase of the order parameter governed by a free energy functional

$$H[\phi] = \tfrac{1}{2}K \int d\underset{\sim}{r} (\nabla\phi)^2 \qquad (3.5)$$

Note that this is precisely the form of the planar rotor model where only small deviations of adjacent spins are taken into account so that both systems can be treated simultaneously with this free energy. The only difference is in the definition of the constant K

$$K = J/kT \quad \text{(planar rotor)}$$

$$= \hbar^2 \rho_o(T)/m^2 kT \quad \text{(superfluid)} \qquad (3.6)$$

where k is Boltzmann's constant, m is the mass of a helium atom and $\rho_0(T)$ is a "bare" superfluid mass per unit area in which effects due to elementary excitations etc. have already been taken into account.

The partition function is

$$Z = \text{Tr} \exp - \int d^2\underset{\sim}{r} \, H[\phi] \qquad (3.7)$$

which is a deceptively simple expression because $\phi(r)$ is an
angular variable so $H[\phi]$ is really periodic.   If, however, we
make a naive first attempt at evaluating eq.(3.7) by ignoring the
periodicity then we can calculate all correlation functions exactly
because we only have to evaluate some gaussian integrals.   The
expectation value of the order parameter is

$$<\exp i\phi(r)>\quad = 0 \tag{3.8}$$

which says that the magnetisation is zero in the planar rotor model,
or there is no Bose Einstein condensation in agreement with the
Mermin-Wagner-Hohenberg theorem.   Two point correlation functions
are also easily calculated

$$<\exp i(\phi(r) - \phi(o))> \quad \sim \quad r^{-\eta(T)} \tag{3.9}$$

where $\eta(T) = 1/2\pi K$.   This power law decay of correlation functions
is characteristic neither of a conventional low temperature phase
with complete long range order nor of a high temperature phase
which has exponential decay of correlations.   This type of power
law decay is normally associated with a critical point.   However,
as we shall see later, the behaviour described by eqs.(3.8) and
(3.9) defines the "ordered" phase of this type of system at its
lower critical dimension.

3.2 Crystalline Order in Two Dimensions

In this section we shall consider only the ideal system
namely an adsorbed monolayer on a smooth substrate whose only
role is to confine the atoms to a plane.   We shall ignore most
of the effects which are important in real systems such as the
periodicity of a graphite substrate etc.   These topics are
covered in other lectures.   The experimental systems to which
this discussion is relevant are adsorbed monolayers with a
structure incommensurate with the substrate, electrons on the
surface of helium and freely suspended films of smectic B
liquid crystals.

The most common structure in two dimensions is a triangular
lattice and suppose the atoms are at positions $\underset{\sim}{r} = \underset{\sim}{R} + \underset{\sim}{u}(\underset{\sim}{R})$
where $\underset{\sim}{R}$ denotes the sites of the triangular lattice and $\underset{\sim}{u}(\underset{\sim}{R})$
small displacements from these positions.   Exactly as in the
previous section, we ignore the fact that the mass density is a
periodic function with the periodicity of the lattice, and we
can write the Hamiltonian describing small displacements as

$$H = \tfrac{1}{2}\int d^2R\{\mu(\nabla\underset{\sim}{u})^2 + (\lambda+\mu)(\underset{\sim}{\nabla}.\underset{\sim}{u})^2\} \tag{3.10}$$

where μ and λ are the Lame coefficients.   Note that up to

quadratic order in the displacements, the elastic energy of a
triangular lattice is the same as for an isotropic system so that
one can just take over well known results of isotropic elasticity
theory.

The absence of long range order (Mermin 1968) shows up clearly
in the structure function defined by

$$S(\underset{\sim}{q}) \quad = \quad <|\rho(\underset{\sim}{q})|^2> \tag{3.11}$$

where $\rho(q)$ is the Fourier transform of the density.   In the
solid phase, we can write

$$S(\underset{\sim}{q}) \quad = \quad \sum_{\underset{\sim}{R}} e^{i\underset{\sim}{q}\cdot\underset{\sim}{R}} <\exp i\underset{\sim}{q}\cdot(\underset{\sim}{u}(R)-\underset{\sim}{u}(o))> \tag{3.12}$$

where the summation is over the lattice sites.   In three dimensions, the thermal average of eq.(3.12) taken over a Boltzmann
factor with the Hamiltonian of eq.(3.10) is essentially unity
because of true long range order.   When $q = G$, a reciprocal
lattice vector, $\exp iG.R = 1$, the structure function has a
δ-function singularity corresponding to the well known Bragg peaks
in the scattering spectrum.   Equation (3.12) and the definition
of the Fourier transform of the density $\rho(q) = \sum \exp(iq.R+iq.u(R))$
gives us a hint as to what the appropriate local positional order
parameter of a crystal should be.   If we define the local
quantities

$$\rho_G(R) \quad = \quad \exp(i\underset{\sim}{G}\cdot\underset{\sim}{u}(R)) \tag{3.13}$$

we see that these are exactly analogous to the local magnetisation
of the planar rotor model.   The only difference is that these are
many order parameters for a crystal, one for each reciprocal lattice
vector.

In three dimensions, because the mean square displacement is
finite, we find $<\rho_G(R)> > 0$, implying true long range order in
the positions of the atoms.   In two dimensions, on the other hand,
$<\rho_G(R)> = 0$, so we have no long range order, but the correlation
function decays with a power law

$$<\rho_G(R)\rho_G{}^*(o)> \quad \sim \quad R^{-\eta_G(T)} \tag{3.14}$$

where $\eta_G(T)$ is given in terms of the Lame coefficients as

$$\eta_G(T) \quad = \quad kTG^2(3\mu+\lambda)/4\pi\mu(2\mu+\lambda) \tag{3.15}$$

This behaviour is analogous to the low temperature behaviour of
the planar rotor model. The slow power law decay of eq. (3.14) is

very different from the exponential decay one expects in a
liquid so that one can say that this behaviour characterises a
two dimensional crystal.   If one could make a sufficiently large
system experimentally, one could investigate this behaviour by
neutron scattering, since the intensity would have a very
characteristic behaviour with smeared out Bragg peaks at a
reciprocal lattice vector

$$S(q) \sim |q - G|^{-2+\eta_G(T)} \tag{3.16}$$

In principle, one should be able to see several peaks for G
small enough so that $\eta_G(T) < 2$.   In practice, owing to the fact
that substrates with a large enough surface area per unit volume
to enable neutron scattering to be carried out consist of a large
number of small crystallites at different orientations, this
characteristic power law smearing of the peak cannot be measured
with sufficient accuracy.

There is true long range order of a different sort even in
two dimensions (Mermin 1968 , Halperin and Nelson 1979).   The
quantity $\psi(r) = \exp(6i\theta(r))$, where $\theta(r)$ is the angle the
"bond" between two neighbouring atoms makes with some fixed axis,
is an orientational order parameter.   In terms of the displacement field, $\theta = \frac{1}{2}$ curl u, so that in the solid phase where the
fluctuations in u are governed by the elastic Hamiltonian (3.12),

$$|<\psi(r)>| \simeq \exp(-9kT\Lambda^2/8\pi\mu) \tag{3.17}$$

where Λ is a suitable large momentum cut-off of the order of
an inverse lattice spacing.   Provided the shear modulus μ is
finite, there is true long range order which is important in
that this allows one to define crystal axes in a simple way, even
without true long range positional order.   The reader with a
fertile imagination will have already asked about the possibility
of a phase with orientational order but exponential decay of
positional order.   The theory of this phase is very similar to
that of the planar rotor model and Halperin and Nelson (1978,79)
have suggested that melting in two dimensions is a two stage
process, first to an anisotropic liquid like a liquid crystal
and at a higher temperature to a true isotropic liquid.

3.3 Topological Defects in Two Dimensions

The simple minded theory of the planar rotor model ignores
the fact that ϕ is an angle and that the Gaussian Hamiltonian
of eq.(3.5) should really be a periodic function of ϕ since the
original order parameter is a single valued function.   This
implies the existence of vortices which are configurations of the
order parameter in which the phase changes by a multiple of 2π

on going round a closed contour

$$\oint d\phi \ = \ 2\pi n \tag{3.18}$$

Of course, this statement has meaning only in a continuum limit
but a physicist will have little difficulty in giving it meaning
on a lattice.   One has to choose a contour sufficiently far
from regions in which spins on adjacent sites point in very
different directions so that one can decide how much the phase
has changed on going from one site to the next by a continuity
argument.

The naive Gaussian approximation includes only slowly varying
configurations with n=0 ignoring singular configurations of the
type shown in fig. 3.1.   To perform the statistical mechanical

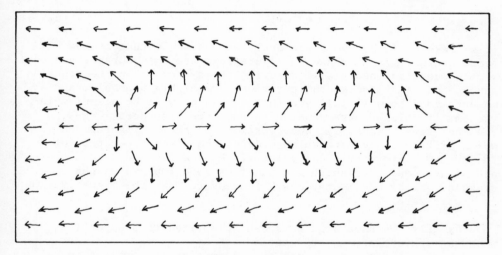

Fig.3.1 A spin configuration with a vortex and antivortex.

calculation, the first step is to compute the energy of a given
configuration.   We can readily find the local energy minimum for
a given set of vortices of vorticity n_i with centres at r_i by
minimising the energy of eq.(3.5), taking into account that there
are a set of vortices.   Minimisation yields

$$\nabla^2\phi \ = \ 0 \tag{3.19}$$

which has a solution

$$\phi(x,y) \ = \ \text{Im} \sum_i n_i \ln(z-z_i) \tag{3.20}$$

where $z = x + iy$.   Note that ϕ is no longer a single valued
function of position.

Before we go any further, let us investigate if such configurations are of any statistical significance. To do this, we must compute the energy associated with them because if it is too large the probability of such a configuration will be negligible and will make no contribution to the partition function. The energy of a single isolated vortex in a system of linear dimension L is easily seen to be

$$\Delta E = kT\pi K\ell nL + O(1) \tag{3.21}$$

where K is defined in eq. (3.6).

Since the centre of the vortex can be anywhere in the system, the entropy is

$$\Delta S = 2k\ell nL + O(1) \tag{3.22}$$

$$\Delta F/kT = (\pi K-2)\ell nL + O(1) \tag{3.23}$$

If we take eq.(3.23) very seriously we can argue that at low temperatures, $\pi K>2$, these configurations are unfavourable and our simple Gaussian theory will describe the system. On the other hand, when $\pi K<2$, such configurations are highly probable, and a moments thought will tell you that the spins will be pointing in random directions corresponding to complete disorder. Thus, we can tentatively identify these vortices as the configurations which differentiate between an ordered and disordered state of the system, with a critical value of $K_c = 2/\pi$. Note that there are always long wavelength fluctuations superimposed on the singular vortex configurations which ensure that the magnetisation is always zero.

Exactly the same arguments can be applied to a superfluid with the appropriate interpretation of K. To make a direct connection with the more familiar description of vortices in a fluid in terms of the superfluid velocity $\underset{\sim}{v}$ we simply use

$$\underset{\sim}{v} = - (i\hbar/2m)(\psi^*\underset{\sim}{\nabla}\psi - \psi\nabla\psi^*) \tag{3.24}$$

where we have taken $|\psi| = 1$ except at isolated points – the vortex cores. Then we immediately see with the aid of eq.(3.20) that curl $\underset{\sim}{v} = (2\pi\hbar/m)\sum n_i\delta(r-r_i)$ which is precisely the standard description of a set of vortices.

This is a convenient point to see the Hamiltonian which describes the interaction of an arbitrary configuration of vortices and which one must use in a real statistical mechanical calculation and to get a picture of the system without going into gory details.

$$H/kT = \tfrac{1}{2}K \int (\nabla\theta)^2 - \pi K \sum_{i\neq j} n_i n_j \ln\left|(\underset{\sim}{r}_i - \underset{\sim}{r}_j)/a\right| - \ell ny \sum_i n_i^2 \qquad (3.25)$$

which is subject to a neutrality condition $\sum n_i = 0$. In eq.(3.25), the first term is our old friend the Gaussian approximation describing smooth fluctuations or the potential flow part of the superfluid velocity. The second term is the interactions between the vortices with a short distance cut off of the order of a lattice spacing or a vortex core size and the last term is a vortex core energy. In the planar rotor case the core energy is of course determined by the interaction J but we consider here a generalisation with y arbitrary. Note that one can adjust y by adding to the Hamiltonian terms like $(\underset{\sim}{s}_i \cdot \underset{\sim}{s}_j)^2$ etc. which preserve the symmetry of the system.

Of course eq.3.25 is only an approximation to the true fixed length spin planar rotor model but it can be shown that it describes quite accurately the long wavelength fluctuations of the real model (Jose et al 1977). Moreover Villain (1975) and Knops (1977) have constructed planar rotor models which correspond exactly to eq. 3.25.

Consider now the situation when y is very small which implies a dilute system of vortices with a density $O(y^2)$. At low temperatures, $\pi K>2$, we immediately see that the vortices must be bound together in pairs of zero total vorticity with mean square separation $<r^2/a^2>\alpha 1/(\pi K-2)$ A physical picture of a superfluid is a few vortex pairs penetrating a superfluid background. When the temperature is raised so that $\pi K<2$ all that happens is that a vortex pair already well separated will completely unbind. There is essentially no change in the mean density of vortices, which implies that there will be at worst a very weak anomaly in the specific heat. A more detailed calculation shows that this anomaly is so weak that it is unobservable.

These simple arguments indicate that the low temperature phase, $\pi K>2$, will consist of a dilute gas of closely bound vortex pairs of zero total vorticity in a superfluid background. In this case, a uniform superfluid flow will be metastable because, in a system with periodic boundary conditions, the only way of reducing the velocity is to create a pair and send them across the system in opposite directions perpendicular to the flow and finally recombine them. Since there is a logarithmic energy barrier to such a process, the flow will be metastable. Note that a finite superfluid velocity is not stable since the velocity couples linearly to the separation (Langer and Reppy 1970). At higher temperatures, $\pi K<2$, the system consists of a uniform distribution unbound vortices still penetrating a superfluid background. A uniform flow is now unstable since it is obvious that there is no energy barrier

to sending a vortex right round the system as it is not necessary
to pull a pair apart first.

These arguments also support the contention that the ordered
phase of a superfluid or planar rotor model is described by a
Gaussian model.   Two tightly bound vortices of equal and opposite
vorticity have little effect because their effect on the velocity
field some distance from the pair falls off as $1/r^2$.

The change from metastability to instability will occur when
the largest pair unbinds as the temperature is raised.   Note that
not all pairs unbind at the same time, which means that the
transition will be very hard to see, and one must look at the
response to a flow.   Mathematically, one must compute a velocity-
velocity correlation function, the transverse part of which is
called the superfluid density $\rho_s(T,k,\omega)$ which is a function of
frequency, wavenumber and temperature.   This is precisely what
the Bishop-Reppy torsional oscillator experiment measures.

A theoretical problem immediately raises its ugly head at this
point.   We have argued that if we want to see this transition, we
ought to look at the superfluid density or the response to an
externally applied velocity.   Equilibrium statistical mechanics
allows one to compute such quantities at zero frequency and (with
difficulty) at constant velocity (Myerson 1978).   Unfortunately,
to measure such a response function is impossible under such
conditions.   One is compelled to measure one of the following:
(i)   the decay of a persistent current - velocity is not constant
(ii) frequency shift and dissipation in oscillating superfluid -
non-zero frequency   (iii) third sound resonance or decay of third
sound pulse - non-zero frequency and wavenumber.   To make contact
between theory and experiment it is necessary to extend the theory
to take into account these effects but this will not be done here
(see however, Ambegaokar et al 1978, Huberman et al 1978).

Melting in Two Dimensions

The problem of melting in two dimensions is very similar to
that of the phase transition in the planar rotor model.   It is
believed that melting (at least on a smooth substrate) is driven
by the topological excitations, namely dislocations (Nabarro 1967,
Friedel 1964).   The harmonic theory described in section 3.2 ignores
such excitations.

Because of the underlying periodic lattice structure we can
imagine inserting an extra half line of atoms so that locally the
crystal looks perfect, except in a small region close to the core
(Fig.3.2).   If one now goes round a contour which would be closed
 in a perfect crystal one finds that, if the contour encloses a
single dislocation, the path fails to close by an elementary lattice

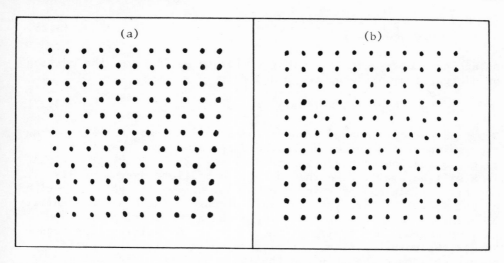

Fig.3.2 Square lattice with (a) single (b) pair of dislocations.

vector.   This is called the Burgers vector of the dislocation.
In continuum language

$$\oint d\underline{u} \;=\; \underline{b}$$

where $\underline{b}$ is the Burger's vector, a multiple of elementary lattice
vectors.   The importance of these defects is that motion of
dislocations permits the crystal to shear.   Consider the configuration of fig. 3.2b.   Application of a shear stress will cause the
two dislocations to move in opposite directions parallel to their
Burgers vectors until they reach the surface when we will have a
distorted crystal which is internally perfect.   Note that this
process of glide does not require the diffusion of vacancies and
interstitials and so the only barrier to such a motion is an
energetic one.   If there are a number of dislocations in the
crystal which are free to move, the response to an applied shear
stress will be liquid-like rather than solid-like.   Thus, the
stability of a two dimensional crystal may be determined by the
properties of dislocations in analogy to vortices in a superfluid.
Note that dislocation motion can also occur via climb, a process
in which the dislocation moves perpendicular to its Burgers vector.
This requires the diffusion of vacancies and interstitials and will
consequently be very slow, but this is a dynamical problem outside
equilibrium statistical mechanics.

One can go through similar arguments which led to eq.(3.23)
for dislocations, and one finds that the energy ΔE of a single
isolated dislocation in an otherwise perfect crystal described by
isotropic elasticity theory

$$\Delta E \quad = \quad \frac{b^2 \mu (\mu + \lambda)}{2\pi (2\mu + \lambda)} \quad \ell n L \quad + \quad O(1) \qquad\qquad (3.26)$$

Balancing energy against entropy yields an estimate of the melting or dislocation unbinding temperature

$$kT_m \quad = \quad b^2 \mu (\mu + \lambda) / 4\pi (2\mu + \lambda) \qquad\qquad (3.27)$$

Provided the core energy of a dislocation is large, the core being the region of the system where the displacement of the atoms is large so that non-linear effects dominate, the density of dislocations will be small. We then have a very dilute system of dislocations floating about in an otherwise harmonic crystal described by eq. (3.10). Note that for this picture to have any possibility of being correct, the mean spacing between dislocations must be large compared to the size of the cores at the melting temperature. Should the cores overlap, our picture based on linear elasticity theory collapses and then the theory is no longer applicable.

With these provisos, the picture is now very similar to that of a two dimensional superfluid. For $T < T_m$, we will have a perfect crystal penetrated by tightly bound pairs of dislocations (and in a triangular lattice, also tightly bound triplets) with zero total Burgers vector. Such a system will be metastable against shearing, just as the superfluid is metastable against an externally applied flow. We can thus say that the system will have a response typical of a solid in that the measured shear modulus $\mu(T)$ will be finite. Above the melting temperature, the dislocations will be free and the system will have a liquid-like response and $\mu(T) = 0$.

Such a system is an experimentalists nightmare because the melting transition has no entropy and hence no specific heat anomaly connected with it, and one must measure the shear modulus. Substrate problems are very important in contrast to the superfluid since the atoms couple directly to the substrate. However, some very recent experiments on freely suspended liquid crystal films by D. Bishop may provide a direct experimental probe of this picture.

Recent theoretical work (Halperin and Nelson 1979) has shown that an ideal system on a smooth substrate will melt to a true liquid with exponential decay of both positional and orientational order by a two stage process. The dislocation unbinding transition does not destroy orientational order so that the phase to which the solid melts is an anisotropic liquid with properties very similar to a liquid crystal. This is because dislocations floating in an otherwise perfect crystal destroy positional but not orientational order. Orientational order is then destroyed by a second type of topological defect which is directly analogous to a vortex - disclination (Nabarro 1967).

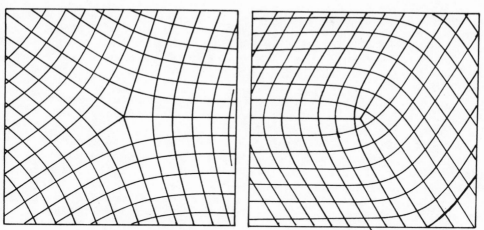

Fig.3.3 Disclinations in a square lattice

It is clear from a few moments contemplation of fig.3.3 that the energy of a disclination in a solid must be proportional to the area of the crystal since the displacement of an atom is proportional to the distance from the disclination. Thus these play no role in the crystalline phase. However, in the phase in which there is a gas of free dislocations, their energy is screened by dislocations and is reduced to log A. From this point all the arguments about vortices can be taken over directly and one has the possibility of a second disclination unbinding transition to a true isotropic liquid.

4 Renormalisation Group Analysis

In this rather technical section we will confirm the picture of a superfluid which we have derived by the very naive arguments of earlier lectures. The free energy argument which led to eq.(3.23) sounds rather convincing but it is only an indication that something important may be happening near $\pi K=2$. It is quite possible that many body effects may render the conclusions invalid. As the argument is presented, it is valid for logarithmically interacting particles in any dimension but it can be shown that many body effects invalidate the conclusions except in two dimensions.

The importance of many body effects can be seen by the following simple argument. Since the problem (Eq.3.25) has been mapped into a Coulomb gas problem of interacting charges with overall electrical neutrality, we can use our knowledge of electro-

statics to see what is going on.   A pair of equal and opposite
charges (vortices) form a dipole whose separation is variable.
The interaction between the charges forming a large dipole will
be modified by the presence of smaller dipoles between the charges
and the interaction between the charges forming these will be
modified by yet smaller dipoles and so on.   This immediately
implies that one could introduce a separation dependent dielectric
constant to describe the modified interaction.

The original treatment of this problem by Kosterlitz and
Thouless (1973) followed these ideas but an unnecessary (and
incorrect) approximation was made which has since been corrected
by Young (1978,79).   He obtains results identical to those
obtained by more sophisticated renormalisation group methods
(Kosterlitz 1974, Jose et al 1977) although these latter methods
yield more information.   In what follows, we shall follow the
treatment of Young (1978) since for those readers unfamiliar with
renormalisation group technology it makes the results at least
plausible.

Consider a widely separated pair of charges and absorb the
effects of all the other smaller pairs into an effective dielectric
constant $\varepsilon(r)$.   Then the force between these charges can be
written as $2\pi K_o/r\varepsilon(r)$ and the interaction energy as

$$2\pi U(r)\ln r/a \;=\; 2\pi K_o \int_a^r dr'/r'\varepsilon(r') \tag{4.1}$$

The density of such pairs of separation r and orientation θ
relative to an arbitrary axis is

$$n(r,\theta) \;=\; y_o^2 (r/a)^{-2\pi U(r)} a^{-2} + O(y_o^4) \tag{4.2}$$

and they have a polarisability $p(r) = \pi K_o r^2/2$.   We can now write
down a self-consistent equation for $K(r)$ using the standard
formula relating dielectric constant to susceptibility

$$\varepsilon(r) \;=\; 1 + 4\pi^3 y_o^2 K_o \int_a^r (r'/a)^{4-2\pi J(r')} dr'/r' \tag{4.3}$$

Finally, defining $r/a = e^\ell$ and $K(\ell) \equiv K_o/\varepsilon(\ell)$ we obtain

$$K^{-1}(\ell) = K_o^{-1} + 4\pi^3 y_o^2 \int_0^\ell d\ell' \; \exp(4\ell'-2\pi \int_0^\ell d\ell''K(\ell'')) \tag{4.4}$$

This rather forbidding integral equation can be unravelled by
defining an auxiliary variable y(ℓ) by

$$y(\ell) \;=\; y_0 \exp\left(2\ell - \pi\int_0^\ell K(\ell')d\ell'\right) \tag{4.5}$$

so that eq. (4.4) is equivalent to the pair of differential
equations

$$dK^{-1}/d\ell \;=\; 4\pi^3 y^2 \;+\; O(y^4)$$

$$dy/d\ell \;=\; (2-\pi K)y \;+\; O(y^3) \tag{4.6}$$

which are the key results of the theory.   Note that these equations
are derived only to lowest order in y and a more sophisticated
theory will yield the right hand side of eq.(4.6) as a power series
expansion in y (Amit et al 1979).   These equations can be derived
by a variety of different methods, mostly based on modern renormalisation group ideas (Kosterlitz 1974, Jose et al 1977, Wiegmann
1978, Amit et al 1979) and the one you find most convincing will
depend on your taste.   The fact that all yield the same results
is good evidence that they are correct.   The extra information
yielded by these techniques is a scaling equation for the free
energy f

$$df/d\ell \;=\; 2f + Ay^2 \tag{4.7}$$

from which one can compute thermal properties such as the specific
heat.   (Kosterlitz 1974, Berker and Nelson 1978).

Before considering the physical predictions of these equations
let us briefly review the approximations which have been made.
The essential one is that y_0, the fugacity of charges, is very
small so that the concentration is low.   This is vital because
it allows us to approximate the concentration of dipoles in
eq.(4.2) by the $O(y_0^2)$ term.   It also permits us to ignore higher
charges in the original vortex problem defined by eq.(3.25).
These approximations allow us to treat vortices as point excitations
in an otherwise uniform superfluid background.   If the density
of vortices were high, we would be in trouble because the
magnitude of the order parameter is reduced in the immediate
vicinity of a vortex (Gross 1961, Pitaevskii 1961), and then it
is possible that the cores start to overlap and the model is no
longer a reasonable description of the physical system.   This
problem is probably not a serious one in a superfluid because the
vortex core size is about 1Å but may be important in two dimensional crystals where dislocations play the role of vortices.
The core size of a dislocation may be several lattice spacings
and overlap becomes a real possibility.   However, from now on we
shall assume that our system of vortices is dilute so that the
possibility of overlap can be ignored and eqs.(4.6) and (4.7) hold.

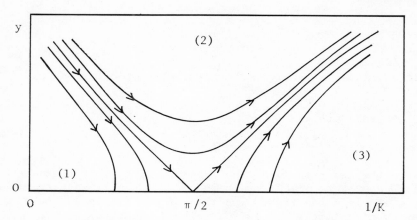

Fig. 4.1   Trajectories of eq.4.6.   Arrows indicate direction
           of increasing ℓ.

The solution of eq.(4.6) is best illustrated by a diagram
(Fig.4.1).   The lines indicate the solution for given
initial conditions and the arrows the direction as ℓ
increases.   These trajectories may be interpreted by going
back to the derivation of eq. 4.6.   The physical values of K
and y are the values in the Hamiltonian (3.25) K_0 and y_0
which correspond to $\ell = 0$.   This Hamiltonian describes the
interactions of all vortices separated by more than the physical
core spacing a.   As ℓ is increased, the effect of vortices
separated by less than ae^{ℓ} is buried in new effective interaction
constants $K(\ell)$ and $y(\ell)$ which are related to the physical ones
by eq. (4.6), so that an effective Hamiltonian parametrised by
$K(\ell)$ and $y(\ell)$ describes the interaction of vortices separated by
a larger minimum spacing ae^{ℓ}.   Thus, a calculation of a physical
property in terms of the physical parameters, $\ell = 0$, can be
related to a calculation at some other ℓ.   This renormalisation
procedure is not a magic prescription for obtaining properties of
a system but merely relates the properties at one set of values
of K and y to those at some other values.   One must perform
a real computation somewhere by some other method.

However, there is one region of Fig. 4.1 where the computations are straightforward and that is the line y = 0 which
corresponds to no vortices, and the system is described by a
trivial Gaussian Hamiltonian which was discussed in section 3.1.
Thus, a calculation involving a Hamiltonian in region 1 can be
related to one in a vortex free environment, and we interpret
this region as being the ordered phase $T < T_c(y)$.   On the other
hand, if the physical values K_0, y_0 lie in regions 2 or 3, as we
go to larger length scales $y(\ell)$ increases and the equivalent

system has a higher density of vortices which we interpret as being in the disordered phase $T > T_c(y)$. The line separating these $T = T_c(y)$ meets the $y = 0$ axis at $K^{-1} = \pi/2$ which is the value obtained by the naive free energy argument. Note that regions 2 and 3, although they appear to be different, become equivalent at large length scales and for our purposes are taken to correspond to the same high temperature phase (Amit et al 1979).

In deriving equation 4.6 and the discussion following it, all sorts of possible interactions have been ignored which in a real physical system must be present. These include the higher order vortices, short range interactions, many vortex interactions, interactions with a substrate etc. However, renormalisation group investigations show that, as far as long wavelength probes of the system are concerned, they have a negligible effect near T_c so that our Hamiltonian of eq.(3.25) provides a good description of the system.

At this stage, we have confirmed by an explicit calculation our earlier claim that there is a phase transition in the two dimensional planar rotor model and superfluid. These phases are distinguished by the presence or otherwise of free vortices and our claim that the low temperature phase is described at long wavelengths by a simple Gaussian Hamiltonian is also confirmed. It is possible to extract more detailed predictions using eq.(4.6) to transform our problem defined by the parameters K_o, y_o to another problem defined by $K(\ell)$ and $y(\ell)$. These can be solved more or less convincingly when $y(\ell) = 0$ (region 1) and when $y(\ell) \approx 1$ (region 2). In the first case, there are no vortices and we are left with a Gaussian problem which we have discussed earlier and in the second we can use a Debye – Huckel approximation familiar from other problems since there is a dense system of vortices. This approximation corresponds to treating the vorticity as a continuous variable and integrating rather than the awkward summations involved in the original problem.

To see what happens near the awkward point $\pi K = 2$ we linearise eq.(4.6) around $\pi K = 2$ and $y = 0$ to obtain

$$dx/d\ell = - z^2; \quad dz/d\ell = - xz \qquad (4.8)$$

where $x = \pi K - 2$ and $z = 4\pi y$. These are easily solved to yield

$$x^2(\ell) - z^2(\ell) = C \qquad (4.9)$$

where $C > 0$ in region 1 and $C < 0$ in region 2. The critical line $\pi K_c - 2 = z$ corresponds to the separatrix of figure 4.1 which goes into the point $K^{-1} = \pi/2$. By evaluating eq.(4.9) at $\ell = 0$ we easily find that

$$C = 4t(t + z) \qquad\qquad (4.10)$$

where $t = T_c(z)/T - 1$.

In region 1 where $C = a^2$, the solution for $\ell \gg 1/2x_0$ is

$$x(\ell) \simeq a(1 + e^{-2a\ell})/(1 - e^{-2a\ell}) \qquad\qquad (4.11)$$

This shows that there are two distinct ranges of ℓ. $\ell < 1/2a$ and $\ell > 1/2a$.   Provided we are sufficiently close to T_c, $t \ll z$, so that $a \ll x_0$, the trajectory follows the separatrix for $\ell < 1/2a$ when $x(\ell) \sim 1/\ell$ and then drops rapidly to its $\ell = \infty$ value of a.   This allows us to define a correlation length below T_c, $t \ll z$

$$\xi_- \sim \exp 1/2a = \exp 1/4\sqrt{zt} \qquad\qquad (4.12)$$

This length scale is not of much relevance for equilibrium properties of the systems under consideration but is important when one studies the dynamics of thin helium films (Ambegaokar et al 1978).

One of the most important results of a more sophisticated renormalisation group analysis (Kosterlitz 1974, Jose et al 1977) is that an effective, renormalised Hamiltonian describing very long wavelength fluctuations below T_c is simply the Gaussian one of eq.(3.5) with K replaced by

$$K(\infty) \simeq 2/\pi (1 + \sqrt{zt}) \qquad\qquad (4.13)$$

for $t \ll z$.   This immediately gives us the effect of the vortices on the order-order parameter correlation function (eq.3.9)

$$\Gamma(r) \sim r^{-\eta(T)} \qquad\qquad (4.14)$$

where $\eta(T)$ is modified from its linear dependence on T of the Gaussian approximation to

$$\eta(T) \simeq (1 - \sqrt{zt})/4 \qquad\qquad (4.15)$$

Thus $\eta(T_c) = 1/4$ which is a universal result.

It is readily seen that the order parameter-order parameter susceptibility which is, unfortunately, not accessible to direct experimental measurement, is infinite in this region.   The specific heat, obtained from the free energy by integration of eq.4.7, is proportional to ξ_-^{-2}, an unobservable essential singularity singularity.

In region 2 above T_c where $C = - a^2$, the solution of eq.4.8 is

$$\tan^{-1}\frac{x(o)}{a} - \tan^{-1}\frac{x(\ell)}{a} = a\ell \qquad\qquad (4.16)$$

These equations cannot be integrated until ℓ is arbitrarily large since $y(\ell)$ then increases and the equations themselves become invalid. However, one can use them until $y(\ell) = 0$ (1) at which stage $\ell \simeq \pi/a$. This implies the existence of a length scale $\xi_+ \sim \exp\pi/2\sqrt{z|t|}$ which is identified as the correlation length of the high temperature phase. This can be interpreted as the maximum separation of a vortex pair which can be considered to be bound. Note that for equal distances in temperature above and below T_c, $\xi_+ = \xi_-^{2\pi}$. Note also that this form of the correlation length is valid only in the narrow temperature range $|t| \ll z$. Further from T_c, this square root behaviour of the exponent crosses over to a $1/|t|$ form. It is intriguing to speculate if this is an explanation of the results of Camp and Van Dyke (1975) who found, by analysing high temperature series for the susceptibility, that the correlation length behaved like $\exp(b|t|^{-\nu})$ with $\nu \simeq .75$.

The evaluation of the order-order parameter correlation function in the high temperature regime may be carried out by first using the scaling relation (Wilson & Kogut 1974)

$$\Gamma(r) \simeq \Gamma'(r/\xi_+), \ r > \xi_+$$

where $\Gamma'(r/\xi)$ is calculated using the effective hamiltonian describing fluctuations of a wavelength larger than ξ_+. The problem is that Γ' must be evaluated for a system with a large number of vortices. This can be done by treating the vorticity as a continuous rather than discrete variable which corresponds to a Debye-Huckel approximation (Berker and Nelson 1979). This yields the expected exponential decay of the correlation function at high temperatures.

The singular part of the specific heat as T_c is approached from above behaves in a similar manner as below T_c, namely as ξ_+^{-2}. A recent numerical calculation (Berker and Nelson 1979) shows that the vortex contribution to the specific heat shows a fairly sharp bump rather above T_c. The sort of curve obtained is shown in Fig. 4.2.

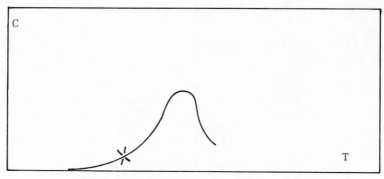

Fig.4.2   Vortex specific heat

Note that the curve is completely smooth at T_c.   Precisely similar
effects are expected in dislocation mediated melting (Halperin
and Nelson 1979) so specific heat experiments are rather uninformative for such transitions.

     Perhaps the most striking result of the static vortex theory
described here is that the response function measuring what one
normally thinks of as the superfluid density has a finite value
below T_c and then drops discontinuously to zero.   This discontinuity is proportional to T_c, the proportionality constant being
a universal number which is given in terms of fundamental quantities
such as Planck's constant, Boltzmann's constant and the bare mass
of a helium atom.   Hohenberg and Martin (1965) have shown that the
superfluid density can be calculated from the current-current
correlation function where the current $j = \rho_s^o \underset{\sim}{v}$.   The observed
superfluid density is given by

$$\rho_s(q,T) \;=\; (C_\ell(q) - C_t(q))/kT \tag{4.17}$$

where C_ℓ and C_t are the longitudinal and transverse parts of
$<j_\mu j_\nu>$.   The Fourier transform of the superfluid velocity is

$$v_\mu(q) \;=\; \frac{i\hbar}{m}\, q_\mu \phi(q) + \frac{2\pi i\hbar}{m}\, \varepsilon_{\mu\nu} q_\nu \, n(q)/q^2 \tag{4.18}$$

where $n(q)$ is the vorticity.   It is easily shown that the
observed quantity

$$K_R(q,T) \;=\; K_o(1 - 4\pi^2 K_o <n(q)n(-q)>/q^2) \tag{4.19}$$

where $K_R(q,T)$ is defined in exact analogy to eq. 3.6

$$K_R(q,T) \;=\; \hbar^2 \rho_s(q,T)/m^2 kT$$

Renormalisation group considerations show that on rescaling the

vortex core size from a to ae^{ℓ}, this expression remains the same
except that K_o is replaced by $K(\ell)$ as determined by the recursion
relations of eq. 4.6 and the vortex-vortex correlation function
<nn> is to be computed with the corresponding effective Hamiltonian.

Below T_c, the rescaling can be carried out to infinite ℓ at
which stage there are no vortices left because $y(\infty) = 0$, which in
turn means that the term in eq. (4.19) containing < nn > vanishes.
Thus, $K_R(0,T)$ tends to a finite limit $K(\infty)$ which implies that
the superfluid density is finite and is equal to $2/\pi$ at T_c. The
ratio $\rho_s(0,T_c)/T_c$ has the universal value of 3.5×10^{-9} $gm/cm^2/^{\circ}K$
(Nelson and Kosterlitz 1977). Note that neither ρ_s nor T_c are
themselves universal, but their ratio is.

Above T_c, the vortex fugacity increases with increasing ℓ, so
that the procedure is to scale up to $\ell_c = \ln \xi_+$ by the recursion
relations and then compute < nn > by integrating over the vorticity
field $n(q)$ to obtain

$$<n(q)n(-q)>/q^2 \simeq 4\pi^2K(\ell_c) \; + \; q^2\xi_+^2 \qquad\qquad (4.20)$$

which is valid for $q\xi_+ \ll 1$. Substituting this in eq.(4.19) we
see that $K_R(q,T)$ vanishes like $q^2\xi_+^2$, so that the superfluid
density $\rho_s(0,T)$ is zero above T_c as expected (Halperin and
Nelson 1979). These results can be summarised in fig. 4.3.

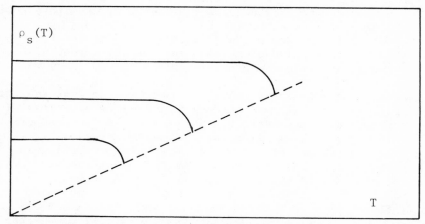

Fig.4.3 Superfluid density at zero wavenumber and frequency

The dashed line in fig.4.3 is the locus of critical points
for various values of ρ_s and it has a slope of $2m^2k/\pi\hbar^2$. The
curves include only the depletion of the superfluid density due to
vortices, all other effects being buried in $\rho_s^{\,o}$. The sort of
effects ignored in this calculation are terms like $(\nabla\phi)^4$ in the

Hamiltonian which, in this classical theory, would lead to a
depletion linear in T at low temperatures, but are irrelevant
near T_c so that the ratio $\rho_s(T_c)/T_c$ is unaffected. The cusps
close to T_c are due to the fact that $K(\infty)$ approaches its
critical value $2/\pi$ as a square root (eqs. 4.10, 4.11)

$$K(\infty) = 2(1 + \sqrt{t})/\pi \qquad t << z \qquad\qquad (4.21)$$

Since ρ_s is the areal superfluid density, the zero temperature
value is adjusted in thin helium films by simply adsorbing more
or less helium. Very similar results hold for the shear modulus
in the melting of monolayers provided it is dislocation mediated.

5   Summary

        I have attempted, in the limited time available, to review
the rapidly developing field of phase transitions in two dimensional
systems. These may be divided into two classes according to
whether or not the system is at its lower critical dimensions, and
I have concentrated mainly on the former case. In such a situation
the theory is now sufficiently well developed to yield quantitative
predictions and agreement with experiment is good. Renormalisation
group considerations show that certain experimentally accessible
quantities such as the superfluid density tend to exactly calculable
universal values, but which are not directly accessible to experiment. I have not discussed the extension of the theory to time
dependent phenomena which is necessary before contact with experiment can be made. This extension has been carried out by
Ambegaokar et al (1978) and by Huberman et al (1978) for the
vortex case. The dynamics of melting and the corresponding
experiments are currently being developed.

        Since systems at a lower critical dimension are very sensitive
to small perturbations it is necessary to choose an experimental
system very carefully indeed. Suitable ones seem to be superfluid onset in thin helium films adsorbed on Mylar or glass,
electrons on the surface of helium and freely suspended liquid
crystal films. To study transitions in such films one must
look at the appropriate dynamic response function such as the
superfluid density or shear modulus. Specific heat measurements
are not informative, so that the necessary experiments are difficult.
On the other hand, systems above their lower critical dimension
are relatively easy to investigate experimentally because the
anomalies in, for example, the specific heat are readily detectable
and they are much less sensitive to deviations from the ideal.
This makes experiments on monolayers adsorbed on a poor substrate
such as grafoil meaningful. There are a variety of transitions
of this type such as the melting of a registered monolayer which may
be described by a Potts model. The disadvantage of such systems

is that theory has much less predictive power although recent work has been able to reproduce the experimentally measured phase diagram of Krypton on graphite.   Since most of the theoretical tools now exist to cope with both types of system, I expect a flood of papers over the next year or so treating more and more complicated systems, but most of the fundamental theory has been done.

An important area of research in two dimensional systems which I have not mentioned is a theoretical one being pursued vigorously by Leo Kadanoff.   This is a study of how various models such as the eight vertex model, Ashkin Teller model, F model etc. are connected.   It seems that, at their critical points, these models can be mapped on to the simple Gaussian model discussed here so that all these models are closely related.   References on this area may be found in the papers by Jose et al (1977), Kadanoff 1978,79, and Kadanoff and Brown 1979.   The interested reader will find food for thought in these papers.

References

Ambegaokar, V., Halperin, B.I., Nelson,D.R. and Siggia, E.D., 1978
        Phys. Rev. Lett.40 783.
Amit, D.J. Goldschmidt, Y.Y., and Grinstein, G., 1979, to be
        published.
Alexander, S., 1975, Phys. Lett. 54A   353
Bak, P., 1979, to be published.
Barker, R.A., Estrup, P.J., Jona, F., and Marcus, P.M., 1978
        Solid State Comm. 25   375
Beasley, M.R., Mooij, J.E. and Orlando, T.P., 1979, Phys.Rev.Lett.
        42 1165.
Berezinskii, V.L., 1970 Sov.Phys. J.E.T.P. 32   493.
Berlin, T.H., and Kac, M., 1952, Phys. Rev. 86   821.
Berker, A.N., Ostlund, S. and Putnam, F.A., 1978, Phys. Rev. B17
                                                            3650.
Berker, A.N. and Nelson, D.R., 1979, Phys.Rev. B19   2488.
Berthold, J.E. Bishop, D.J. and Reppy, J.D., 1977, Phys. Rev.Lett.
        39   348.
Bishop, J.D. and Reppy, J.D., 1978, Phys. Rev. Lett. 40   1727.
Blakeley, D.W. and Somorjai, G.A., 1977, Surf.Sci. 65,   419.
Bretz, M., 1973, Phys. Rev. Lett. 31   1447.
Brewer, D.F. and Mendelssohn,K., 1961, Proc.Roy.Soc. A260   1.
Brezin, E. and Zinn-Justin, J., 1976, Phys.Rev.Lett. 36   691;
        Phys. Rev. B14   3110.
Camp, W.J., and Van Dyke, J.P., 1975, J.Phys. C8   336.
Chan, C.M., Luke, K.L., Van Hove, M.A., Weinberg, W.H. and
        Withrow, S.P., 1978, Surf.Sci 78   386.
Chester, M. and Yang, L.C. 1973, Phys. Rev. Lett. 31   1377.

Chester, M. and Yang, L.C., 1974, Phys. Rev. A9 1475.
Chui, S.T. and Weeks, J.D., 1976, Phys. Rev. B14 4978.
Dash, J.G.,1975, Films on Solid Surfaces (Academic Press,New York).
Debe, M.K. and King, D.A., 1977, Phys. Rev. Lett. 39 708.
Doniach, S. and Huberman, B.A., 1979, Phys. Rev. Lett. 42 1169.
Felter, T.E., Barker R.A. and Estrup, P.J., 1977, Phys.Rev. Lett.
        38 1138.
Fisher, D.S. Halperin, B.I., and Platzman, P.M., Phys. Rev. Lett.
        42, 798.
Friedel, J., 1964, Dislocations (Pergamon Press, London).
Grimes, C.C. and Adams, G., 1979, Phys.Rev.Lett. 42 795.
Gross, E.P., 1961, Nuovo Cimento, 20, 454.
Halperin,B.I. and Nelson, D.R. 1978, Phys.Rev.Lett.41, 121.
        1979, Phys.Rev. B19 2457.
Halperin, B.I. and Nelson, D.R., 1979, J.Low Temp.Phys. in press.
Henkel, P.P., Smith, E.N. and Reppy, J.D., 1969, Phys.Rev.Lett,
        23, 1276.
Herb, J.A. and Dash, J.G., 1972, Phys.Rev.Lett. 29 846.
Hohenberg, P.C. 1967, Phys. Rev. 158 383.
Hohenberg, P.C. and Halperin, B.I., 1977, Rev.Mod.Phys. 49 435.
Hohenberg, P.C. and Martin, P.C., 1965, Ann.Phys. 34 291.
Huberman, B.A. and Doniach, S., 1979, Phys.Rev.Lett. 42, 1169.
Huberman, B.A., Myerson, R.J. and Doniach, S., 1978, Phys.Rev.
        Lett. 40, 780.
Huff, G.B and Dash, J.G., 1976, J.Low.Temp.Phys. 24, 155.
Imry, Y. and Gunther, L., 1971, Phys.Rev. B3, 3939.
Jancovici, B., 1967, Phys.Rev. Lett. 19 20.
de Jongh, L.J., 1976, Physica 82B, 247.
de Jongh, L.J. and Miedema, A.R., 1974, Adv. Phys. 23, 1.
Jose, J.V., Kadanoff, L.P., Kirkpatrick S. and Nelson, D.R., 1977,
        Phys. Rev. B16 1217.
Josephson, B.D., 1966. Phys.Lett. 21, 608.
Joyce, G.S. 1972 in Phase Transitions and Critical Phenomena
        (C.Domb and M.S.Green, Eds.)(Academic Press, London
        and New York) Vol.II. Ch.10.
Kac, M., 1971, Phys. Norveg. 5, 163.
Kadanoff, L.P., 1978, J.Phys. A11, 1399.
Kadanoff, L.P., 1979, Ann.Phys. 120, 39.
Kadanoff, L.P. and Brown, A., 1979, to be published.
Knops, H.J.F., 1977, Phys. Rev. Lett. 39 766.
Kosterlitz, J.M., 1974, J.Phys. C7 1046.
Kosterlitz, J.M. and Santos, M.A., 1978, J.Phys. C11 2835.
Kosterlitz, J.M. and Thouless, D.J., 1973, J.Phys. C6. 1181.
Kosterlitz, J.M. and Thouless, D.J. 1978.Prog.Low Temp.Phys.
        (D.F.Brewer, ed.)(NorthHolland) Vol.7B 373.
Langer, J.S. and Reppy, J.D., 1970. Prog.Low Temp.Phys.
        (C.J.Gorter,ed.)(North Holland)Vol.6.
Langmuir, I., 1918, J. Amer. Chem.Soc. 40 1361.
Lieb, E.H. and Wu, F.Y., 1972, in Phase Transitions and Critical

Phenomena (C.Domb and M.S.Green.eds)(Academic Press, London and New York). Vol.1, Ch.8.

McTague, J.P. and Nielsen, M., 1976, Phys. Rev. Lett. 37 596.

Mermin, N.D. 1967, J. Math.Phys. 8 1061.

Mermin, N.D. 1968, Phys.Rev. 176, 250.

Mermin,N.D. and Wagner, H., 1966, Phys.Rev. Lett. 17 1133.

Migdal, A.A., 1975, Sov.Phys. J.E.T.P. 42 413; 42 743.

Mikeska, M.J. and Schmidt, H., 1970, J.Low Temp.Phys. 2 371.

Myerson, R.J., 1978 Phys.Rev. B18 3204.

Nabarro, F.R.N.,1967, Theory of Dislocations.(Clarendon Press, Oxford)

Navarro, R. and de Jongh, L.J., 1976, Physica 84B 229.

Nelson, D.R., 1978, Phys. Rev. B18, 2318.

Nelson, D.R. and Kosterlitz, J.M., 1977, Phys. Rev. Lett. 39 1201.

Onsager, L., 1944, Phys. Rev. 65, 117.

Ostlund, S. and Berker, A.N., 1979, Phys. Rev. Lett. 42 843.

Peierls, R.E., 1934, Helv. Phys. Acta 7, Suppl.II 81.

Peierls, R.E., 1935, Ann.Inst. Henri Poincare 5 177.

Pelcovitz, R.A., 1978, Ph.D. Thesis (unpublished).

Pelcovitz, R.A. and Nelson, D.R., 1976, Phys. Lett. 57A, 23.

Pelcovitz, R.A. and Nelson, D.R., 1977, Phys. Rev., B16, 2191.

Pitaevskii, L.P., 1961, Sov. Phys. J.E.T.P. 13, 451.

Polyakov, A.M., 1975, Phys. Lett. 59B 79.

Potts, R.B., 1952, Proc. Camb. Phil. Soc. 48, 106.

Richards, MG., 1979, this summer school.

Rudnick, I., 1978, Phys.Rev. Lett. 40,1454.

Scholtz, J.H., McLean, E.O. and Rudnick, I., 1974, Phys. Rev. Lett. 32 147.

Siddon, R.L. and Schick, M., 1974, Phys. Rev. A9, 907; 1753.

Stanley, H.E., 1968, Phys.Rev. 176, 718.

Stewart, G.A. and Dash, J.G., 1970. Phys.Rev. A2, 918.

Taub, H., Carneiro, K., Kjems, J.K., Passell, L. and McTague, J.P., 1977, Phys. Rev. B16 4551.

Telschow, K.L. and Hallock, R.B. 1976, Phys.Rev. Lett. 37 1484.

Thomy, A. and Duval, X., 1969, J. Chim.Phys. 66. 1966.

Thomy,A. and Duval, X., 1970, J.Chim.Phys. 67 286; 1101.

Villain, J., 1975, J. Physique 36 581.

Webster, E., Webster, G and Chester, M.,1979, Phys.Rev.Lett. 42 243.

Wiegmann, P.B., 1978, J.Phys. C11, 1583.

Wilson, K.G. and Kogut, J., 1974, Phys. Rept. C12, 75.

Yonehara, K and Schmidt, L.D., 1971, Surf.Sci. 25 238.

Young, A.P., 1978, J. phys. C11 L453.

Young, A.P., 1979. Phys. Rev. B19 1855.

SUPERFLUIDITY IN THIN HELIUM FOUR FILMS

John D. Reppy

Laboratory of Atomic and Solid State Physics and
Materials Science Center
Cornell University, Ithaca, New York 14853

INTRODUCTION

I wish to discuss in these lectures the experimental aspects
of superfluidity in thin films of adsorbed helium. I will be concerned mainly with those films ranging from about five atomic
layers down to the thinnest which show evidence of superfluidity.
I shall begin with a brief review of the superfluid properties of
bulk helium as these ideas are well understood and form the conceptual basis for our understanding of superfluidity in films. The
experimental methods used for observation of the superfluid properties of films will then be surveyed. In the final lecture, I
will discuss the nature of the superfluid transition in films, and,
in particular, the evidence for both two and three dimensional
superfluid phase transitions in films, depending on the topology
of the adsorption substrate.

PROPERTIES OF BULK SUPERFLUID HELIUM FOUR

It is useful to review the properties of bulk liquid ^{4}He, as
they are well known and form the basis for much of our conceptual
understanding of the superfluid properties of films. In addition,
it is particularly interesting to determine which properties of
the bulk superfluid carry over to the film and the ways in which
they are modified.

A schematic view of the phase diagram for bulk ^{4}He is shown
in figure 1A. Its dominant features are the existence of the
liquid phase down to zero temperature and the liquid-liquid phase
transition λ-line, separating the high temperature normal phase

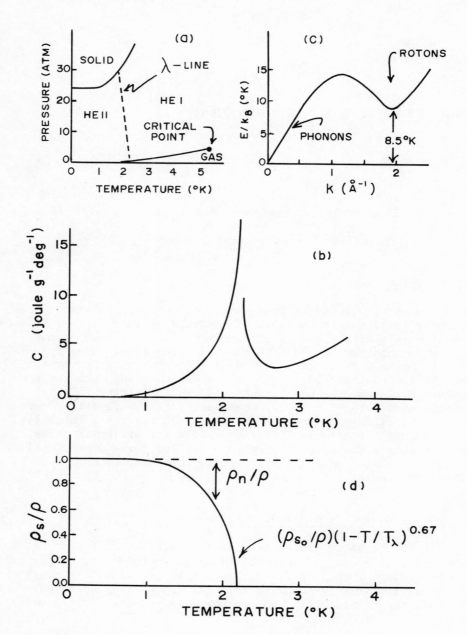

Figure 1.Properties of Bulk ^{4}He

(Helium I) from the lower temperature superfluid phase (Helium II).
A pressure of about 25 atmospheres is required to solidify liquid
^{4}He. In an adsorbed film, the van der Waals attraction to the wall
can provide effective pressures greater than this value, so one might
expect to find a layer of solid next to any surface. The heat capacity for liquid helium at saturated vapor pressure is indicated in
figure 1B. The normal-to-superfluid transition is second order; the
heat capacity displays the famous λ form at the transition. The
actual heat capacity anomaly at T_λ closely approaches a logarithmic
singularity. Between one and two degrees Kelvin, the heat capacity
is dominated by a contribution from roton excitations. For temperatures well below one degree Kelvin, the rotons are frozen out and
the heat capacity has a T^3 form due to phonons. Figure 1C shows the
Landau excitation spectrum for bulk helium. This spectrum, which
has been determined from neutron scattering measurements[1], is in
good agreement with the measured values of heat capacity and determinations of the superfluid density in bulk helium.

Figure 1D shows the temperature variation of the superfluid
density, ρ_s, as a function of temperature at saturated vapor pressure. Rotons provide the dominant contribution to the normal fluid
density, $\rho_n = \rho - \rho_s$, except where critical fluctuations become important. Although phonons contribute to the normal fluid density,
the fraction ρ_n/ρ is always small compared with unity. Near the
transition, where critical fluctuations are important, the super-
fluid density follows a power law, $\rho_s = \rho_{s_0}(1 - T/T_\lambda)^{.67}$. For a
more complete discussion of the properties of ^{4}He in the critical
region, one is referred to the lecture notes of Dr. G. Ahlers from
the previous Erice Summer School on Quantum Liquids.[2]

The principal experimental property that distinguishes a superfluid is the ability to flow without dissipation. In a discussion
of superflow, there are two points of view which prove useful. The
first of these is the two-fluid model of superfluid helium. In this
model, the total density of the system is divided into ρ_s, the
superfluid density, and ρ_n, the normal fluid density. Associated
with each of these densities is a separate velocity field. The
superfluid is presumed to flow without friction and to carry no
entropy. In figure 2, a channel formed by two parallel plates and
filled with superfluid ^{4}He is pictured. The roton and phonon
excitations which constitute the normal fluid are schematically
pictured as dots. These excitations interact with the wall. When
the spacing between the walls is sufficiently small, the excitations
are effectively locked to the wall. The density of the normal fluid
can be easily measured by oscillating the plates and detecting the
effective mass of the normal fluid which is carried along in this
oscillation. This technique is that due to Andronikashvili[3] and can
be employed equally well for bulk helium or for measurements of the
normal fluid mass in thin films.

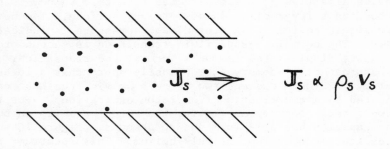

Figure 2.Superfluid mass transport between parallel plates

It is also possible to have a dissipationless supercurrent, $J_s \propto \rho s v_s$, existing between the plates. This current can also be detected in a number of ways. Probably the most sensitive techniques are the doppler shift technique utilizing 4th or 3rd sound or the detection of the angular momentum carried by the persistent current flowing between the plates.

In addition to the two-fluid model, another useful representation of the superfluid is given in terms of a "macroscopic order parameter", ψ, which is taken to be complex. The superfluid velocity field is a potential flow given by $v_s = (\hbar/m)\nabla\varphi$, where φ is a macroscopic phase. The conventional representation of the order parameter is $\psi = \psi_o e^{i\varphi}$. The kinetic energy in the flow field of the superfluid is $E_{kin} \propto \frac{1}{2}\rho_s v_s^2$ or $\frac{1}{2}\rho_s \int ((\hbar/m)\nabla\varphi)^2 d\tau$. (This expression is similar to the expression for the elastic strain energy in a crystal, where ρ_s plays the role of a stiffness constant which determines how much energy can be accumulated in the superfluid state for a given gradient in the phase, φ.)

In this representation of the superfluid order parameter, the phase must be single valued, modulo 2π. By analogy to the properties of a single particle wave function, we then expect

$\oint v_s \cdot d\ell$ to be equal to nh/m, where n = 0, $\pm$ 1, $\pm$ 2,...

This relation was originally suggested by arguments that the bulk superfluid exists in some sort of macroscopic quantum state. Fortunately, we do not have to depend on theoretical argument for this quantization condition. It is an experimental fact that has been demonstrated in bulk helium in a number of different ways.[4] At the present time, however, a direct demonstration of the quantization of superflow has not yet been possible.

SUBSTRATES AND SUPERFLUID FILMS

The superfluid contained between two smooth parallel plates, as indicated in Fig. 2, forms our idealization of the adsorbed helium film. Only one plate is necessary in the film case for the adsorbed atoms are held on the wall by van der Waals attraction. For film thickness ranging from a few tens of angstroms downward, the normal fluid excitations are firmly locked to the substrate at typical experimental frequencies.

In reality, this idealization of the film with a smooth, featureless substrate is far from the truth. One of the interesting questions in the study of superfluid films is the question of to what degree the properties of the real substrate determine the behavior of the superfluid film.

In Fig. 3, I have sketched a rough representation of the real situation. The substrate is far from smooth, it has atomic structure, crystalline structure, and, in most cases, contamination of adsorbed dirt of various sorts. Immediately above the substrate, there will usually be a region or layer of localized helium atoms with effective thickness d_o. These localized atoms are often spoken of as the "solid layer". For most substrates used in studies of superfluidity in films, this region is not a well-defined crystalline layer but is a highly disorganized region of atoms localized to some degree on the substrate.

Above the localized layer, there are the mobile atoms, which may form a superfluid at sufficiently low temperature. The density in this mobile region for thick films is similar to bulk liquid helium densities. However, for low coverages the density, in terms of mobile atoms per unit area, can be much less than that for mono-

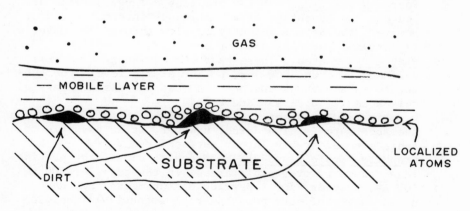

Figure 3. Conceptual view of "real" substrate

layer coverage at equivalent bulk densities. At very low coverage, the mobile atoms may possibly constitute "dilute" surface gas, moving above the region of localized atoms. Above the film, there exists a region of gas. At temperatures below one degree, or for very thin films, the atoms in the gas phase can often be neglected. However, for thicker films, especially at temperatures above 1 degree, a large fraction of the helium atoms in the system may, in fact, be in the gas phase. This situation can be a considerable experimental nuisance. Often, to reduce the influence of the gas phase, the experimenter will choose a substrate which offers a large surface to free volume ratio. A large surface area is also desirable in measurements such as heat capacity studies in order to increase the total signal size.

Superflow experiments in films impose special requirements which should be emphasized. Superflow is a long-range phenomenon, and most superflow experiments will require a continuous substrate. This requirement often goes contrary to the requirement of large surface area, and accordingly, many superfluid film experiments have been performed on less than ideal geometries.

A few examples of substrates used for superfluid experiments are listed below:

1. Fine Powders (lampblack, rouge, aluminum oxide). Powders have the advantage of extremely large surface areas. Their main disadvantages are that the grains may be poorly connected together and the substrate lacks the continuity which is desirable for superfluid flow experiments.
2. Porous Vycor Glass. Vycor provides a large surface area, and has the advantage of a continuous internal surface. However, the actual geometry of flow through the material is extremely complex and tortuous.
3. Flat Glass Plates. These have the advantage of a well-defined geometry, however, on the atomic level, the substrate is amorphous and one also has the disadvantage of only a small total surface area.
4. Plastic Films. Plastic Films are similar to glass in that the substrate is disordered, however, it is possible to achieve considerably larger surface areas.
5. Graphite (Grafoil and ZYX). Graphite substrates provide ideal geometry over small areas, on the order of $100 \times 100 \, \text{Å}^2$, however, for superflow measurements they have the disadvantage of very poor internal connectivity between the crystallites of the graphite.

THIN FILM EXPERIMENTS

In the remainder of these lectures, I wish to discuss a number of different types of superfluid film experiments, with an emphasis on the information gathered from each type and the experimental technique employed. The categories of experiments I will discuss are: 1) Heat Capacity measurements, using powders and graphite as the substrates, 2) Mass Transport measurements, 3) Third Sound, 4) Persistent Currents, 5) Oscillating Substrate methods.

HEAT CAPACITY MEASUREMENTS

In bulk liquid helium, the heat capacity anomaly, the λ point, occurs at the same temperature as the onset of superfluidity. In a helium film, one has the possibility of studying the evolution of the heat capacity as the thickness of the film is reduced. In 1949, Frederikse published his important results for the heat capacity of ^{4}He films adsorbed on a powder sample.[5] The heat capacity helium confined to the pores of Vycor glass was also measured by the Univ. of Sussex group, Brewer et. al.[6], and most recently with considerably higher resolution by Joseph and Gasparini.[7]

Figure 4 shows the general features of Frederikse's data. As the film is made thinner, the sharp λ point seen in bulk helium becomes rounded and the temperature of the specific heat maximum moves to lower temperatures. When the film thickness has been reduced to the order of three or four layers, very little evidence of the maximum remains.

In addition to full pore heat capacity measurements made on Vycor by Brewer, a series of heat capacity measurements has been made for films adsorbed on this substrate by Symonds and Evanson at the University of Sussex.[8] These measurements show much the same qualitative character as seen in Frederickse's work. As the film thickness is reduced, the heat capacity maximum broadens and shifts to lower temperatures until at coverages on the order of three layers, the evidence of a maximum has almost disappeared.

The data shown in Fig. 5 is in marked contrast to the heat capacity results on Vycor and powder samples. Here, we show a sketch of the heat capacity measurements made by M. Bretz (1973) for ^{4}He films adsorbed on a Grafoil substrate. In this work, the heat capacity maximum, instead of broadening as the film thickness is reduced, develops a truncated form with reasonably sharp corners. One point of similarity remains, however. When the film thickness is reduced to the order of three layers, the heat capacity maximum essentially disappears.

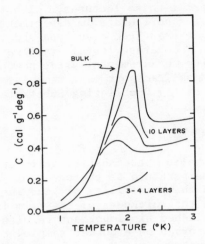

Figure 4. Heat capacity measurements of
He films on powder substrate

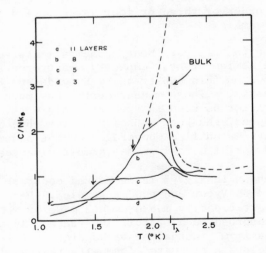

Figure 5. Heat capacity measurements of He films on
Grafoil substrate

When these data were first obtained, it was suggested that the relatively sharp features seen were due to the more ideal nature of the graphite substrate. Now, however, the interpretation is less clear. For instance, the corner which remains visible near the bulk λ transition temperature is thought to be probably due to capillary condensation between the platelets of graphite.[33]

Fortunately, in these measurements, Bretz was also able to make an important correspondence between the onset of superflow and features of the heat capacity. The onset of superfluidity was detected through a study of the thermal relaxation time of the sample. In Fig. 5, the superfluid onset is indicated by arrows, and it is seen that the onset temperature occurs very near the sharp lower temperature corner in the heat capacity data.

DETECTION OF SUPERFLUID ONSET BY DRIVEN FLOW

The superfluid transition in bulk helium is easily detected since drastic changes occur in certain of the transport properties, such as the effect of thermal conductance and the ability of the superfluid to flow with zero viscosity through very small channels. The sharp change at the λ point in one of these properties is often used as a signal of the superfluid phase transition. The same approach can be applied for the detection of the onset of superfluidity in adsorbed films. There are two major variations depending on whether the superflow in the film is driven by a temperature or pressure gradient.

In Fig. 6, I've schematically illustrated the temperature gradient method. The apparatus consists of a low conductivity box with a heater and thermometer at one end and a heat sink at the other. The experiment consists of making a thermal resistance measurement on this system. When power is supplied by the heater, evaporation takes place at the hot end of the box and matter is transported by gas flow to the cold end, where recondensation takes place. When the temperature is below the superfluid onset, mass may be returned through superflow in the film. This configuration is similar in concept to the countercurrent flow which occurs during heat transport in bulk helium, only in this case, the flow of gas takes the place of the normal fluid transport. The temperature is raised above the superfluid onset temperature for the film, then the thermal resistance of the apparatus increases rapidly. Fokkens, Taconis and deBruyn Ouboter[10] have used this method extensively. The other version of the driven flow experiment involves the use of a pressure gradient applied across a superleak. In this case, the superfluid onset is signalled by a large increase in the mass flow through the superleak as the temperature is lowered below the transition. Herb and Dash[11] have used this technique to observe the onset for helium films adsorbed on Grafoil.

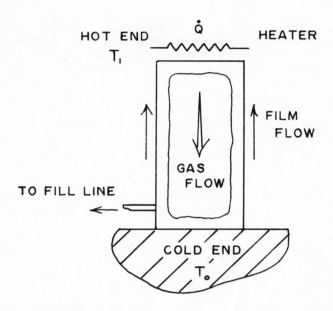

Figure 6. Schematic view of temperature gradient method.

In many of these experiments, the film thickness is controlled by varying the vapor pressure above the unsaturated film. If one knows the van der Waals attraction, it is possible to estimate the film thickness from the vapor pressure data, however, it has been customary to plot the observed onset temperatures as a function of the measured parameter P/P_o, P being the vapor pressure and P_o the saturated vapor pressure. Fig. 7 shows a collection of such data, obtained by a number of workers using the driven flow technique. A conspicuous feature of this plot is the fact that a large number of the experiments scatter about an S shaped curve, which one might draw through the data. The scatter probably arises from variations in the van der Waals attraction and, consequently, in the thickness of the films for the different substrates. In particular, it seems that the Grafoil onset temperatures lie considerably above those of most of the rest of the experiments for the same relative vapor pressure. Bretz has pointed out[12] that this at first apparent disparity is, in fact, mostly a reflection of the much stronger van der Waals forces which are in action above a graphite substrate. If the comparison is made on the basis of the thickness of the superfluid portion of the film, one finds that the onset temperatures are in much better agreement.[12]

Thermal and pressure driven flow measurements provide simple techniques for the detection of superflow in films and work well,

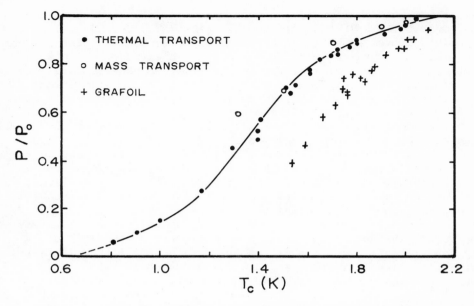

Figure 7. Vapor pressure fraction at superfluid onset

provided that the transition is sharp. An important disadvantage
of the temperature gradient method is the requirement of a return
gas flow. This requirement makes the temperature gradient method
progressively less useful as the temperature is lowered below 1°K,
as the vapor pressure then approaches zero. In addition, the driven
flow technique is only capable of demonstrating high surface mobility and does not prove in any sense the existence of a superflow
with zero dissipation.

THIRD SOUND MEASUREMENTS

        Third sound is the name given to a unique propagating surface
mode in helium films. For a comprehensive treatment of the subject
of third sound, I would recommend the excellent review by Atkins
and Rudnick.[13] Third sound modes in superfluid films are closely
analogous to ordinary sound. A variety of experimental configurations, some familiar from acoustics, have been used for third
sound investigations. Particularly useful among these have been
resonant cavity techniques, including Helmholtz resonators and
time of flight measurements.

        An interesting geometry has been employed by K. Mountfield
in his investigations of mass transport in helium films.[14] This
apparatus spans the driven flow regime discussed in the previous

section and the regime of the more conventional third sound measurement. The apparatus is shown schematically in Fig. 8B. It consists of two chambers filled with a porous medium (filter paper) with a large surface area (about 20 m^2). Electrodes are placed between the layers of filter paper to form a capacitor used for sensing the film thickness. If an increase in the film thickness is created on one side by the admission of a small dose of helium gas, one can then observe the migration of helium atoms to the other chamber along the connecting tube. The experiment was usually run at temperatures well below one degree, where there would be no transport through the connecting tube by gas flow. The operation of this system is analogous to the U tube configuration, as shown in Fig. 8A. If a height difference exists between levels in the two tubes, then a gravitationally-driven flow will take place through the connecting channel. When the damping of the flow through the channel is sufficiently small, oscillations may be seen. The undamped oscillations will have a frequency $\omega^2 = (\rho_s/\rho)(a/A)(g/L)$, where a is the area of the small tube, A is the standpipe area, L is the length of the connecting tube, and g is the acceleration due to gravity. In Mountfield's film experiment, the van der Waals attraction plays the role of gravity. and the difference in height between the two standpipes corresponds to the difference in thin film thickness in the two chambers.

The experiment was then performed by increasing the thickness of the film in one chamber and watching the relaxation of the system. The behavior seen at a temperature of 0.5 K with several different film thicknesses is shown in Fig. 9. At this temperature, and for coverages on the order of two layers, a small increase in the film thickness results in a slow exponential relaxation as the helium atoms migrate or diffuse to the other chamber. When the thickness is further increased to the range of four layers, a constant flow velocity is observed at first. When equilibrium is reached, there is a sharp break in the curve. This behavior is reminiscent of critical velocity flow as seen for bulk helium passing through narrow channels. When the film thickness is again increased, free oscillations of the film can be seen. These oscillations are the third sound Helmholtz oscillations of the system. The value of this approach lies in the possibility of studying the dynamics of the film ranging from the diffusive mobility regime to the nearly undamped third sound oscillations.

The two-chamber arrangement of Mountfield's experiment, with an interconnecting tube, represents a rudimentary third sound oscillator. The more conventional configuration for third sound

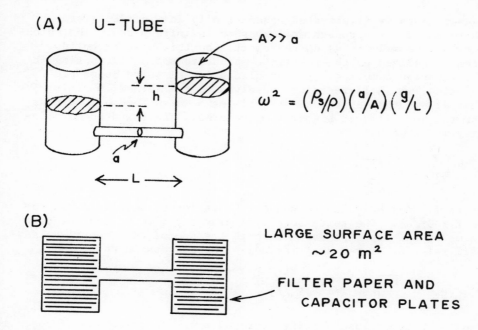

$$\omega^2 = \left(\rho_s/\rho\right)\left(a/A\right)\left(g/L\right)$$

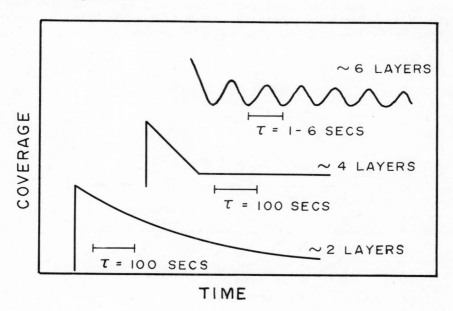

Figure 8. U-tube oscillation apparatus and variation used in Mountfield experiment

Figure 9. Character of thickness relaxation in Mountfield experiment

measurements is illustrated schematically in Fig. 10. Here, a third sound wave propagates across an infinite plane. The thickness of the film is d; d_o represents the thickness of helium atoms localized on the substrate which are not participating in the superfluid motion. In our discussion of third sound propagation, we shall restrict ourselves to the thin film limit; this allows the neglect of thermal terms which become important for thicker films at higher temperatures. The third sound velocity is given by

$$C_3{}^2 = df\left(\frac{\overline{\rho_s}}{\rho}\right)$$

where $f = [\partial\mu/\partial d]_T$ is the van der Waals force, usually taken to be $f = 3\alpha/d^4$. The third sound velocity may also be expressed in terms of the vapor pressure through the use of the Frenkel-Halsey-Hill relation, $P = P_o\exp[-\alpha/kTd^3]$. The expression obtained is

$$C_3{}^2 = \left(\frac{\overline{\rho_s}}{\rho}\right)\left(3k_BT/m\right)\ell n[P_o/P] \qquad .$$

Thus, the superfluid density averaged over the thickness of the film, $[\overline{\rho_s/\rho}]$, can be obtained from the measured third sound velocity and vapor pressure. At zero temperature, where the normal fluid excitations have disappeared, the third sound velocity has a simple dependence on thickness, $C_3{}^2 \propto (d-d_o)/d^4$. This behavior is illustrated in Fig. 11. The velocity increases from zero at a

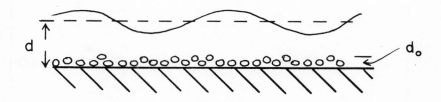

Figure 10. Conceptual representation of third sound wave

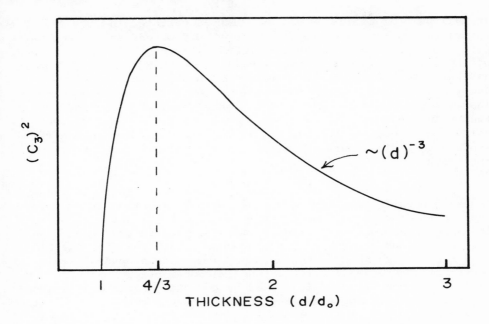

Figure 11.Thickness dependence of third sound velocity

thickness of $d/d_O = 1$ to a maximum at a thickness of $d/d_o = 4/3$, and then decreases monotonically as the film thickness is further increased. The majority of third sound measurements have been made in this region above the maximum in the third sound velocity. Fig. 12 illustrates the most commonly used third sound configuration. The third sound is generated by a thin film heater evaporated on a glass or quartz substrate and the third sound signal is detected by a sensitive superconducting bolometer, also evaporated on the substrate. This configuration has been developed primarily by I. Rudnick and his group at UCLA. Recently, R. Hallock at U. Mass has successfully observed the propagation of a third sound signal with

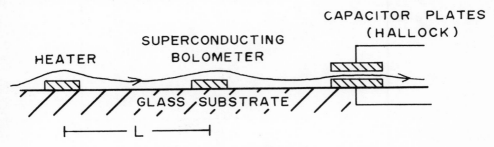

Figure 12.Third sound measurement techniques

a small pair of capacitor plates, also illustrated in Fig. 12. The
determination of the third sound velocity proceeds by the generation
of a heat pulse which produces a small disturbance in the thickness
of the film. This disturbance propagates outward from the heater
and, at a time T, later is seen by the superconducting bolometer
located a distance L from the heater. The third sound velocity,
then, is simply L/T. In Fig. 13, I show some data obtained by Rudnick using this method. As the film thickness is reduced, the
third sound velocity increases. The expected maximum in the third
sound signal disappears, however, in the region indicated by the
hatchmarks.

     In addition to the open geometry employed by the UCLA group,
several investigators have developed true resonant cavity methods
for the study of thin superfluid films. With the cavity approach,
it has proved possible to reduce signal loss into the vapor considerably, thus achieving a greater sensitivity than is possible
with the open geometry. Ratman and Mochel[15] developed a simple
cavity configuration for third sound measurements. The third sound
cavity shown in Fig. 14, is formed by two thin quartz plates fused
together on their edges forming an envelope structure. In addition
to a small amount of helium gas, the cavity also contains argon,
which preplates the substrate as the cavity is cooled to helium temperatures. Third sound resonances are excited in the cavity by an
ac heater and the resonances are detected by a sensitive thermometer. At low temperatures, the resonances observed are quite sharp.

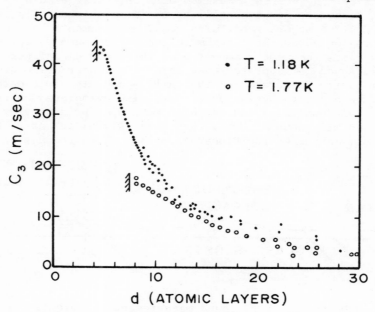

Figure 13. Thickness dependence of third sound velocity

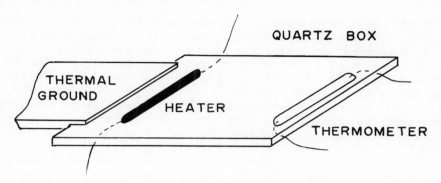

Figure 14. Cavity configuration for thermally excited third sound measurements

Q values as large as 10^4 are observed. As the temperature is
raised, the Q decreases as the onset temperature is approached.
Measurements of the resonant frequency and the dimensions of the
cavity are combined to obtain the third sound velocity. Values of
the third sound velocity near the onset temperature are shown in
Fig. 15. The third sound velocity is seen to decrease as the onset
temperature is approached. In this region, the Q is also dropping
rapidly. The large increase leading to the loss of the third sound
signal is characteristic of the behavior of third sound near onset
for two dimensional films.

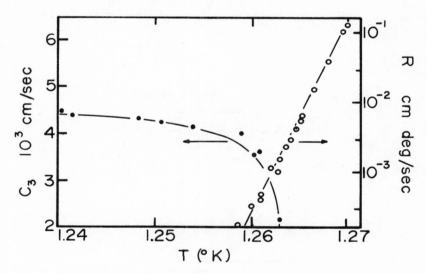

Figure 15. Data from Mochel cavity experiments

In addition, Ratman and Mochel were also able to make a thermal resistance measurement on their cell, thus obtaining an entirely independent determination of the onset temperature. They
find that the onset temperature determined by signal loss in the
third sound experiment and that obtained from the thermal resistance method are in good agreement. In Fig. 15, their measurements of the thermal resistance are also plotted. It is seen that
the thermal resistance, R, varies by over three orders of magnitude in a 10 mk range of temperature.

An entirely different approach which can be applied to the
creation and detection of third sound resonances has been developed
by H.E. Hall and J.D. Reppy at the University of Manchester and
Cornell University. This technique has the advantage that it can
be applied to the detection of any of the resonant sound modes.
In fact, it was originally developed for the observation of fourth
sound in porous media.[16] The first application of this method to
unsaturated helium films was made by A. Yanof[17], who observed third
sound resonances for helium adsorbed on lampblack. The apparatus
used is sketched in Fig. 16. It consists of a cavity containing
the porous medium, such as Vycor glass or pressed lampblack on
which is a helium film. The cavity is mounted on a spring and
driven by a periodic force, $F = F_0 \cos \omega t$. The amplitude response
of the cavity is observed as a function of the frequency, ω. The
drive frequency matches the resonant frequency of a half wave
standing mode within the cavity; the amplitude, A, has the minimum
value. At a slightly higher frequency, the center of mass of the
fluid and the cavity move 2π out of phase with each other, and the
amplitude goes through a maximum. The difference in frequency be-

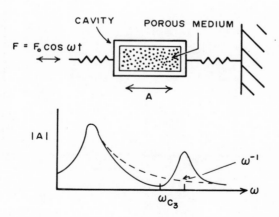

Figure 16. Oscillating cavity technique for third sound measurements

tween the minimum and maximum allows a determination of the actual superfluid mass within the cavity. For details of this
method, see Yanof and Reppy.[18]

   At temperatures below [1]K it has been possible to make precise
measurements of the third sound velocity for films of constant
thickness. In this case, $C_3{}^2 \propto \rho_s/\rho$, and one can obtain the temperature dependence of ρ_s with high resolution. Fig. 17 shows an
example of the temperature variation of the third sound velocity
taken from the work of Rutledge et. al.[19] The curve drawn through
the data is a fit based on a normal fluid density calculated
assuming two dimensional phonons and rotons with an energy gap in
the range of 5K. The presence of a roton or gap-type excitation
with an energy near 5 to 6 K seems to be characteristic of thin
films and helium confined to small channels, such as those of
porous Vycor glass. This is in contrast to the 8.5 K gap for
rotons in bulk liquid helium.[1] The values for the roton gap obtained for a number of different third sound measurements are
shown in Fig. 18. In this figure, data from the work of Mochel
et. al.[19] and values obtained by D.J. Bishop[20] from third sound
measurements on Vycor glass are shown. The agreement in the magnitude of the gap obtained from measurements on these different
substrates is remarkable considering the difference in substrate
geometry. When the pores of Vycor are filled, fourth sound can be
observed and again, a roton gap in the range of 5-6 K is obtained.

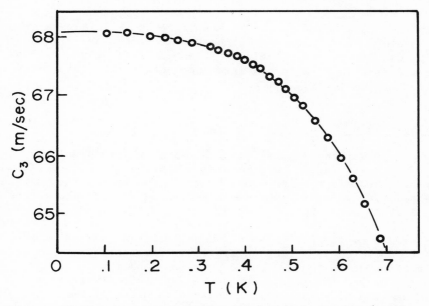

Figure 17.Temperature dependence of third sound velocity

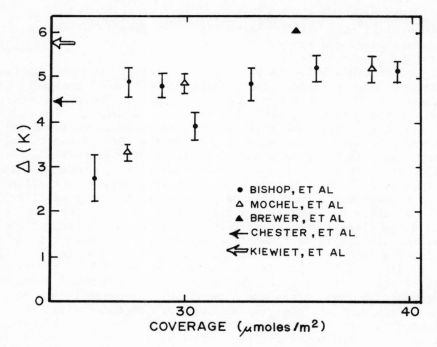

Figure 18.Values of roton gap for thin He films

One may then conclude that the free surface which is always present
in a two dimensional film experiment probably does not play an important role for this excitation. It seems likely that these 5K
rotons are associated with the interface between the liquid and the
localized helium layer.

PERSISTENT CURRENTS IN SUPERFLUID FILMS

The observation of persistent currents in liquid helium provides the most stringent experimental demonstration of superfluidity.
Dissipative processes which would escape detection in the most sensitive oscillating and driven experiments can easily lead to the
rapid decay of a persistent current.

A simple configuration in which a persistent current might
exist is indicated as a ring in Fig. 19. Imagine that a superfluid
film is adsorbed on the surface of the ring and that a supercurrent
is flowing around the ring. The superfluid velocity is $v_s =$
$(\hbar/m)\nabla\varphi$ and the circulation is quantized in units of h/m. $\oint v_s \cdot d\ell =$
$n(h/m)$, where n is an integer. Lack of experimental sensitivity
places a lower restriction on the magnitude of persistent currents

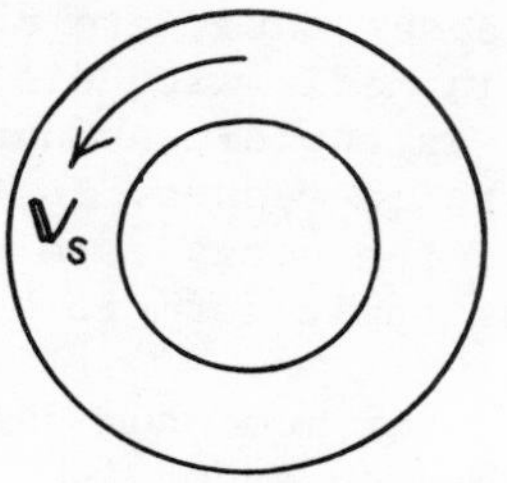

$$V_s = (\hbar/m)\nabla\phi$$

$$\oint V_s \cdot d\vec{l} = n(h/m)$$

Figure 19. Quantized circulation for superfluid He

which can actually be observed. In practice, persistent currents which can be easily seen will have quantization numbers, n, in the range of 10^2 to 10^4.

The angular momentum resulting from a persistent superflow will be given by $L = \rho_s v_s G$, where G is a geometric constant. The angular momentum can be measured either directly by gyroscopic technique or indirectly by third sound doppler shift. This latter technique is discussed by R. Hallock at this summer school.

The superfluid persistent current is a metastable state, but, under suitable conditions, it can have an effectively infinite lifetime. For systems capable of supporting a long-lived persistent current, there exists a region in velocity and temperature space where persistent currents are stable. In the region of stability, the temperature can be changed slowly and the angular momentum of the persistent current simply tracks the temperature variation of ρ_s in a reversible fashion. However, if the temperature is raised above a certain value, the current can become unstable and may decay rapidly. Thus, the study of the temperature dependence and stability of persistent currents yields considerable information about the superfluid state.

The decay of the metastable persistent current involves a change in the quantized circulation around the ring. This can be achieved by the motion of a quantized vortex line across the ring. Thermal activation often plays a major role in this process. The thermal activation model for the decay of superfluid persistent currents has enjoyed considerable success. For an early review of these ideas and their application to persistent current decay, see Langer and Reppy.[21]

When the decay rate of the persistent current is sufficiently slow, the thermal activation model predicts a particularly simple form for the time dependence of the superfluid velocity. In this case, the velocity is expected to decrease linearly with the loga-

rithm of time. In Fig. 20, fractional decay data, are shown for a
number of persistent currents created in media with different characteristic dimensions. It is seen that as the dimensions of the
flow channel are reduced, the decay rate increases. In fact, for
thin films of only a few atomic layers, the decay rate can be very
large and one does not expect the logarithmic form to hold.

In Fig. 21, persistent current angular momentum data obtained
by Chan et. al.[22] for thin films adsorbed on Vycor is shown for a
series of films with different thicknesses where the persistent
currents were all formed with the same initial flow velocity.
Therefore, the low temperature angular momenta reflect the mass of
the superfluid in each film. For each thickness, there exists a
stable, reversible region (indicated by closed circles in the
figure). However, if the temperature is raised high enough, a
rapid, irreversible decay takes place (open circles). As the film
thickness is progressively reduced, the temperature at which decay
becomes rapid also decreases, and, for the thinnest films shown
here, something on the order of two and one half layers, the decay
temperature is less than 0.3K.

In Fig. 22, a plot of the decay temperature and the initial
angular momenta is shown as a function of film thickness. The in-

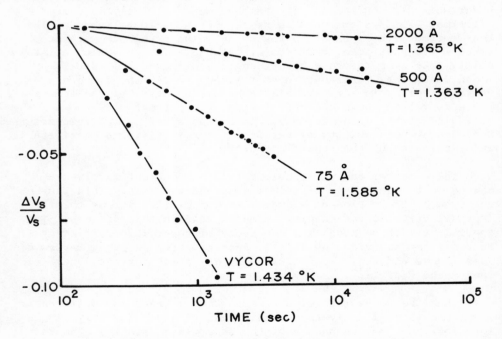

Figure 20. Fractional decay of persistent currents

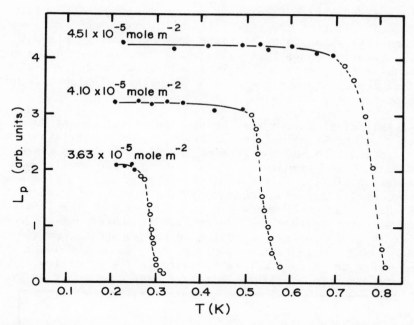

Figure 21.Persistent currents on Vycor. Solid dots show reversible
region. Open circles represent **irreversible** decay

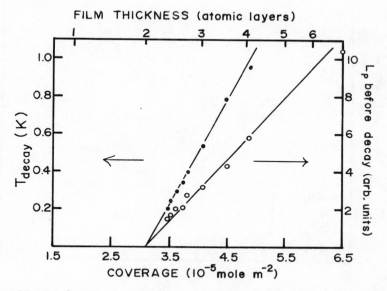

Figure 22.Thickness dependence of persistent current behavior

teresting feature of this plot is the convergence of both sets of
data toward a single coverage. It would appear that for coverages
less than this value, superfluidity does not exist. Also, it is
only the coverage above this value which contributes to the superfluid mass. This data is then in accord with the concept of a
localized layer adsorbed directly on a substrate which does not
participate in the superflow.

    A similar effect can also be seen in an experiment due to R. P.
Henkel,[23] in which the angular momentum, L, of a persistent current
of an adsorbed film was observed as the film thickness was slowly
increased with the temperature held constant. The angular momentum
as a function of film thickness is shown in Fig. 23. It is seen
that the angular momentum increases linearly with the amount of
helium added, for coverages above a certain value. The intercept,
D_0, has some temperature dependence, but it approaches a non-zero
value of about $1\frac{1}{2}$ layers, depending on the substrate, as $T \rightarrow 0$.
This data again suggests the existence of a localized layer next to
the substrate which does not participate in the superflow. Rudnick
has constructed plots of the superfluid mass as a function of film

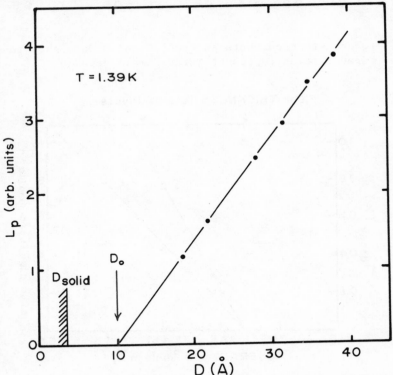

Figure 23. Angular momentum of persistent current as thickness is
varied at constant temperature

thickness from third sound data, and a similar linear dependence on thickness is observed.[24]

OSCILLATING SUBSTRATE METHODS

The oscillating substrate technique, as applied to films, is an adaptation of the classic method of Adronikashvili. This approach was first successfully applied to thin films by Chester and Yang[25], using a quartz microbalance, and, more recently, has been developed in the torsional oscillator configuration by Berthold, Bishop and Reppy at Cornell.

Fig. 24 gives a schematic representation of the experimental arrangement. The substrate, with mass M, is mounted on a spring with spring constant k; the system is driven at its resonant frequency by a force, $F_o \cos \omega t$. In the case of the quartz microbalance, a thin slab of quartz is driven in a resonant shear mode with the motion of the crystal parallel to the major part of the surface area.

Helium gas is allowed to adsorb on the surface of such an oscillator; for temperatures above the onset temperature the helium merely sticks to the surface and adds to the effective mass of the oscillator. In view of the very large (about 10^6) ratio between M and the mass of the adsorbed film, the resonant frequency must be determined with high precision. The success of this approach has been possible through the development of oscillators with frequency stabilities in the range of 1 part in 10^8 or even 10^9. Usually, the low temperature mechanical oscillator is incorporated as a frequency-determining element in a feedback oscillator. The amplitude of oscillation is usually kept low in order to achieve the high Q's necessary for greatest frequency stability.

Fig. 25 shows the variation of the period of such an oscillator as the areal density, ρ, of the adsorbed film is increased. In

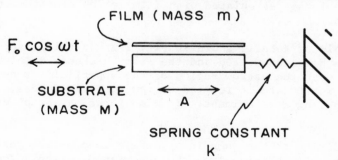

Figure 24. Oscillating substrate technique.

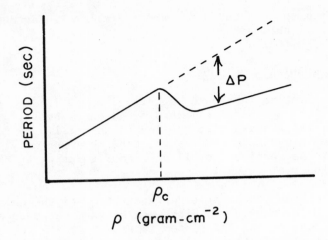

Figure ·25. Characteristic form of oscillating substrate data

this type of measurement, the temperature is held constant. The
areal density necessary for superfluid onset is exceeded; there is
an abrupt change in the period. The period deficit, Δp, is proportional to the superfluid density, ρ_s (g/cm2).

One set of data from Chester and Yang's measurements is shown
in Fig. 26. The superfluid density (here called σ_s) is plotted as
a function of film coverage, σ. The striking feature of this data
is the jump in σ_s at the critical coverage.

Bishop and Reppy[26] have made somewhat similar measurements
using a torsional oscillator. The oscillator contains a spiral
roll of Mylar film wound around the axis of oscillation. With
this substrate, it was possible to obtain a surface area of approximately 0.4 m2. It was then possible to achieve considerably
greater resolution of the superfluid mass than possible with the
quartz microbalance technique. They were also able to study the
dissipative losses associated with film motion near the onset
temperature.

A sample of data is shown in Fig. 27. In this case, the
temperature has been varied and the film thickness remains nearly
constant   over the narrow region of the superfluid transition.
The period shift data is related to the superfluid mass, as in
Chester and Yang's experiment. It also shows an abrupt change at
onset.

In addition to the jump in ρ_s, there is a marked peak in the
dissipation, Q^{-1}, arising from the superfluid motion. The curves
passing through the data are a fit taken from the dynamic theory

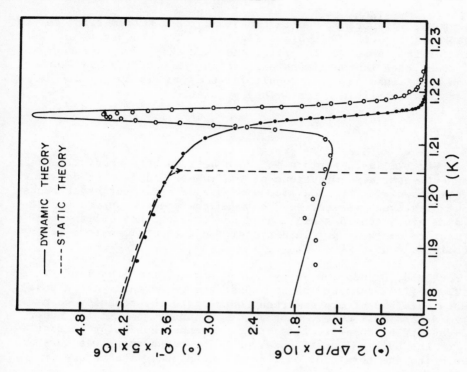

Figure 27. Dissipation and period shift at superflow onset on mylar substrate using tor-sional oscillator technique.

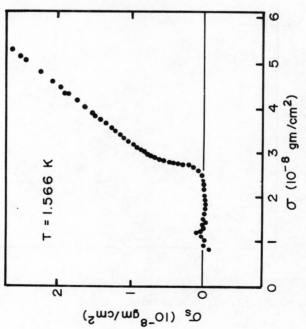

Figure 26. Data from quartz microbalance experiment.

of the superfluid transition in two dimensional helium films due to
Ambegaokar, Halperin, Nelson and Siggia.[27] This theory is an extension of the ideas of Kosterlitz and Thouless[28] for the superfluid transition in two dimensional films. Kosterlitz and Thouless have obtained an exact result for the jump in superfluid density at the superfluid onset. They predict,

$$\rho_s(T_c^-) = 8\pi k_B (m/h)^2 T_c \quad .$$

In Fig. 28, a comparison is shown between this prediction and
the experimental determinations of the superfluid jump at onset.
The data shown was obtained from a number of different experiments.
The quantitative agreement seen between the prediction and the actual data gives considerable confidence that the Kosterlitz Thouless picture of the superfluid transition in two dimensions is
basically correct.

Berthold, Bishop and Reppy[29] have studied the behavior of
superfluid films adsorbed on the substrate of porous Vycor glass.
The data obtained offers a marked contrast to the behavior seen for
the two dimensional case. Instead of an abrupt jump in the superfluid density at onset, a temperature dependence similar to that
seen for the bulk helium is observed. The measurement was performed using the torsional oscillator technique. The change in

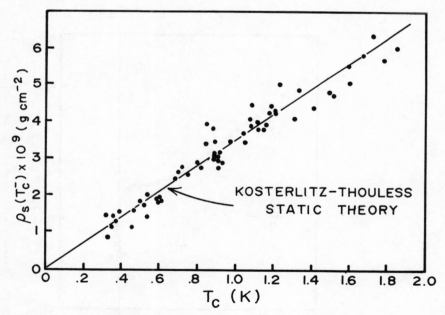

Figure 28. Superfluid density jumps at T_c compared with theory

period seen below the transition, ΔP, is shown in Fig. 29 for a number of films of differing thicknesses. The data near the transition can be fit to the power law form:

$$\Delta P = \Delta P_0 (1 - T/T_c)^{\zeta} \quad .$$

It was found that the best fit was given for a value of $\zeta = 0.63 \pm .05$. Another important difference between the Vycor film measurements and those made on two dimensional substrates is the absence of any excess dissipation associated with the transition. The fact that the superfluid density exponent, ζ, has a value near 0.67, the value for bulk helium, strongly suggests that we are dealing with a three dimensional phase transition.

Josephson[30] has pointed out that the coherence length, ξ, is related to ρ_s by

$$\xi = \left(\frac{m}{\hbar}\right)^2 \frac{k_B T}{\rho_s} \quad .$$

It is necessary that ξ be considerably larger than to 50-100 Å pore size of the Vycor in order that three dimensional behavior be

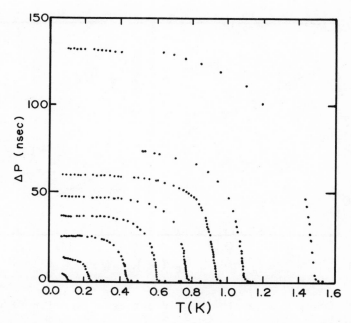

Figure 29. Superfluid density measurements on Vycor substrate

seen. For purposes of comparison with experiment, it is convenient to divide the temperature dependence of ξ into a singular part and a weakly varying coefficient ξ_0,

$$\xi = \xi_0 \left(1 - \frac{T}{T_c} \right)^{-\zeta} \quad .$$

Using Josephson's relation, we can obtain the coherence length coefficient, ξ_0, from the experimentally determined ρ_s and T_c for each film. The values calculated from the experimental data are shown in Fig. 30. The values of ξ_0 obtained are sufficiently large to insure an ample temperature region near the transition where the coherence length is much larger than the pore size.

P.C. Hohenberg[31] has pointed out that the large value of ξ obtained for superfluid helium in Vycor has implications for the amplitude of the heat capacity anomaly expected at the transition. Following the arguments of Hohenberg, Halperin, Aharony and Siggia[32], the strength of the anomaly is expected to scale as ξ^{-3}, and since ξ, for Vycor, is at least 50 times larger than for bulk helium, one expects the anomaly at the transition to be unobservable in accordance with the recent high resolution experiments of Joseph and Gasparini.[7]

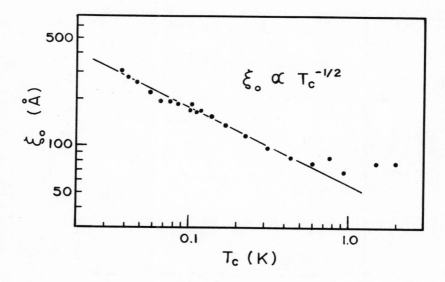

Figure 30. Coherence length coefficient as a function of onset temperature.

ACKNOWLEDGEMENTS

        I would like to thank Glenn Agnolet and Hillary Rettig for
their very considerable help in preparing this manuscript. The
work discussed here was supported in part by the National Science
Foundation through Grant No. DMR 77-24221 and through the Cornell
Materials Science Center Grant No. DMR 76-81083. Technical Report
#4171.

REFERENCES

1. P.J. Bendt, R.D. Cowan, and J.L. Yarnell, Phys. Rev. 113, 1386
    (1959). A.D.B. Woods and R.A. Cowley, Rep. Prog. Phys. 36,
    1135 (1973).

2. G. Ahlers, "Quantum Liquids", J. Ruvalds and T. Regge, Editors.
    North-Holland, New York, 1978; also G. Ahlers, "The Physics of
    Liquid and Solid Helium", K.H. Benneman and J.B. Ketterson,
    Editors, John Wiley & Son, New York, 1976.

3. E.L. Andronikashvili, Zh. Eksp. Teor. Fiz. 16, 780 (1946).

4. W.F. Vinen, Proc. Roy. Soc. (London) A260, 218 (1961).;
    G.W. Rayfield and F. Reif, Phys. Rev. Lett. 11, 305 (1963).,
    also Phys. Rev. 136, No.5A, 1194 (1964).

5. H.P.R. Frederickse, Physica 15, 860 (1949).

6. For a review of the Vycor heat capacity measurements, see
    D.F. Brewer, J. Low Temp. Phys. 3, 205 (1970).

7. R.A. Joseph and F.M. Gasparini, Journal de Physique 39, C-6,
    310 (1978).

8. A.J. Symonds, Thesis, University of Sussex, 1965 (unpublished).
    A. Evenson, Thesis, University of Sussex, 1968 (unpublished).

9. M. Bretz, Phys. Rev. Lett. 31, 1447 (1973).

10. K. Fokkens, W.K. Taconis and R. deBruyn Ouboter, Physica 32,
    2129 (1966).

11. J.A. Herb and J.G. Dash, Phys. Rev. Lett. 29, 846 (1972).

12. M. Bretz, remarks at this summer school.

13. K.R. Atkins and I. Rudnick, "Progress in Low Temperature Phy-
    sics", Vol. VI, C.J. Gorter, editor, North-Holland (1970).

14. K.R. Mountfield, Thesis, Cornell Univ., 1973, (unpublished).

15. B. Ratnam and J. Mochel, Phys. Rev. Lett. 25, 711 (1970).

16. For a description of this method see, C.W. Kienriet, H.E. Hall and J.D. Reppy, Phys. Rev. Lett. 35, 1286 (1975).

17. A.W. Yanof, to be published.

18. A.W. Yanof and J.D. Reppy, Phys. Rev. Lett. 33, 631 (1974) and A.W. Yanof, Thesis, Cornell University (1975).

19. J.E. Rutledge, W.L. McMillan, J.M. Mochel and T.E. Washburn, Phys. Rev. B18, 2155 (1978).

20. D.J. Bishop, J.M. Parpia and J.D. Reppy, Proc. of 14th Int. Conf. on Low Temp. Phys., M. Krusius and M. Vuorio, editors, North-Holland/American Elsevier Pub. Comp., Amsterdam, New York (1975). Also D.J. Bishop, Thesis, Cornell University, (1978) unpublished.

21. J.S. Langer and J.D. Reppy, Progress in Low Temp. Phys., Vol. 6, C.J. Gorter, Editor, North-Holland, Amsterdam, 1970.

22. M.H.W. Chan, A.W. Yanof and J.D. Reppy, Phys. Rev. Lett., 32, 1347 (1974). M.H.W. Chan, Thesis, Cornell University, (unpublished).

23. R.P. Henkel, E.N. Smith and J.D. Reppy, Phys. Rev. Lett., 23 1276 (1969).

24. I. Rudnick, Proc. of 12th Int. Conf. on Low Temp. Phys., E. Kanda, Editor, Academic Press of Japan, (1971).

25. M. Chester and L.C. Yang, Phys. Rev. Lett. 31, 1377 (1973).

26. D.J. Bishop and J.D. Reppy, Phys. Rev. Lett. 40, 1727 (1978).

27. V. Ambegaokar, B.I. Halperin, D.R. Nelson and E.D. Siggia, Phys. Rev. Lett. 40, 783 (1978).

28. See the review by J.M. Kosterlitz in this volume.

29. J.E. Berthold, D.J. Bishop and J.D. Reppy, Phys. Rev. Lett. 39, 348 (1977).

30. B.D. Josephson, Phy. Lett. 21, 608 (1966).

31. P.C. Hohenberg, private communication.

32. P.C. Hohenberg, A. Aharony, B.I. Halperin and E.D. Siggia, Phys. Rev. B13, 2986 (1976).

33. T.P. Chen and F.M. Gasparini, Phys. Lett. 62A, 231 (1977).

CHEMISORBED PHASES

E. Bauer

Physikalisches Institut
3392 Clausthal-Zellerfeld
and SFB 126, Göttingen - Clausthal

INTRODUCTION

This chapter deals with phases, phase diagrams and phase transitions in chemisorption systems. Chemisorbed phases are phases in which the bonding energy E_O of the adsorbate to the substrate is higher than about 1eV per atom or molecule. The lateral adsorbate-adsorbate interaction energy E_{aa} is usually more than an order of magnitude smaller. Therefore, phase transitions within the layer which are determined to a large extent by E_{aa} can be studied conveniently over a wide temperature range without loss of material to the environment. A phase transition is defined here as a discontinuous or continuous change of the pair correlation function as evidenced, for example, by changes of the long range order ("structure") or of quantities which are sensitive to the short range order. Only transitions in the adsorbate alone or in the adsorbate and the topmost layer of the substrate are considered here. Transitions in which an exchange with the bulk of the substrate occurs will be excluded as well as transitions from 2- to 3-dimensional phases.

Phase diagrams of chemisorbed phases are usually presented as T-Θ diagrams as illustrated in fig.1, Θ being the coverage in monolayers. Two monolayer definitions are used: a) the monolayer (ML) is a layer which contains the same number N_a of atoms (molecules) as the topmost layer of the substrate (Θ_a); b) it is the layer

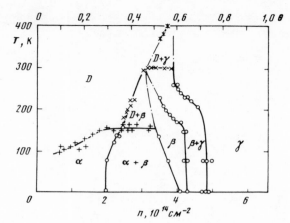

Fig.1: Phase diagram of Sr on W(110).
$\Theta=1$ corresponds to the close-
packed monolayer with $n_{sat}=$
6.6×10^{14} atoms cm^{-2} $(\Theta=\Theta_b)$.
α, β and γ are ordered phases,
D is a disordered phase. [82]

with the highest number N_b of atoms (molecules) which
can be incorporated in the first layer (Θ_b). Depending
upon the relative size of the atoms, $N_a \lessgtr N_b$. Both definitions will be used here.

     Phase transitions are usually classified in first,
second and higher order, discontinuous or continuous,
on the basis of the change of the energy of the system
with temperature T or density (here Θ). Such a classification is useful if the characteristic thermodynamic
quantities can be measured. These measurements are
difficult in most chemisorption systems, consisting of
a thin single crystal with a surface area of the order
1 cm^2 and have not been done, therefore. However, some
other classification systems are obvious:
1. According to the atomic layers involved: a) adsorption
   layer only, b) topmost substrate layer (without exchange with the adsorbate) and c) both layers (with
   exchange of atoms)
2. According to reversibility: a) reversible and
   b) irreversible
3. According to the variable of state which is held
   constant: a) T, b) Θ
The understanding of phase formation and phase transitions requires knowledge of
a) the lateral interactions in the adsorbate. The may

be short range but usually are long range; they may
be repulsive, attractive or oscilling in sign with
changing distance;

b) the place exchange, such as activation energies for
lateral place exchange (surface diffusion) or normal
place exchange which occurs in reconstruction;

c) the processes in which the chemical species is changing such as dissociation during adsorption or asso-
ciation (molecule formation) during desorption.

On the basis of these considerations the paper will be
organized as follows. First (Sect.2) the experimental
techniques used in the study of 2-d phases and phase
transitions will be discussed briefly and illustrated
with application examples. Next (Sect.2) some of the few
phase diagrams and transitions studied in some detail
will be reviewed together with some less well examined
systems as examples for the various kinds of phase
transitions mentioned before. Sect.4 will be concerned
with the elementary processes whose knowledge is nesec-
sary for understanding phase transitions. A brief outlook will conclude the paper.

2. EXPERIMENTAL METHODS

2.1. Diffraction techniques: LEED

Of the various diffraction techniques (atomic beam,
neutron, X-ray, reflection high energy electron diffraction (RHEED) and low energy electron diffraction (LEED))
only the latter is generally useful. The low penetration
depth of the 10-200eV electrons used in LEED ensures a
high surface sensitivity so that a fraction of a monolayer can be detected provided it shows long range order.
The method is so well known that the reader unfamiliar
with it must be referred to books and review articles
on the subject.[1-5] We will discuss its application to
the study of phase transitions extensively in Sect.3 and
mention here only some limitations of the method. A complete structure analysis - i.e. determination of the
atomic positions - is presently possible only for simple
surface structures such as (1x1), (2x1), p(2x2), c(2x2),
$\sqrt{3}x\sqrt{3}R(30°)$ and other high symmetry unit meshes. In
most other cases only the dimensions of the unit mesh
can be determined and even this is not always unambiguous
due to the complicating effect of double- and multiple
scattering between adsorbate and substrate. In spite of
these limitations a great amount of information on phase
transitions can be extracted from LEED: the spot size

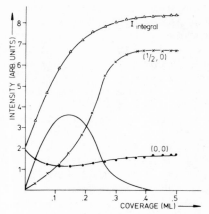

Fig.2: LEED intensities as
a function of coverage of
oxygen on W(110). Shown
are the total intensity on
the LEED screen, the intensities of the (0,0) and
($\frac{1}{2}$,0) beams and the background intensity.[7]

and shape allows conclusions about size and shape of
ordered regions[6], the spot intensity as a function of
T or Θ is a measure for the degree of order in the
ordered regions and for the coverage by ordered regions.
Even the background gives some information about the
amount of disordered adsorbate. This is indicated in fig.2
which shows the variation with coverage of the background intensity together with the intensity of a
reflex characteristic of ordered two-dimensional oxygen
islands with p(2x1) structure on a W(110) surface.[7]
This chemisorption system will be discussed in detail
in Sect.3. Here it should only be mentioned that the
phase diagram in Fig.1 was also obtained by LEED from
the regions of existence of the periodic structures
characteristic for the various phases.

2.2. Ion scattering techniques
2.2.1. High energy ion scattering

   The Rutherford backscattering (RBS) spectrum of
MeV ions from a single crystal surface when measured
in a "channeling" direction[8] - i.e. a direction of high

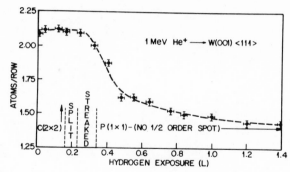

Fig.3: Intensity of the RBS surface peak from a W(100) surface for <111> channeling as a function of hydrogen exposure. The corresponding LEED structures are indicated at the bottom of the figure.[9]

transparency to the incident ion beam - has a high energy peak, the "surface peak", whose intensity is a measure for the number of atoms directly visible to the incident beam. Change of the surface peak intensity with T or - in case of negligible scattering by the adsorbate - with Θ can be attributed to displacements of the topmost atoms which change the visibility of the atoms in the second layer. Fig.3 illustrates the appli- cation of this technique to the study of the "re-recon- struction" of the W(100) surface by H_2 adsorption[9] (type 1b phase transition). At H_2 exposures up to about .3L (=.3x10^{-6}torr.sec) the surface is "reconstructed", i.e. the topmost W atoms are displaced laterally from their "normal" positions, so that an abnormally large number a W atoms is visible. With further increasing H adsorption this reconstruction is increasingly suppressed until a high H coverage the W atoms assume thier "normal" lateral positions. This phase transi- tion as a function of Θ is completely reversible.

2.2.2. Low energy ion scattering

     The basis of the application of low energy (100- 1000eV) ion scattering spectroscopy (ISS)[10] to the study of phase transitions is similar to that in 2.2.1: if two different kinds of atoms change their relative position during the transition then their scattering behaviour changes, in part due to changes in the mutual

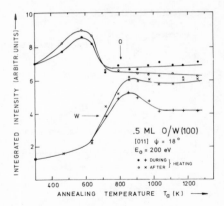

Fig.4: Intensity of the O
and W ISS signals from a
W(100) surface with .5 monolayers of oxygen as a function
of annealing temperature. The
200eV He$^+$ ions are incident
at a glancing angle of 18° in
the [011] azimuth.[11]

"shadowing" and in the double and multiple scattering
conditions, in part due to changes in the neutraliza-
tion probability which is very high for low energy
ions, in particular for He. Due to the difficulties in
describing ion scattering at low energies quantitatively,
ISS signal changes serve presently predominantly as
indicators of a phase transition rather than as a tool
for detailed structure analysis. Fig.4 shows an example,
the irreversible and reversible phase transitions of a
W(100) surface covered with half a monolayer of oxygen
as a function of T.[11] The irreversible reconstruction,
i.e. mixing of O and W atoms in the topmost layer (type
1c phase transition) begins at about 600K. At approximately 800K a reversible phase transition of the reconstructed surface begins. Both transitions will be
discussed together with other results in Sect. 3.

2.3. Ion emission techniques

2.3.1. Secondary ion mass spectroscopy (SIMS)

A surface which is bombarded with primary ions
- usually Ar$^+$ with 500-5000eV energy - emitts secondary
ions whose mass spectrum is determined by the composition
of the topmost few atomic layers.[12] The absolute ion

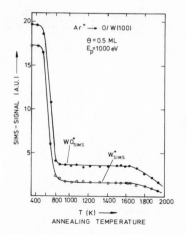

Fig.5: Intensity of the WO$^+$ and
W$^+$ secondary ion signals from a
W(100) surface with .5 monolayers
of oxygen as a function of annealing
temperature.[13]

yields and the relative magnitude of the various mass
peaks depend, however, not only upon the composition,
but also on the distribution of the various atoms in
the topmost layers and upon the work function. Therefore, phase transitions which cause changes in these
parameters can be followed by SIMS, as illustrated in
fig.5.[13] Here,the same irreversible reconstructive phase
transition which was shown in fig.4 causes a dramatic
decrease of the WO$^+$ and W$^+$ ion yield - as well as of
all other observed secondary ions - and a considerable
change of the relative intensities. Again, as in the
case of ISS (2.2.2.) it is still difficult to extract
detailed structural information from the signal changes.
The application of energy and angular distribution
measurementes to structure analysis is still in its
early stages.

2.3.2. Electron stimulated desorption (ESD)

     Many chemisorbed atoms or molecules can be desorbed
by electron bombardment in form of neutrals or - usually
positive - ions. The energy and angular distribution of
the ions which can be detected much easier than that of
the neutrals is - similar to SIMS - very sensitive to

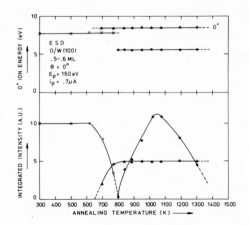

Fig.6: Energy (upper part)
and intensity (lower part)
of the O$^+$ ions emitted
normal to the surface from a
W(100) surface with .5-.6
monolayers of oxygen during
bombardment with 160eV
electrons as a function of
annealing temperature.[15]

structural and work function changes[14] and is, therefore, a useful tool for the study of phase transitions.
As an example, fig.6 shows the change of the ESD ion
energy and intensity during the reconstruction of the
W(100) surface covered with somewhat more than half a
monolayer of oxygen at 300K.[15] In this irreversible
phase transition the 7.8eV ESD peak is replaced by a
8.3eV peak. In addition, a peak with significantly
lower energy (5.5eV) forms which is caused by O atoms
in excess of .5ML which cannot be incorporated into the
reconstructed half monolayer. ESD discriminates strongly
between atoms in different bonding configurations: the
large 5.5eV ESD signal in fig.6 is caused by a few
percent of a monolayer while, for example, no ESD signal
at all is seen from O on Mo(100) up to 1/3 monolayer.[16]
Therefore, it is a very selective technique which has to
be used with considerable caution.

2.4. Work function change ($\Delta\phi$) measurements

      In contrast to ESD, $\Delta\phi$ averages over the whole
surface seen by the $\Delta\phi$-measuring probe which usually is

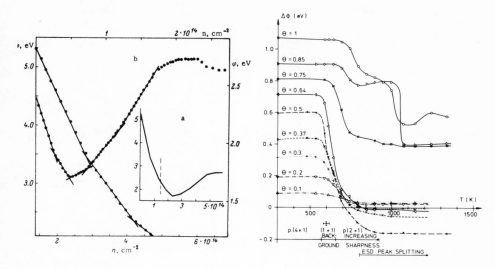

Fig.7:a)Dependence of the work function of a W(110)
surface upon Ba coverage.[18] Upper scale: lower coverage
range (with ϕ starting at 5.3eV); lower scale: high
coverage range; insert: complete $\Delta\phi(n)$ curve. b) Work
function change with temperature of annealing of W(100)
surfaces with various coverages of O.[19]

an electron beam in the retarding field method or a
vibrating electrode in the Kelvin (vibrating capacitor)
method.[17] ϕ is determined by the dipoles of the surface
atomic configurations which depend on monoatomic surfaces on the atomic roughness (e.g. steps, single atoms),
on surfaces consisting of more than one kind of atoms,
upon the distribution of these atoms and on the electro-
negativity difference. Therefore, $\Delta\phi$ measurements are
good indicators of phase transitions if these parameters
change during the transition. Figs. 7a and 7b illustrate
this for phase changes with Θ and T respectively. At
each break in fig.7a a new phase emerges while fig.7b
indicates that the phase changes in the surface system
O/W(100) depend in a complicated manner upon O coverage.
$\Delta\phi$ measurements belong to the most powerful methods for
studying phase transitions.

2.5. Adsorption-desorption measurements

2.5.1. Adsorption-desorption equilibrium

It is usually difficult to achieve adsorption-

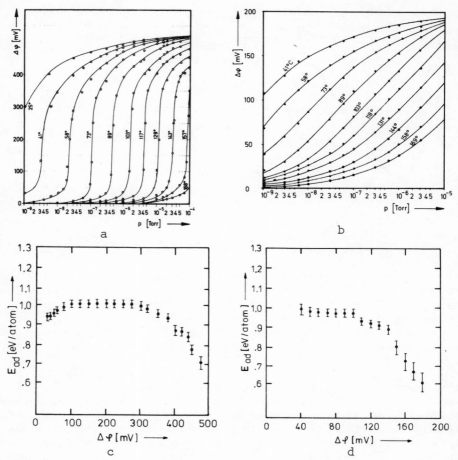

Fig.8: Adsorption isotherms for H on Ni(110) (a)
and on Ni(111) (b) and isosteric heats of
adsorption derived from them (c,d).[22]

desorption equilibrium in chemisorbed layers in a temperature range in which the order induced in the layer by
the weak lateral interactions (E_{aa}) is not already
destroyed by thermal agitation. This is due to the fact
that the adsorbate-substrate interaction (E_o) is much
stronger than the adsorbate-adsorbate interaction. A
good though extreme example is the system O/W(110) in
which temperatures between 2100K and 2400K are necessary
to maintain an oxygen coverage of $\Theta=.5$ in the pressure
range from 10^{-7}torr to 10^{-5}torr[20]. This is to be compared with a order-disorder transition temperature of
700K.[21] No two-dimensional phase-transition can be

expected, therefore, under experimentally accessible
equilibrium conditions. Nevertheless, some chemisorption
systems are known to show such a transition, e.g.
H/Ni(110)[22] (Fig.8a). Here, the work function change
which is a monotonic function of coverage in this system
is taken as a measure of the coverage. Other measurable
quantities which frequently serve as a measure for Θ are
the Auger electron, XPS, SIMS or ISS signals of the
adsorbate, only to name a few of them. For comparison,
a more usual adsorption isotherm is shown in fig.8b.
The isosteric heats of adsorption derived from the two
sets of curves via the Clausius-Clapeyron equation
$q_{st}=-R(\partial \ln p/\partial(1/T))_{\Theta}$ are shown in figs. 8c and d.
Recent atomic beam scattering data suggest that the
first-order phase transition indicated in fig.8a is not
a simple one involving only the adsorption layer but
probably a reconstructive or displacive one in which
incorporation of H into the topmost Ni layer or a
H-induced displacement of the Ni atoms occurs.[22a]

2.5.2. Adsorption kinetics

The adsorption behaviour of a gas on a surface is
usually described by the sticking coefficient $s(\Theta)$ which
is the fraction of the arriving atoms (molecules) which
stick to the surface. The Θ dependence of s is determined by the detailed adsorption mechanism[23] and will
be illustrated here only for two special though frequent
cases: atomic adsorption with a commensurate-incommensurate transition and b) dissociate adsorption via a
precursor - usually molecular - state. An example of
the first case is the temporary drop of s during the
transition from the commensurate c(2x2) structure to
an incommensurate structure during Pb adsorption on a
Au(100) surface as illustrated in fig.9a by the reduction of the increase of the Pb Auger electron signal.[23a]
If in the second case α is the trapping probability in
the precursor state and m is the number of hops of the
presursor over occupied sites then the simplest model
predicts $s(0)=\alpha(1-\Theta^{m-1})$. If m is large, then $s(\Theta)=\alpha$ up
to close to saturation where $s(\Theta)$ rapidly goes to zero.
A similar result is obtained in more sophisticated
models if the desorption from the precursor state is
very small. If several phases are formed as a function
of Θ and if Θ_{sat} is taken to be the saturation coverage
of each phase then $s(\Theta)$ will be essentially constant
until near to completion of each phase where it drops
rapidly to a new lower constant value. An example for

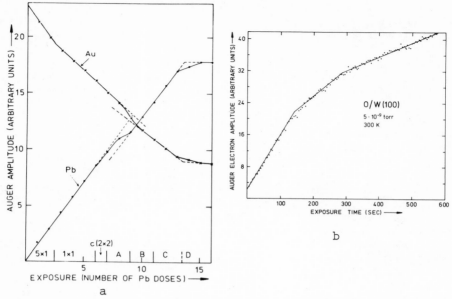

Fig.9: a) Pb and Au Auger electron signal on Au(100) as
a function of Pb exposure. The LEED structures are indicated at the bottom. A to D are incommensurate structures.[23a] b) Oxygen Auger electron signal on W(100) as
a function of exposure time to 5×10^{-9} torr O_2.[24] Slope
changes correspond to changes of the sticking coefficient.

this is shown schematically in fig.9b for the system
O/W(100). Thus s(Θ) measurements allow determination
of phase transitions with increasing Θ.

2.5.3. Desorption kinetics[23,25]

Thermal desorption has developed rapidly into one
of the major methods for the study of lateral interactions in the last few years. The reason for this is
that the desorption energy E_d, though determined predominantly by the adsorbate-substrate bonding energy
E_o, contains also the contributions E_{aa} from adsorbate-
adsorbate interactions. Two techniques are used: isothermal desorption and temperature-programmed desorption,
usually with a linear temperature rise ("thermal desorption spectroscopy", TDS). The analysis is based on the
assumption that the desorption rate can be described by
$d\Theta/dt = -f(\Theta,T)\exp(-E_d(\Theta)/kT)$ and that the temperature
dependence of the pre-exponential $f(\Theta,T)$ is negligible

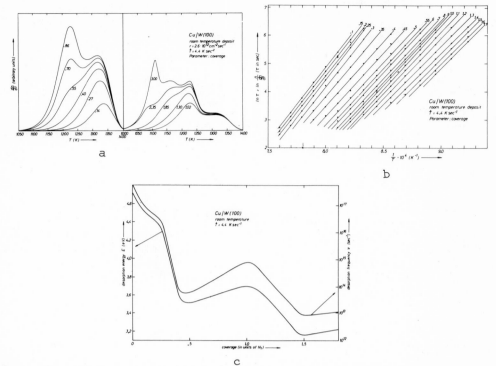

Fig.10: Thermal desorption of Cu from W(100). a)Desorp-
tion spectra for a heating rate $\dot{T}$ = 4.4Ks^{-1}. The initial
coverage Θ_o is the curve parameter. b) Arrhenius plots
for constant Θ derived from a). c) Desorption energy E_d
and pre-exponential as a function of Θ obtained from
a).[27]

as compared to that of the exponential. This is only
approximately true because even in the simplest models
of the desorption kinetics $f(\Theta,T)$ varies linearily or
quadratically with T^{25} and a more detailed analysis[26]
gives even higher powers of $T(T^3)$. The Θ- and T-dependence in TDS is usually separated by taking TDS spectra
either with different initial coverage Θ_o at the same
rate of temperature rise $\dot{T}^{25,27,28}$ or with different
heating rates $\dot{T}$ at the same initial coverage Θ_o.[29]
Experimentally either the momentary coverage Θ is monitored by $\Delta\phi$, Auger electron spectroscopy or other
Θ-sensitive methods, or the desorption flux $\dot{n}\sim$-dΘ/dt is
measured with a mass spectrometer. Fig.10 gives an
example of the first procedure. Except for the range
$\Theta <$ .2ML the variation of E_d with Θ is attributed to the
changes in lateral interactions E_{aa}. If a first order

phase transition occurs in the temperature range of
desorption the Arrhenius plots (Fig.10b) show a break
at the transition temperature corresponding to the
sudden change of the pair distribution function and the
associated change of the lateral interaction. This has
been observed in Cu^{25} and Pd^{30} on W(110).

2.6. Vibration spectroscopy techniques

2.6.1. High resolution electron energy loss spectroscopy (EELS)[31]

    Electrons interacting with an adsorbate-covered
surface suffer energy losses due to excitation of
vibrations of the adsorbed species. The frequencies ω_v
of the vibrations depend upon the adsorption site (type
of bonding) and the effective bond strength. Different
surface phases usually differ in both parameters so
that phase changes are reflected in changes of the
energy loss values. Fig.11a illustrates this for the
adsorption of CO on Ni(111).[32] The initial linear
increase of ω_v in the disordered phase is caused by the

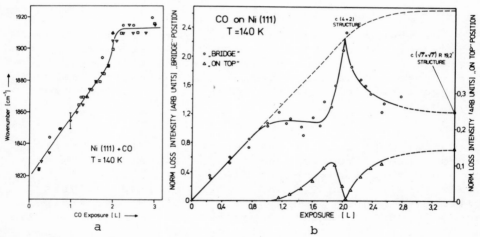

a   b

Fig.11: High resolution electron energy loss results
for CO on Ni(111). a) Change of the dominant energy
loss (expressed in $cm^{-1} \approx 1/8meV$) with exposure to CO
at 140K. b) Change of the intensity of the dominant
loss ("bridge" position, circles) and of the high
energy (2050 cm^{-1}) loss ("on top" position, triangles)
with exposure to CO at 140K. Energy and angle of
incidence of primary electrons: 5eV, 70°.[32]

change of the effective bond strength with coverage due
to lateral interactions, the sudden increase at about 2L
is associated with ordering (sharp LEED pattern!). A
deeper insight into the phase changes may be obtained
by following the intensities of the major loss peaks
(Fig.11b): between about 1L and 2L exposure all additional CO is adsorbed in "on top" positions. With the
formation of the well-ordered structure all CO is converted in "bridge" positions. With further increasing
coverage a mixture of both sites forms again.

2.6.2. Infrared Reflection-Absorption Spectroscopy[33]

Adsorbate vibrations can also be excited by infra-
red light which causes absorption bands in the light
reflected from an adsorbate-covered surface. An example
is shown in fig.12 for CO on Pd[34]. Similar to fig.11a
the loss value sudden increases when a well-ordered
structure is formed at a coverage of 1/2 on the (100)
surface (a) and of 1/3 on the (111) surface (b), respectively. The further increase at higher coverage is
connected with increasing lateral interactions in the
compression region. Increasing lateral interactions are

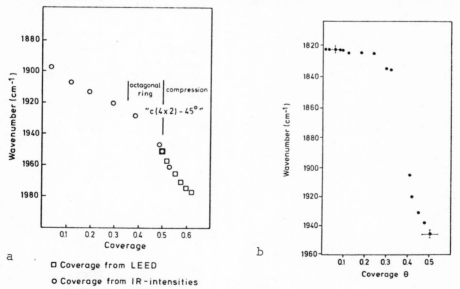

Fig.12: IR reflection-absorption spectroscopy results
for CO on Pd. Change of C-O stretch frequency with CO
coverage. a) (100), b)(111) surface.[34]

also responsible for the increase of ω_v in the disorder
region on the (100) surface below $\Theta = 1/2$ while the
constant ω_v below $\Theta = 1/3$ on the (111) surface is
attributed to constant lateral interactions within
islands of constant packing density.

2.7. Surface diffusion measurements

The diffusion behaviour of adsorbed atoms across
the surface is expected to be different in phases
differing in atomic density or distribution. As a
consequence, measurements of the diffusion constants
should give some information on surface phases. Indi-
cations that this is true are provided by microscopic[35]
and macroscopic[36] measurements. The microscopic method
is based on the field emission current fluctuations
associated with the in- and out-diffusion of adsorbed
atoms from a small surface region whose emission current
is measured. Fig.13 shows results obtained this way for
diffusion of O on W(110). E_{sd} and D_O determine the
diffusion constant $D = D_O \exp(-E_{sd}/kT)$. The changes in
E_{sd} and D_O at $\Theta=.3$ are attributed to the formation of
the p(2x1) structure.[35] Macroscopic methods can use
any high resolution probe for the adsorbed atoms which
allows to measure diffusion profiles with sufficient
precision. Results for O on W(110) obtained with a high
resolution work function change probe are shown in
Fig.14a for two different diffusion temperatures as a
function of O coverage.[36] They may be compared with a

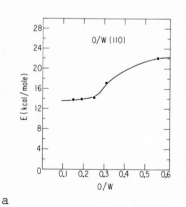
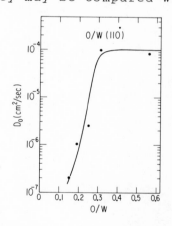

a   b

Fig.13:Activation energy for surface diffusion E_{sd} and
surface diffusion coefficient D_O as obtained from field
emission current fluctuations. Diffusion temperature:
600K.[35]

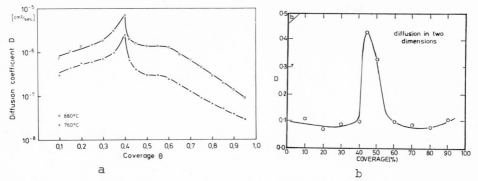

Fig.14: a) Surface diffusion coefficient D of O on
W(110) at 760°C and 880°C as obtained from diffusion
profiles by measuring the coverage-dependent work
function change with a high resolution vibrating
electrode.[36] b) Model calculation of D taking into
account the influence of nearest neighbours on E_{sd}.[37]

model calculation which takes into account the influence
of nearest neighbours on E_{sd} (see fig.10b).[37] Both show
a peak at or near $\theta = .5$ where usually the highest order
is obtained.

3. Phase diagrams and phase transitions

3.1. Clean surfaces: "Selfadsorbed layers"

Although most clean surfaces have the lateral
periodicity which would be expected on the basis of
the bulk crystal structure some do have abnormal structures ("superstructures") which show reversible or
irreversible transitions to the normal structure or to
other superstructures with temperature. Examples are
the surfaces of the elemental semiconductors Si and Ge,
the (100) surfaces of Au, Pt and Ir and the (100)
surfaces of Mo and W. The vacuum-cleaved Si and Ge(111)
surfaces have a (2x1) structure, which transforms irreversibly into a more complicated superstructure - a
(7x7) structure in the case of Si and a (2x8) structure
in the case of Ge - in the temperature range 200°C -
300°C[38] (see fig.15a) and 20°C - 120°C.[39] A further but
reversible surface phase transition to the normal (1x1)
structure occurs between 865°C and 890°C in Si[40] and
between 200°C and 400°C in Ge[41]. None of the transitions has been studied in sufficient detail and too

many models of the superstructures exist as to justify further discussion of these transitions.

Not much more is known about the phase changes on the (100) surfaces of Au, Pt and Ir. It is now generally accepted that the superstructure is caused by a over-layer of a (somewhat distorted) (111) plane of these materials. This close-packed "self-adsorbed" layer represents the stable surface configuration for Pt and Ir over the whole temperature range observable with LEED while in the case of Au a rapid and reversible transition to the "normal" (1x1) structure occurs at $806\pm5°C$.[42] This transition temperature T_C has been observed to depend on the O_2 partial pressure: at $p_{O2} \approx 1\times10^{-7}$torr $T_C=820-830°C$, at $p_{O2}<1\times10^{-9}$torr $T_C=770-790°C$.[43] On Pt and Ir metastable "normal" (1x1) surfaces can be produced by proper gas adsorption-reaction cycles.[44-46]

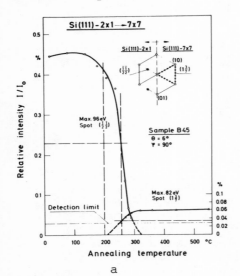

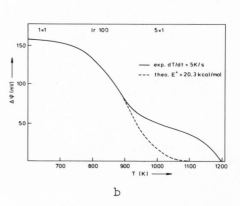

Fig.15: Irreversible surface phase transitions on clean surfaces. a) Intensities of selected LEED spots charac-teristic for the (2x1) and (7x7) structures of the Si(111) surface as a function of the temperature of 10 min anneals. Intensities are referenced to the incident current.[38] b) Work function change of an Ir(100) surface with (1x1) structure as a function of T during heating with a linear temperature rise of 5Ks^{-1}. The dashed line is a theoretical fit assuming a first order kinetic law suggested by isothermal measurements.[46]

These metastable surfaces transform irreversibly to the
superstructure state above 400K (Pt)[44] and 800K (Ir)[45,46],
respectively. The transformation has been studied in
some detail for Ir using $\Delta\phi$ measurements as a probe of
the state of the surface.[46] Fig.15b shows the $\Delta\phi$ change
during heating with a heating rate of $5Ks^{-1}$. The transition occurs in two stages of which the first one can
be described by a first order kinetic law with an
apparent activation energy of about .9eV. Although the
superstructure reflections develop already clearly
during the first stage, the strong background is eliminated only during the second stage which ends at 1200K.

The surface phase transition on a clean surface
which has been studied in most detail up to now is the
reversible "displacive" or "reconstructive" transition
on W(100)[47-49] which is similar but nor identical to
that on Mo(100).[47] It has been well known for some time
that "clean" W(100) surfaces show a diffuse c(2x2) LEED
pattern at room temperature[50,51] whose sharpness increases with decreasing temperature.[50] This pattern was
tentatively attributed to H dissolved in W above room
temperature which precipitates at the surface upon
cooling[50] or to a clean surface with every second W
atom removed so that a "microfacetted" surface is formed
which disorders above room temperature.[51] The second
model was supported by field ion images of W(100) surfaces produced either by field evaporation at 78K or
by annealing in a high field at 390K.[52,53] This indicates
that the "normal" (1x1) structure of the clean W(100)
surface is not very stable. The field ion images show
that in the (1x1) structure all sites are occupied
while in the c(2x2) structure every second atom is
missing. It is difficult to understand how a phase
transition between these two structures could occur
at low temperature.

Recent models based on extensive LEED evidence
therefore assume that the number of W atoms in the
complete topmost layer does not change during the
transition and that the superstructure is due to periodic
displacements of the atoms in the topmost layer as
indicated in fig.16a.[49] The transition from the (1x1)
structure to the c(2x2) structure which can be monitored by measuring the intensity of superstructure
spots (see fig.16b) is an order-order transition in
which with decreasing temperature islands with c(2x2)
structure grow in a sea with (1x1) structure. The
order-order nature of the transition is deduced from

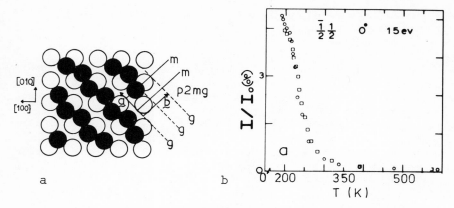

Fig.16: Reversible surface phase transition on the
clean W(100) surface. a) Periodic displacement model
of the c(2x2) structure. The atoms in the topmost layer
(full circles) are displaced horizontally .324Å from
the "normal" position above the center between 4 atoms
in the second layer (open circles) alternately right
upward and left downward; their vertical displacement
is .380Å. b) Temperature dependence of the intensity
of a superstructure spot as a measure of the order-
order transition.[49]

the fact that the c(2x2) diffraction spots grow at the
expense of the (1x1) diffraction spots instead of the
background intensity, the island growth follows from
the decreasing width of the c(2x2) spots with decreasing
temperature.

     Furthermore, on the basis of the change of a
surface-sensitive Auger electron feature with T, it is
concluded that the islands grow with an enthalpy and
entropy change of -.04eV/atom and -1x10⁻⁴eV/atom K.
The enthalpy change is of the same order of magnitude
as in polymorphic transitions in bulk metals, the
entropy decrease indicates that in the c(2x2) structure
fewer surface phonon modes are excited than in the (1x1)
structure. The driving mechanism is believed to be a
periodic lattice distortion caused by a charge density
wave which forms if the electronic energy in the
surface region is reduced this way.[54,55] Most of these
conclusions are likely not to be influenced by recent RBS
results[56] (see 2.2.1) which indicate that the structure
model of fig.16a has to be modified: only 1/2 ML of
atoms is displaced laterally by about .23Å instead of

all atoms by about .32Å as assumed in fig.16a.
Undoubtedly, the W(100) surface will be a major testing
ground for the understanding of phase transitions on
clean surfaces for the next few years.

3.2. Phase transition in the adsorbate only

3.2.1. One ordered phase only: H/Ni(111)[57,58]

     This system is probably the simplest one studied
to date with respect to phase transitions. Except for
the (1x1) structure at $\Theta=1$ it forms only one ordered
structure, a(2x2) structure which is best developed at
$\Theta=.5$. In this structure the H atoms form a network like
the C atoms in the basal plane of graphite. The structure
shows a continuous order-disorder transition with a
coverage-dependent transition temperature (see fig.17).
The half-order spot intensity $I_{1/2,1/2}$ depends quadratically on Θ up to the saturation coverage of the (2x2)
structure which shows that no (2x2) structure islands
are formed – in this case $I_{1/2,1/2} \sim \Theta$ – but that holes
in the already existing imperfect (2x2) structure are
filled with increasing coverage. Island formation can
also be excluded on the basis of the strong Θ-dependence
of T_c (see fig.17b). Therefore the order is attributed
to lateral repulsive forces whose magnitude is certainly
very small: the desorption energy E_d is constant up to
$\Theta=.5$ ($E_d=.93\pm.05eV$) where is drops suddenly by .1eV.
The E_d values agree well with the isosteric heats of
adsorption E_a (fig.8d).[22] Atoms in the completed (2x2)
structure have 3 nearest neighbours, additional atoms 6
nearest neighbours. The upper limit of the lateral interaction energy E_{aa} per neighbour is therefore .02eV. A
similar value is obtained from $T_{c,max}=270K$. The asymmetry
of the phase diagram (fig.17b) is not incompatible with
a simple lattice gas model with coverage-independent
interaction energy which would produce a _symmetric_ phase
diagram if the adsorbate would show particle-hole symmetry.
The coverage scale used in fig.17b is, however, referred
to the number of substrate atoms (Θ_a). If all sites which
have to be assumed as being equivalent in the graphitic
(2x2) structure are occupied then the maximum coverage
would be $2\Theta_a$. $\Theta=.5$ in fig.17b corresponds, therefore,
to $\Theta'=.25$ referred to the number of equivalent sites so that
there is no particle-hole symmetry and, consequently,
no symmetric phase diagram is to be expected.[60] In
contrast to the (111) plane, the (110) plane adsorbs
hydrogen with at least initially attractive interactions
leading to island formation (see fig.8a,c). A detailed

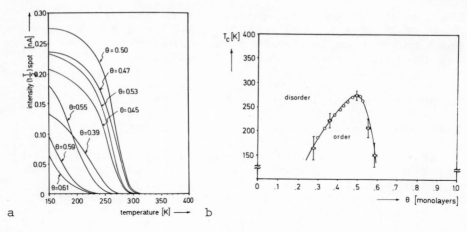

Fig.17: Order-disorder transition of H on Ni(111).
a) Intensity of a half-order spot - which is a measure
of the order - as a function of temperature for various
coverages. b) Phase diagram (order-disorder transition
temperature T_c as a function of coverage).[57,58]

comparative study of phase transitions on these two
surfaces would be interesting.

3.2.2. Several (discrete) ordered phases

3.2.2.1. O/W(110)

In this system six different structures have been
reported: p(2x1), p(2x2), (1x1) and three superstructures
with large unit meshes.[7,60,61] The second and third
structure transform irreversibly to the complex superstructures upon heating above 800K. The p(2x1) structure
and the complex structure resulting from heating the
p(2x2) structure show order-disorder transitions. That
of the (2x1) structure which is best developed at $\theta = .5$
has been the subject of extensive investigations.[62-68]
Fig.18a shows the normalized intensity $I_{1/2}$ of a superstructure (half order) LEED spot corrected with the
instrument function as a function of temperature for
$\theta = .14$ and $\theta = .5$.[67] The transition temperature T_t clearly
is different for the two coverages and at $\theta = .14$ the
intensity does not go to zero at high T; at intermediate coverages a more complex T-dependence is seen.
The coverage dependence of T_t is shown in fig.18b.[67]
From $I_{1/2}(T,\theta)$ and the variation of the halfwidth of
the half order LEED spots the conclusion may be drawn

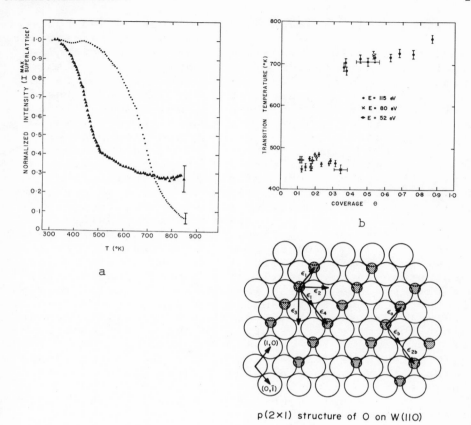

p(2×1) structure of O on W(110)

Fig.18: Phase transitions of the p(2x1) structure of
O/W(110). a) Temperature dependence of normalized
intensity $I_{1/2}(T)$ of half order LEED spots for $\Theta=.14$
(triangles) and $\Theta=.5$ (dots).[67] b) Transition temperature
T_t (inflection point of $I_{1/2}(T)$ as a function of coverage.[67] c) Pairwise interactions taken into account in
various Monte Carlo calculations.[65]

that the thermal disordering of the p(2x1) structure
is due to an order-disorder transition at high coverage
($\Theta \geq .4$) and to p(2x1) island dissolution in the low-
coverage limit.[67] That island formation occurs already
at very low coverage was also deduced from the coverage
dependence of the LEED background and $I_{1/2}(\Theta)$ (see fig.2).
Island formation implies that the lateral interactions
must at least in part be attractive.

The lateral interactions can be determined by treating the adsorbate as a two-dimensional lattice gas with a finite number of pairwise interaction energies (fig.18c). A calculation of the configurations with minimum total interaction energy with the Monte Carlo method and of the kinematic LEED intensities of these configurations for the high- and low-Θ disorder transition with a large number of ε_i-combinations allows to select the best set of values by comparison with experiment. In earlier calculations[63],[64] the parameters $\varepsilon_a = \varepsilon_{2b} \neq \varepsilon_b$ (see fig.18c) were chosen which, however, violate the symmetry properties of the substrate ($\varepsilon_a = \varepsilon_b$).[63] More recent calculations[65],[66] take the symmetry and more neighbours such as the set ε_1 to ε_4 in fig.18c into account, with the following results: $\varepsilon_1 = -.09eV$, $\varepsilon_2 = \varepsilon_3 = +.075$, $\varepsilon_4 = -.03eV$[65],[66] and $\varepsilon_1 = -.072eV$, $\varepsilon_2 = +.080eV$ and $\varepsilon_4 = -.049eV$ if ε_3 is set zero.[66] These lateral interaction energies have to be compared with a desorption energy[7] and isosteric heat of adsorption[20] of about 6eV! The lateral interaction is oscillatory as expected in view of the small dipole moment - $\Delta\phi(\Theta=1/4)\approx0$, $\Delta\phi(\Theta=1/2)\approx.2eV$ - which leaves the indirect interaction via the substrate (see Sect.4) as the most probable one.

The pairwise interaction energy calculations have one flaw: they produce a phase diagram symmetric with respect to $\Theta=1/2$ - which is an immediate consequence of the particle-hole symmetry of pairwise interactions - while experiment shows a clear asymmetry: a p(2x2) structure at $\Theta=3/4$, but not at $\Theta=1/4$. The asymmetry can be accounted for if many-body (non-pairwise) interactions - which are expected to be important anyway[69] - are included. A first step in this direction is the inclusion of three-body ("trio") interactions.[70] If to the four pairwise interaction energies ε_1 to ε_4 shown in fig.18c trio interactions ε_{Tp} between the atoms involving the two smallest atom triangles are added then the following values give excellent agreement with the observed $I_{1/2}(T)$ for $\Theta=1/2$: $\varepsilon_1 = -.072eV$, $\varepsilon_2 = +.056eV$, $\varepsilon_3 = 0$, $\varepsilon_4 = -.049$, $\varepsilon_{Tp} = .052eV$, indicating that trio interactions are indead comparable in magnitude with pairwise interactions. Calculations[70] confirm this but show simultaneously that trio interactions involving larger atomic triangles are of comparable magnitude and may not be neglected.

A second difficulty of the pairwise interaction potential calculations is the fact that they predict islands elongated in one direction[65] while experiment shows approximately equilateral islands.[7],[62] This

discrepancy may be eliminated by the inclusion of trio
interactions but can also be attributed to the influence
of surface imperfections such as steps which are known
to have a strong influence on the growth of p(2x1)
islands.[71] A careful analysis of the influence of the
instrument response function on the LEED pattern reveals
that the surface contains monoatomic steps with a
density of about 2% under the conditions of the experiments analyzed above with regard to lateral interactions.[72]
If these steps act as nucleation sites of the p(2x1)
islands - as suggested by studies on unidirectionally
stepped surfaces[71] - then their small size and high
density - which are needed to get quantitative agreement with experimental spot widths and with $I_{1/2}(T)$ -
would be understandable. Unfortunately, this would
limit the possibility of extracting lateral long-range
interactions from island shapes.[73],[74]

    Closely related to the question of island formation
and lateral interactions is the coverage dependence of
the diffusion coefficient discussed in sect.2.7. (see
figs.13 and 14). D_O in fig.13b increases by 3 orders of
magnitude in about the same coverage range in which T_t
rises from 470K to 700K in fig.18b (which was attributed
to the change from island dissolution to order-disorder
transition). The results of fig.14a were obtained above
T_t so that the maximum of D is not related to the
optimum p(2x1) structure. Nevertheless, the delicate
balance between attractive (ε_a) and repulsive (ε_b)
interactions which stabilizes the p(2x1) structure also
causes the maximum of D(Θ) as shown by a Monte Carlo
computer simulation of the diffusion process using the
$\varepsilon_a,\varepsilon_b$ parameters mentioned above (see fig.14b).[37] More
work, both experimental and by computer simulation,
needs to be done before the influence of the various
parameters on island growth and dissolution, order -
disorder transition and surface diffusion will be
fully understood.

3.2.2.2. Na/W(110)

    Although this system is not the alkali/W system
experimentally[61] studied in most detail it is the
best analyzed one. The system shows several (discrete)
phases corresponding to Na:W atom ratios of 1:6, 1:4,
1:3, 3:8 and 2:5 and a close-packed Na atom layer
(see fig.19a). Except for the last phase which is out
of registry all phases form superstructures whose unit

meshes are related to that of the substrate in a simple
manner. In the first three phases all Na atoms are in
equivalent positions with respect to the substrate, in
the next two they are not unless unequal distances
between the atoms are allowed. The phase transitions
with increasing coverage appear to be continuous except
for the transition between the 1:4 and the 1:3 structure.
The intensity of LEED spots characteristic for the 1:3
structure increases linearly with coverage up to close
to optimum coverage and decreases thereafter linearly
indicating island growth and dissolution. The temperature
dependence of characteristic LEED spots of the various
phases (fig.19b) suggests simple order-disorder transitions for the 1:6 and 1:4 structures and more complicated
ones for the other coincidence structures while the
close-packed layer shows an abnormal T dependence. The
heat of adsorption as obtained from adsorption-desorption
equilibrium decreases monotonically with coverage to a
poorly pronounced plateau at the completion of the 1:3
structure. Thereafter, when the Na atoms are displaced
from the potential minima as indicated in fig.19a
(structures e,f) in order to maintain equal interatomic
distance, the heat of adsorption decreases again. Nearly
equal interatomic distances are caused by the repulsive
forces between the dipoles formed by the partly charged
Na atoms and their image charges which can be deduced
from the work function.

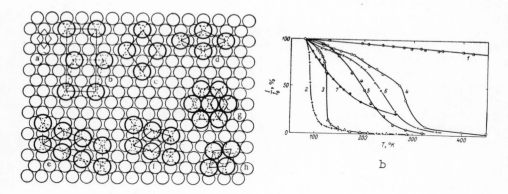

a

Fig.19: Phases of Na on W(110). a) The observed phases
corresponding to Na:W ratios of 1:6(b), 1:4(c), 1:3(d),
3:8(e), 2:5(f) and to a close packed hexagonal layer (g);
(h) is from a multilayer. b) Temperature dependence of
LEED spots characteristic for the clean surface (1) and
for the following phases: 1:6(2), 1:4(3), 1:3(4), 3:8(5),
2:5(6), hexagonal layer (7).[75]

The nature of interatomic forces can be deduced to
some extent from the presence and absence of discrete
structures. For example no discrete structures with
Na:W ratios of 1:7 or 1:5 are observed in the sequence
of ordered phases. This implies that certain inequalities
must be fulfilled between the interaction energies J_k
between an adatom and adatoms on k-th neighbour sites. [76]
The sequence of structures as a function of Θ is
determined by the lowest value E_{tot}^{min} (Θ) of the total
interaction energy of the adsorbate $E_{tot} = \sum_i \sum_k J_k n_i n_{i+k}$
where the n's are 1 or 0 depending upon whether a site
i or i+k is occupied by an adatom or not. $E_{tot}^{min}(\Theta)$ has
been calculated for arbitrary values of the J_k's taking
up to 7-th neighbours into account and J_k-inequalities
were derived which have to be fulfilled for the appearance of the 1:6, 1:4 and 1:3 structures and the absence
of 1:7 and 1:5 structures. For example the condition
$J_6 < J_7$ must be fulfilled if the 1:7 structure is not
allowed as observed. This condition clearly cannot be
fulfilled by a pure dipole-dipole interaction - which is
monotonic - but requires additional force contributions
such as indirect interactions through the substrate
(see Sect.4).

Nevertheless calculations based solely an dipole-
dipole interactions have been quite successful in describing the order-disorder transition of the Na:W=1:4
structure with increasing T (curve 3 in fig.19b). [77]
The interaction energy E_i of the dipole i with all
other dipoles was divided into a sum over all N neighbours within a circle with radius R$\gg$ a (a = W lattice
constant) and an integral over the remaining atoms
which were assumed to form a continuous dipole layer
with the concentration n of the adatoms:

$$E_i = E_1 + E_2 = \sum_i^{r_i \leq R} \frac{p_i^2}{r_i^3} + 2\pi^3 \frac{n^3/2}{N^{1/2}} p^2$$

The total interaction energy of the adsorbate is then
$U = \frac{1}{2} \Sigma E_i$. All adsorbed atoms are allowed to displace
in a Monte Carlo program with a transition probability
from one adsorption site to another determined by the
difference between the E_i's in these two sites. The
dependence of the total energy U on the number of
diplacements is shown for various temperatures in fig.20a.
Typical distributions are seen in fig.20b. From these
distributions the temperature dependence of a LEED spot
characteristic for the Na:W = 1:4 structure can be
obtained by a kinematical LEED calculation. The result
agrees well with experiment (fig.20c) if the dipole

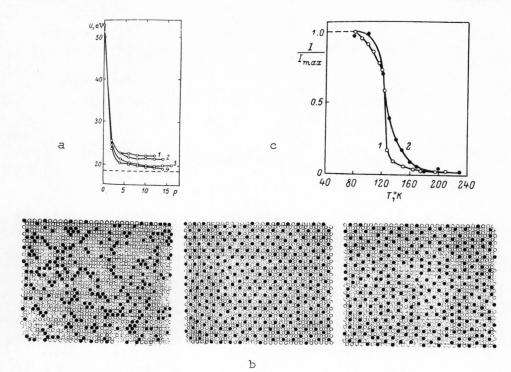

b

Fig.20: Modeling of order-disorder transition in the Na
(2x2) phase on W(110) with a ratio Na:W = 1:4.
a) Convergence of the total energy with number of
displacements starting with a random distribution. The
dashed line is the energy of an ideal (2x2) structure.
b) Typical atomic distributions, from left to right:
initial random distribution, equilibrium distribution
at 80K, equilibrium distribution at 160K, c) Intensity
of (1/2, 1/2) LEED reflex as a function of substrate
temperature: 1) calculated with p=2.5 Debye, 2) from
experiment (ref. 75).

moment p=2.5 Debye obtained from the work function
measurements is used.

3.2.2.3. Other alkali and alkaline earth adsorbates on
         W(110) and Mo(110)

     The phase diagrams of Li, K, Cs, Ba and Sr on
W(110) and of Ba and Sr on Mo(110) have been studied
experimentally in more or less detail (for references

see refs.78,79), with emphasis on Ba and Sr. Here only
some general results concerning the disordering behaviour
of different phases and phase transitions with increasing coverage will be discussed.

It has been observed[80] that there are apparently
two distinct cases: one in which the temperature dependence of the superstructure LEED reflexes is typical
for a order-disorder phase transition as shown by curves
2 and 3 in fig.19b. This occurs typically in commensurate
layers. In the second case which is typical for phases
not in registry with the substrate, i.e. incommensurate
layers,the reflex intensity decreases monotonically
over a wide temperature range as in curve 7 of fig.19b.
This difference between commensurate and incommensurate
films is attributed to differences in the frequency
spectra of the two types of films: the spectrum of the
former begins with a nonzero frequency ω_i determined by
the vibration of a single atom in its potential well;
the spectrum of the latter starts at $\omega = 0$. Therefore
disordering of the commensurate film begins at a critical temperature T_c determined by ω_i while the incommensurate film begins to disorder already at $T>0$. Such a
film does not have strict long-range order but only an
extended short-range order which extends over a
sufficiently wide distance so as to produce reasonably
sharp diffraction spots.

The coverage dependence usually is as follows
(see e.g. refs.81,82): at low coverages ($\Theta_a<.08,\Theta_b<.2$)
- and sufficiently low temperature so that ordering
can occur (see fig.1) - an approximately hexagonal
adsorbate structure with continuously decreasing atomic
distances forms. In this coverage range the work function
decreases linearly with Θ, i.e. the dipole moment per
atom is constant[18] (see fig.7a). This range is followed
by a series of commensurate structures i corresponding
to specific coverages Θ_i. Between Θ_i and Θ_{i+1} the
structures i and i+1 coexist, with structure i+1
growing linearly with coverage at the expense of
structure i. Each structure has a well-defined workfunction ϕ so that ϕ changes linearly between Θ_i and
Θ_{i+1} (see fig.7a).[18,81,82] The next stage seems to
vary from system to system. In some, e.g. Sr/W(110),[81]
a first-order phase transition to a denser commensurate
structure ("β-phase") occurs (α+β-region in fig.1).
This has initially a large number of vacancies which
are filled after the surface is completely covered by
this phase (β-region in fig.1). In other systems, e.g.

Ba/W(110)[82], the region of co-existing structures
("α-phase") is followed by one in which uniaxial com-
pressions occur producing a continuous series of
incommensurate structures until a hexagonal packing is
reached. From here on all systems seem to behave the
same way: continuous homogeneous compression of the
hexagonal structure occurs until saturation coverage
(Θ_b=1) is reached which may be considerably higher than
a normal hexagonal packing of adsorbate atoms, e.g.10%
more in the case of Sr on W(110).[81]

The lateral interactions in such strongly electropositive adsorption systems are primarily direct repulsive dipole-dipole interactions, at least at low coverages and secondarily "indirect" interactions via the
substrate which are oscillatory (see Sect.4). With
repulsive interactions dominating it appears unlikely
that a firstorder phase transition to a denser phase
should occur. However, with increasing coverage the
dipoles become increasingly depolarized so that a denser
phase becomes energetically more favorable inspite of
the smaller distance between the dipoles. This has been
shown theoretically both classically[83] and quantum-
mechanically.[84]

3.2.3. Continuous phase transitions

3.2.3.1. One-dimensional

Furrowed surfaces consisting of densly packed rows
of atoms separated by troughs frequently show continuous
(quasi)one-dimensional transitions. Examples are f.c.c.
{110} and b.c.c. {112} surfaces with furrows parallel
to $[1\bar{1}0]$ and $[11\bar{1}]$, respectively. The system O/Pd(110)
was the first chemisorption system in which a phase
transition with increasing Θ was modeled with the Monte
Carlo method.[85] Fig.21a shows the LEED intensity
profiles in the $[100]$ direction obtained from the
calculated adatom distributions for 1/3<Θ<1/2 using
the following pairwise interaction energies:$\varepsilon^1[1\bar{1}0]$=-.07eV
(attractive), $\varepsilon^1[001]$=.25eV and $\varepsilon^2[100]$=.07eV, the
superscripts referring to 1st and 2nd nearest neighbours.
These calculated intensity profiles agreed well with
the measured transition from a(1x3) to (1x2) structure.
In the same way phase changes with temperature can be
treated. An example is the system O/Ni(110) which, at
Θ=1/3, shows a reversible transition from a(3x1) to
a(2x1) structure at 300°C. (Note that the superstructure

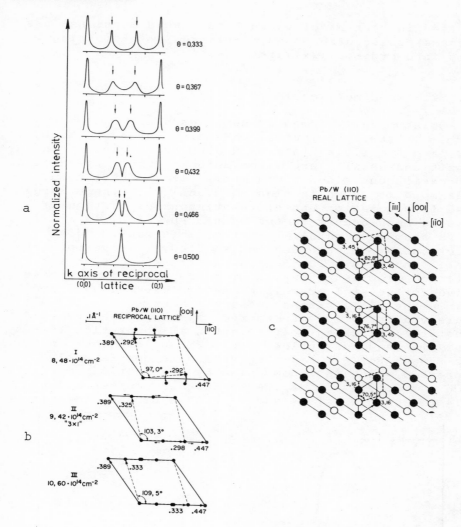

Fig.21: a) Calculated LEED intensity distributions
between (0,0) and (0,1) reflex in the system O/Pd(110)
as a function of coverage[85], b) and c) reciprocal and
real lattice of Pb/W(110) at high coverages.[86]

unit meshes are rotated here 90° with respect to
Pd(110)! Nevertheless is the Θ- and T-dependence of
the structures very similar). Here $\varepsilon^1 [001] = -.165eV$,
$\varepsilon^1 [110] \to \infty$, $\varepsilon^2 [1\bar{1}0] = 0$ and $\varepsilon^3 [1\bar{1}0] = .016eV$ can explain the
phase transition qualitatively. The physical picture

proposed is that at elevated T the two-dimensional gas pressure of isolated atoms compresses the remaining atoms into ordered (2x1) regions.[86]

3.2.3.2. Two-dimensional

On atomically "smooth" surfaces which do not possess a strong azimuthal anisotropy such as the b.c.c. {110} and the f.c.c. {111} and {100} surfaces continuous transitions usually are accompanied not only be uniaxial compression but also by distortion of the unit mesh. An example is shown in figs. 21b and c. In the Θ range close to saturation shown a continuous distortion of the unit mesh occurs in the system Pb/W(110) in which the Pb atom rows parallel to the close-packed W[$1\bar{1}\bar{1}$] directions are compressed and shifted relative to each other.[51] An example for continuous transitions on f.c.c. {111} and {100} surfaces is CO on Pd which is representative for several transition metals (see figs. 12 and 22). On the {111} surface adsorption up to $\Theta = 1/3$ occurs via formation of ordered islands (with ($\sqrt{3}$x$\sqrt{3}$)R30° structure); sticking coefficients, IR absorption frequency ω_v and desorption energy E_d are constant in this range. For $\Theta > 1/3$ the lattice is continuously compressed which is accompanied by a slight increase of ω_V and decrease of E_D. With further compression at $\Theta > .4$ the changes of ω_V and E_D with Θ become dramatic

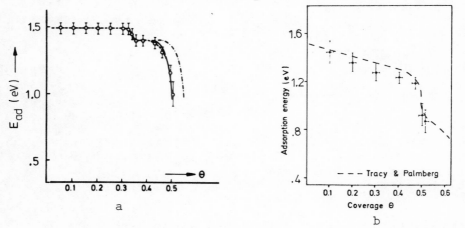

Fig.22: Heats of adsorption of CO on Pd as obtained from adsorption-desorption equilibrium a) {111} using $\Delta\phi$ as a probe, b) {100} using $\Delta\phi$ (dashed line) and IR absorption as a probe.[34]

signaling saturation at $\Theta \approx .5$. On the $\{100\}$ surface
the ordering begins at $\Theta \approx .4$ and is associated with a
smaller increase (decrease) of $\omega_v(E_d)$. After completion
of the ordered structure at $\Theta = .5$ compression occurs and
simultaneously a strong increase (decrease) of $\omega_v(E_d)$.[34]

3.3. Adsorbate-induced phase transition in the substrate

        The idea that adsorption may induce a reconstruc-
tive or more precisely displacive phase transition of
the substrate has been put forward some time ago.[87]
Recently[88] it was invoked as a possible explanation for
the p(2x1) structure of O/W(110) discussed in sect.
3.2.3.3. although models without substrate atom displacements satisfactorily explain the LEED intensities.[89,90]
There are, however, two adsorption systems, H/W(100)
and H/Mo(100), in which the adsorbate induces substrate
atom displacements without doubt. The experimental
evidence comes from RBS (Sect.2.2.1, fig.3)[9], EELS
(Sect.2.6.1)[91] and in particular from LEED (Sect.2.1.)
and has been reviewed recently.[92] The RBS measurements
of H on W(100) show that the transition from the
H-induced c(2x2) structure at low coverages to the
H-induced (1x1) structure at high coverages is accompanied by a displacement of W atoms from displaced
positions to their normal positions.[9] The EELS measurements of vibrational excitations of H on W(100) indicate that the W-H-W bond angle varies with H coverage
in a such manner that the W atoms must be displaced in
the low coverage c(2x2) structure similar to the displacements in the low temperature c(2x2) structure of
the clean surface (fig.16a).[91] Finally, in the order-
order (or order-disorder) transition of the H-induced
c(2x2) structure the change of the half order LEED
reflexes with temperature is similar to that observed
in the corresponding transition on the clean surface
(fig.16b) except for a shift to higher temperatures.
This phase transition was the first one subjected to a
thourough analysis though within the framework of the
now outdated model of a H order-disorder transition
on an ideal square W(100) substrate.[93] This analysis
showed for the first time for a chemisorption system
the dependence of the T-dependence of a superstructure
spot on the size of the coherently scattering domain
(fig.23a), on the deviation from the ideal c(2x2)
coverage (fig.23b) and gave excellent agreement with
experiment (fig.23c). In view of the now well-established
substrate reconstruction this agreement must be taken

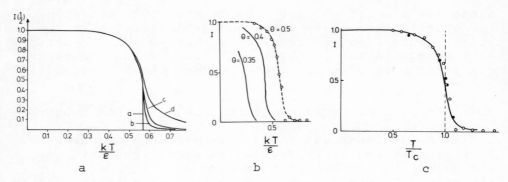

Fig.23: Temperature dependence of the intensity $I_{1/2}$(T) of half order LEED spots of a c(2x2) structure on a square substrate with repulsive nearest neighbour inter-action energy ε. a) Influence of the size M^2 of the coherently scattering region: M=∞, 25,15,5 for a,b,c,d respectively. b) Influence of the coverage Θ on $I_{1/2}$(T). Broken line: analytical solution, circles: Monte Carlo results. c) Comparison of experiment (full circles) with Monte Carlo calculations (open circles) and ana-lytical solution (full line) for M=30 and Θ=.5. From a fit of the temperature scale to the experiment with kT_c/ε=1 the nearest neighbour repulsive energy ε=.076eV is obtained.[93]

as a warning that a structure model should not be considered as well established if predictions derived from it agree well with experiment.

    Although the system H/W(100) has been restudied too recently in some detail more LEED information is available for the system H/Mo(100).[92] The phase diagram of this system is much more complex than that of H/W(100) and is shown in fig.24. With increasing coverage a series of temperature dependent order-order transition occurs which are attributed to various periodic lattice distortions (PLD) induced in the relatively unstable Mo(100) surface by H adsorption. They are similar to the PLD of the clean surface at low temperature and connected with the electronic instability of this surface. Thus the system H/Mo(100) is likely to become not only a major testing ground for phase transitions but also for the coupling between electronic and ionic subsystems in systems with electronic instabilities.

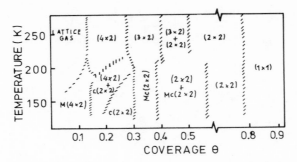

Fig.24: Phase diagram of the system
H/Mo(100).[93]

3.4. Phase transitions involving adsorbate and substrate

On many surfaces which are not close-packed chemisorption is accompanied by an interchange of atoms
between adsorbate and the topmost substrate layer. This
interchange may occur with increasing θ or increasing
T and may be reversible or irreversible. A good example
is the system O/W(100).[24,95] As shown in figs.4-6 and
7b the properties of this surface undergo a dramatic
change upon heating to T>600K, most clearly at θ=.5.
This change is caused by the incorporation of oxygen
originally adsorbed on top of the W surface into the
topmost W layer which is an <u>irreversible</u> order-order
transition from a p(4x1) structure to a p(2x1) structure.
Upon heating this p(2x1) structure undergoes a <u>rever-
sible</u> order-disorder transition with a transition
temperature $T_t \approx 880K$ (fig.25a). This disordering not
only involves lateral displacements but also normal
displacements of O atoms as indicated by the simultaneous
increase of the workfunction and the appearance and
rapid growth of a low energy O^+ESD peak.[95] The normal
O atom displacements are also responsible for the
decrease of the W ISS signal in fig.4 (+signs) which
is due to the shadowing of the W atoms by the O atoms.

Here only the irreversible phase transition (x and
o signs in fig.4) will be discussed. It is accompanied
by the strong drop of the secondary ion yield in fig.5,
the replacement of the medium yield 7.8eV O^+ ESD peak
by the low yield 8.3eV O^+ESD peak in fig.6 and in particular by the sudden decrease of the work function
(fig.7b). This work function change is shown once more
in fig.25b, together with the LEED patterns observed

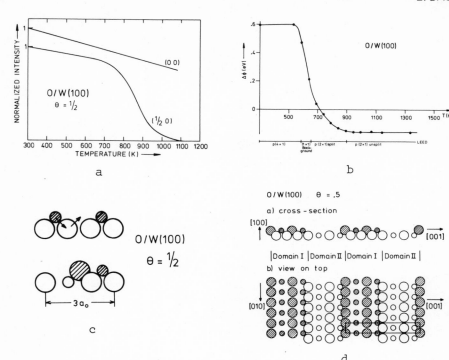

Fig.25: Phase transitions in the system O/W(100) at
Θ=.5. a) Temperature dependence of an integral and
halforder spot during the reversible phase transition
of the p(2x1) structure. b) Work function change and
LEED structures during irreversible reconstructive
phase transition from p(4x1) to p(2x1) structure.
c) Smallest unit of p(2x1) structure, d) Second **smallest**
units of p(2x1) structure.[96]

at 300K after heating to the indicated temperatures,
for exactly half monolayer coverage. It is seen that
the order-order transition p(4x1)→p(2x1) proceeds via
a totally disordered and a partially ordered state
("split"p(2x1)). In the "split" p(2x1) pattern the
diffraction spots are split by an amount Δ (in a_o^{-1})
which depends upon annealing temperature T_a and
annealing time t_a but not on the energy E_o of the
electrons. From the E_o-dependence of the intensity
$I_\Delta(E_o)$ of the split spots it can be concluded that the
surface consists of small regions with p(2x1) structure
separated by monoatomic steps. The smallest unit of
this kind can be formed simply by a place exchange
between O and W atoms resulting in a unit mesh length
$a_s^O = 3a_o$ (fig.25c). The next-largest unit compatible

with the LEED pattern has the unit mesh length $a_s^1 = 7a_0$
(fig.25d). Its formation requires diffusion in addition
to the place exchange. With increasing annealing time
and temperature some of the p(2x1) regions grow at the
expense of others leading to a growth law $\bar{a}_s^2 - \bar{a}_s^{o2} = D(T_a) \cdot t_a$.
The activation energy $E_a = 1.87 \pm .09eV$ derived from the T_a
dependence of $D(T_a)$ can be attributed to the transitions
of W atoms from their positions in the p(2x1) patches
into adsorbed positions on top of the patches where
they can diffuse easily, or to diffusion within the
p(2x1) patches which is expected to have a high activation energy. The activation energy E_a for O atom
displacements normal to the surface as obtained from
the temperature dependence of the O^+ ESD peak seen
during the reversible order-disorder transition mentioned
above is much smaller ($E_a^o = .35 \pm .01eV$) and can, therefore
not be the rate limiting step. More work is certainly
still necessary before this complicated phase transition
can be understood completely.

3.5. Chemisorbed phases on non-metals

Many gases and vapors can be chemisorbed on semiconductor and insulator surfaces but phases and phase
transitions have hardly been studied in such systems,
at least not with the methods discussed in Sect.2. Most
of the work is concerned with semiconductor surfaces,
notably with the Si(111) surface. An example is the
system Ag/Si(111) which has been the subject of numerous
studies (for references see ref.96). The system has three
commensurate phases: a(6x1), a(3x1) and a($\sqrt{3}$x$\sqrt{3}$)R(30°)
structure. Unfortunately no agreement exists about the
saturation coverage of the $\sqrt{3}$-structure which is
essential for the choice of the structure model: most
authors attribute it to $\Theta = 1$, i.e. $N_{Ag} = N_{Si}$ and assume
the Lander model for the $\sqrt{3}$-structure (fig.26a) while
others[97,98], on the basis of at least equally careful
measurements obtain $\Theta = 2/3$ and prefer the graphitic Ag
atom arrangement shown in the right lower half of
fig.26b. Because the latter authors have studied the
phase transition $\sqrt{3} \rightarrow$ (3x1) with decreasing coverage in
most detail their model will be discussed here. In
isothermal desorption measurements - the Ag coverage
was monitored by AES - they found zero order desorption
kinetics - i.e. coverage-independent desorption rates -
both for the $\sqrt{3}$ and the (3x1) phase with desorption
energies of 2.83eV and 2.91eV, respectively. They attribute the order zero to islands with $\sqrt{3}$ and (3x1) structure acting as unexhaustable sources of atoms for the

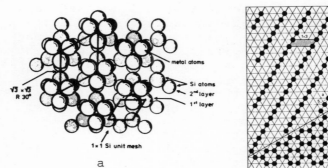
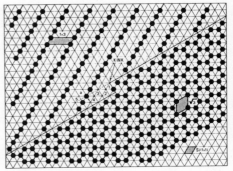

a

b

Fig.26: Phases in the system Ag/Si(111). a) ($\sqrt{3}\times\sqrt{3}$)R(30°)
        structure for $\theta=1$.[97] b) (3x1) and ($\sqrt{3}\times\sqrt{3}$)R(30°)
        for $\theta=1/3$ and $\theta=2/3$, respectively.[98]

lower coverage phase from which desorption occurs at
effectively constant coverage. The reverse transition
(3x1)→$\sqrt{3}$ with increasing coverage can easily be visualized as indicated in fig.26b: the $\sqrt{3}$ phase grows
partially by addition of Ag atoms to the (3x1) structure
(sites 1,2,3), partially by displacements of atoms of
the (3x1) structure (atom a into site 4). Equally
easily the nucleation of the $\sqrt{3}$ islands in the (3x1)
sea can be accounted for. The small difference between
the desorption energies of the two phases makes it
likely that the transition (3x1)→$\sqrt{3}$ occurs easily. It
would be interesting to study this transition and the
T dependence at constant coverage in more detail.

3.6.   Multilayer adsorption

       Multilayer adsorption is a frequently observed
phenomenon in the adsorption of metals on metals. Again
W is particularily suitable as adsorbant because of its
high surface energy. A multilayer is called "adsorbed"
if it is stable up to the desorption temperature.
Multilayer formation is best studied by AES which gives
more or less sharp breaks in the AES signal as a function
of coverage: each additional layer attenuates the Auger
electrons from the lower layers due to the short excape
depth of the Auger electrons. Fig.27a gives an example
for close-packed multilayer formation of Cu on W(110),[27]
fig.27b for multilayer formation on a more open plane,
W(100).[99]   The number of stable adsorbed layers is
difficult to predict because the stability is strongly
effected by still poorly understood electronic effects.

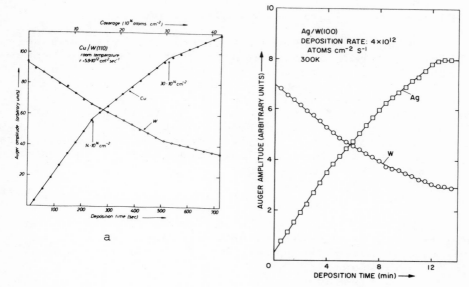

Fig.27: Multilayer formation in metal adsorption on W.
AES signal as function of deposition time.
a) Cu/W(110)[27], b) Ag/W(100).[99]

Thus Cu forms 2 stable layers on W(110)[27] while Pd
forms only one[30] inspite of comparable binding energy
to the substrate. On the more open (100) plane there
are usually more layers adsorbed than on the close-
packed (110) planes but the total amount of adsorbed
material is usually about the same on the two planes.
The structure of the layers also varies considerably.
Thus, the first layer of Cu on W(110) is pseudomorphous
and the second one is incommensurate in one direction,
commensurate in the other.[27] In Ag, Pd and Au[30,99]
the first layer already shows the behaviour which Cu
does in the second layer. The range of existence of
the one-dimensionally incommensurate monolayers agrees
well with the predictions of the Frank-van der Merwe
theory of misfitting monolayers.[100] On W(100) the
layers are usually pseudomorphous, Au being an exception
by forming complex superstructures with increasing
coverage and temperature.[99] On the (110) surface
islands exist already at very low coverage and are
stable up to high temperatures as deduced from TDS[25,27,30]
and LEED[30] at high temperature indicating relatively
strong attractive lateral interactions while on the (100)
surface repulsive forces are present at low coverages.
These question will be discussed in the next section.

4. Fundamental quantities in chemisorbed phases

Although a symmetry analysis of the lateral interaction energy E_{aa} allows to predict the possible types
of transitions of ordered phases (for references see
ref. 101), the question what phases occur in nature
cannot be answered without the knowledge of the specific
form of E_{aa} and of the activation energy E_{sd} for surface
diffusion which must occur in order to reach the minimum
energy configuration at a given coverage and temperature.
This section will briefly deal with these two quantities.

4.1. Lateral interactions between chemisorbed atoms[70],[79]

The lateral interaction energy E_{aa} is in general
anisotropic, distance- and coverage-dependent: $E_{aa}=E_{aa}$
$(\vec{r}_{ij},\theta)$. Starting from the type of chemisorption bond
- ionic (workfunction ϕ > ionization potential I
(electropositive adsorbate) or ϕ < electron affinity A
(electronegative adsorbate)), covalent (localized
electrons) and metallic (delocalized electrons, both
for A < ϕ < I) - three dominant types of interactions
can occur between atoms: dipole-dipole (direct, through
vacuum half space), electron-electron (indirect via
substrate electrons or direct at short distance) and
elastic (via substrate ions). If the work function
change $\Delta\phi$ upon adsorption is very large than the
dipole-dipole interaction is dominating, e.g. at low
to medium coverages of alkali and alkaline earth atoms
on W(110) (Sect.3.2.2.). If $\Delta\phi$ is small and the adsorption energy E_a small compared to the sublimation
energy E_s of the substrate - as for example in the
system H/Ni(111) (Sect.3.2.1) - the indirect electron
interaction is expected to be the most important one.
Finally, if $\Delta\phi$ is small but E_a comparable to E_s as in
the case of Pt adsorption on W(110) elastic interactions
are expected to play a major role.

The elastic interaction is always repulsive for
two identical atoms on an elastically isotropic surface
and falls off at large distances r like r^{-3} just as
dipole-dipole repulsion does.[102],[103] The interaction
can become attractive in certain directions if the
substrate is elastically sufficiently anisotropic[103]
or when the atoms differ.[102] Experimentally - as determined by field ion microscopy (FIM) - the atom most
strongly bound to the W(110) surface, Re, shows the
smallest attractive interaction energy (E_{aa}<.04eV) with

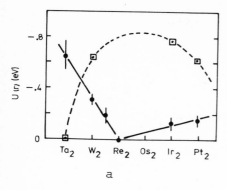 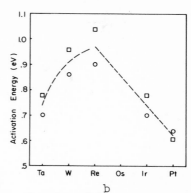

Fig.28: a) Me$_2$ binding energies of adsorbed atoms on
neighbouring lattice sites on W(110) (dots with
error bars).[104] Squares: interaction energies
from Morse potentials. b) Activation energy for
surface diffusion of individual atoms on W(110).
Circles: from single atom data; squares: from
simultaneous diffusion of several atoms.[115]

another Re atom on a neighbouring adsorption site (see
fig.28a)[104] as would be expected because they should
cause the strongest lattice distortion. The fact that
the interaction is not strongly repulsive is to be
attributed to the direct electron-electron interactions
which try to form a Re$_2$ molecule. With increasing
distance the overlap of the electrons on the two atoms
decreases rapidly so that only the indirect electron
interactions via the substrate are left provided the
dipole moments of the adsorbed atoms are sufficiently
small. That the atoms shown in fig.28a have small
dipole moments is well established by field emission
measurements of W(110) surfaces covered with individual
adsorbed atoms.[105] The indirect interaction via substrate electrons has been the subject of much discussion.[70]
The indirect interaction is oscillatory (attractive/
repulsive) with extrema falling off at large distances
with an inverse power law whose exponent can vary from
7[106] to 1[107] depending on the distance and upon the
assumptions made about the electronic structure of the
surface. In the distance range of interest in chemisorption a r^{-3} - r^{-5} dependence[108] or $r^{-2.5}$ dependence[107]
appears most appropriate in the case of a spherical
Fermi surface. A $r^{-2.5}$ dependence actually is compatible
with the coverage dependence of $E_d(\Theta)$ of a two-dimensional gas of Pd atoms on W(110) obtained from desorption
data in the high temperature range.[30]

The influence of the surface orientation on the adatom interaction was studied for W atom pairs on W(110) and (100) surfaces.[109] Inspite of the short-comings of these calculations (see ref.70) the results are of interest for the comparison of the lateral interactions in chemisorbed layers on this surfaces as deduced from TDS measurements.[25,27,99] The coverage dependence of $E_{aa}(\Theta)$ deduced from experiment suggests attractive nearest neighbour interactions on W(110) and repulsive nearest neighbour interactions on W(100) up to $\Theta=.5$ (fig.10c). Theory[109] predicts a repulsive indirect interaction on both surfaces which, however, on the (110) surface is overcompensated by the attrac-tive direct interaction. To what extent this is true for atoms other than W atoms is not clear, however, because the interaction depends critically on the Fermi energy.

One major problem of indirect interactions is the (potentially) slow decrease with distance. With in-creasing coverage the (pairwise) interaction between two atoms is expected to be increasingly influenced by the other atoms similar to the depolarization effect in the case of dipole-dipole interactions. Calculations, already referred to sect. 3.2.2.1, have indeed shown that three-body interactions are comparable to two-body interactions but also that they may be sufficient to account for the modification of the interaction energy at intermediate coverages.[69,70] At high coverages ($\Theta > .5$) the direct interactions should become dominant which has been invoked to explain the increase in $E_d(\Theta)$ of Cu on W(100) (fig.10c). Future work will have to show to what extent other first principle[110] or pairwise interaction potential calculations[111] - which both do not distinguish between direct and indirect interactions - are able to give a quantitative descrip-tion of interaction energies between metal atoms on metal surfaces as a function of coverage.

In concluding the discussion of indirect inter-actions between two adatoms it should be mentioned that there is no direct experimental proof to date. The experimental evidence put forward some time ago[112] has been repudiated.[113] The nonmonotonic interaction between a Pd and a W atom on a W(110) surface[114] does not show the behaviour expected on the basis of the existing calculations and needs to be examined with other different atom pairs. There is, however, mounting indirect evidence for the existence of oscillatory

interactions: the phase transitions in the system
O/W(110) call for them (Sect.3.2.2.1) as does the phase
diagram of the system Na/W(110) which can not be understood in terms of repulsive dipole-dipole forces only
(Sect. 3.2.2.2.).

The dipole-dipole interactions are believed to be
purely repulsive and, therefore, should not allow two-
dimensional condensation. This is not true because with
decreaseing dipole-dipole distance strong depolarization
occurs so that it can become energetically more favorable
to form a higher density phase with lower dipole moment
per adatom than a dilute phase with high dipole moment.
This is believed to occur in alkali and alkaline earth
atoms on W(110) (Sect.3.2.2.3) which finally condense
into metallic monolayers. In electronegative adsorbates,
e.g. O on W(100) the depolarization with increasing
coverage is frequently small but the resulting dipole
layer is unstable and is easily destroyed by reconstruction (Sect.3.4).

4.2. Surface diffusion[115]

Surface diffusion is essential for the establishment of the equilibrium configuration. The mean square
displacement of a diffusing atom within the time inter-
val τ is given by $<r^2>=4\tau D_o \exp(-E_{sd}/kT)$. The diffusivity
D_o and activation energy for surface diffusion E_{sd} can
by determined directly by FIM studies provided that the
atoms are not desorbed by the imaging field. Results
for E_{sd} on W(110) are shown in fig.28b for comparison
with the corresponding lateral interaction energies.
For adatoms with lower adsorption energies E_a such as
Cu, Ag, Au or Pd the E_d values are correspondingly
smaller e.g. .50eV for Pd, for O atoms at low coverages
.6eV[35], only to give a few examples. Formation of small
islands under the influence of attractive interactions
at T>200K is therefore no problem[116] but order over
larger distances is only reached after heating to
higher temperatures. This is quite evident in LEED
studies in which Pd or Au adsorbates had to be annealed
at 800K or higher to obtain the one-dimensionally
incommensurate films with "long range order" alluded
to earlier (sect.3.6). For the W(100) surface no FIM
data for diffusion processes are available but the
difficulty of obtaining order during adsorption of
metal atoms below 600-800K suggests that E_{sd} is significantly higher on this surface which has much deeper

potential minima. Another cause for the lack of order
could of course be the instability of this surface
(Sect.3.1). The establishment of order in other adsorbate systems also frequently requires annealing.
This may not simply be necessary to activate surface
diffusion but more so to rearrange the substrate.
Ag/Si(111) could be a system of this type. Temperatures
of about 500K are necessary to form the sequence of
phases discussed in sect.3.5. In multilayer adsorption
diffusion on top of the second layer is usually much
faster than on the substrate as illustrated by the
example of Pd/W(110).[117] This fast diffusion is certainly a prerequisite for pronounced monolayer by
monolayer adsorption.

5. Outlook

        Chemisorption provides a large zoo of two-dimen-
sional phases and transitions between them, of which
only a few were presented here as examples. Many
experimental techniques - in part also used in physisorption studies - are available for exploring this
zoo and have already been used. Nevertheless the understanding of phases and phase transitions in chemisorbed
layers is far below that in physisorbed layers. This
has several reasons: 1) many of the experimental
techniques are not understood well enough with regard
to the quantitative information content about the
sample analyzed. 2) Chemisorption provides so many
challenging other research directions such as catalysis,
corrosion or electronic structure that the efforts of
experimentalists and theoreticians in the field has
been spread over a wide area. 3) Lack of sufficient
guidance and stimulation from the phase transition
community. Reason 1 should be no barrier to a rapid
development of the study of chemisorbed phases. Even
if only well-established methods such as LEED and AES
are used in careful measurements significant progress
can be made. Reasons 2 and 3 are closely coupled.
If chemisorption researchers are made aware sufficiently
of the significance of the problem of phase transitions
in chemisorbed layers, at least some interest would be
diverted from other areas.

Acknowledgements and apologies

        The author whishes to thank several colleagues, in
particular A.M. Bradshaw, T.L. Einstein, G. Ertl,
P.J. Estrup, D.A. King, J. Küppers, M.G. Lagally and
I. Stensgaard for making (p)reprints available which
were very helpful in the preparation of this review.
He apologizes to many others for not including their
results, in part because of not being aware of them,
in part for lack of space. Some important relevant
references of general nature or to very recent work
are added at the end (refs. 118-124).

References

1. W.B. Webb and M.G. Lagally, in Solid State Physics
   28(1973)302
2. J.B. Pendry, Low Energy Electron Diffraction,
   Academic Press, London 1974
3. E. Bauer, in R. Gomer: Interactions on Metal Surfaces,
   Springer, Berlin 1975, p.227
4. F. Jona, J. Phys. C: Solid State Phys. 11(1978)4271
5. M.A. van Hove and S.Y. Tong, Surface Crystallography
   by LEED, Springer, Berlin 1979
6. M. Henzler, in H. Ibach: Electron Spectroscopy for
   Surface Analysis, Springer, Berlin 1977, p.117
7. T. Engel, H. Niehus and E. Bauer, Surface Sci. 52
   (1975) 237
8. D.V. Morgan, edit., Channeling, John Wiley, London
   1973
9. I. Stensgaard, L.C. Feldman and P.J. Silverman,
   Phys. Rev. Letters 42 (1979) 247
10. W. Heiland and E. Taglauer, Surface Sci. 68(1977)96
11. S. Prigge, H. Niehus and E. Bauer, Surface Sci.65
    (1977)141
12. A. Benninghoven, CRC Critical Reviews in Solid State
    Physics 6 (1976)291
13. S. Prigge, M.S. Thesis, Clausthal 1975
14. E. Bauer, J. Electron Spectrosc. Related Phenom.15
    (1979) 119
15. S. Prigge, H. Niehus and E. Bauer, Surface Sci.75
    (1978) 635
16. E. Bauer and H. Poppa, Surface Sci., in print
17. J. Hölzl and F.K. Schulte, in Springer Tracts in
    Modern Physics, Vol.85, p.1 Springer, Berlin,1979
18. D.A. Gorodetsky and Yu.P. Melnik, Sov.Phys.-Solid
    State 16 (1975) 1805

19. M. Kramer, unpublished
20. E.A. Bas and U. Bänninger, Surface Sci.41 (1974)1
21. G.-C. Wang, T.-M. Lu and M.G. Lagally, J. Chem.Phys.
    69 (1978) 479
22. K. Christmann, O. Schober, G. Ertl and M. Neumann,
    J. Chem. Phys. 60 (1974) 4528
22a.T. Engel, private communication
23. D.A. King, CRC Critical Reviews in Solid State
    Mater. Sci.7 (1978)
23a.A.K. Green, S. Prigge and E. Bauer, Thin Solid Films
    52(1978) 163
24. E. Bauer, H. Poppa and Y. Viswanath, Surface Sci.
    58 (1976) 517
25. E. Bauer, F. Bonczek, H. Poppa and G. Todd, Surface
    Sci. 53 (1975) 87
26. H. Ibach, W. Erley and H. Wagner, preprint
27. E. Bauer, H. Poppa, G. Todd, F. Bonczek, J. Appl.
    Phys. 45 (1974) 5164
28. D.A. King, Surface Sci. 47 (1975) 384
29. J.L. Taylor and W.H. Weinberg, Surface Sci.78(1978)
    259
30. W. Schlenk and E. Bauer, in preparation
31. H. Froitzheim, in Ibach: Electron Spectroscopy for
    Surface Analysis, Springer, Berlin 1977, p.205
32. E. Erley, H. Wagner and H. Ibach, Surface Sci.80
    (1979)612
33. J. Pritchard, in H. Ibach and S. Lehwald, edit.,
    Vibrations in adsorbed layers, KFA Jülich 1978,
    p.114
34. A.M. Bradshaw and F.M. Hoffmann, Surface Sci.72
    (1978) 513
35. J.-R. Chen and R. Gomer , Surface Sci.79 (1979)413
36. R. Butz and H. Wagner, Surface Sci.63 (1977) 448
37. M. Bowker and D.A. King, Surface Sci. 72 (1978) 208
38. P.P. Auer and W. Mönch, Surface Sci. 80 (1979) 45
39. M. Henzler, J. Appl. Phys. 40 (1969) 3758
40. J.V. Florio and W.D. Robertson, Surface Sci. 24
    (1971) 173
41. P.W. Palmberg, Surface Sci. 11 (1968) 153
42. D.G. Fedak and N.A. Gjostein, Phys. Rev. Letters
    16 (1966) 171
43. E. Bauer, in Structure et Propiétés des Surfaces
    des Solides, CNRS Colloqu. No.187, Paris 1970,
    p.111
44. H.P. Bonzel, C.R. Helms and S. Kelemen, Phys. Rev.
    Letters 35 (1975) 1237
45. T.N. Rhodin and G. Broden, Surface Sci.60 (1976)466
46. J. Küppers and H. Michel, Appl. Surface Sci,in press
47. T.E. Felter, R.A. Barker and P.J. Estrup, Phys.Rev.
    Letters 38 (1977) 1138

48. M.K. Debe and D.A.King, J. Phys.C: Solid State
    Phys. 10(1977), L303; Phys. Rev. Letters 39 (1977)
    708
49. M.K. Debe and D.A. King, Surface Sci. 81 (1979)193
50. K. Yonehara and L.F. Schmidt, Surface Sci.25 (1971)
    238
51. E. Bauer, H. Poppa and G. Todd, Thin Solid Films 28
    (1975) 19
52. S. Nishigaki and S. Nakamura, Jap. J. Appl. Phys.15
    (1976) 19
53. T. Adachi and S. Nakamura, Surface Sci. 59 (1976)61
54. E. Tosatti, Solid State Commun. 25 (1978) 637
55. J.E. Inglesfield, J. Phys. C11 (1978) L69
56. L.C. Feldman, P.J. Silverman and I. Stensgaard,
    preprint
57. J. Behm, K. Christmann and G. Ertl, Solid State
    Commun. 25 (1978) 763
58. K. Christmann, R.J. Behm, G. Ertl, M.A. van Hove
    and W.H. Weinberg, J. Chem. Phys., in print
59. M. Schick, private communication
60. L.H. Germer and J.M. May, Surface Sci.4 (1966) 452
61. E. Bauer and T. Engel, Surface Sci.71 (1978) 695
62. J.C. Buchholz and M.G. Lagally, Phys. Rev. Letters
    35 (1975) 442
63. G. Ertl and D. Schillinger, J. Chem. Phys. 66 (1977)
    2569
64. T.-M. Lu, G.-C. Wang and M.G. Lagally, Phys. Rev.
    Letters 39 (1977) 411
65. E.D. Williams, S.L. Cunningham and W.H. Weinberg,
    J. Vac. Sci. Technol. 15 (1978) 417; J. Chem. Phys.
    68 (1978) 4688
66. W. Y. Ching, D.L. Huber, M. Fishkis and M.G.Lagally
    J. Vac. Sci. Technol. 15 (1978) 653
67. G.-C. Wang, T.-M. Lu and M.G. Lagally, J. Chem.
    Phys. 69 (1978) 479
68. W.Y. Ching, D.L. Huber, M.G. Lagally and G.-C.Wang,
    Surface Sci. 77 (1978) 550
69. T.L. Einstein, CRC Critical Reviews in Solid State
    Materials Sci.7 (1978) 260
70. T.L. Einstein, preprint
71. T. Engel, T.v.d.Hagen and E.Bauer, Surface Sci.62
    (1977) 361
72. G.-C. Wang and M.G. Lagally, Surface Sci.81 (1979)69
73. M.G. Lagally, G.-C. Wang and T.-M. Lu,CRC Critical
    Reviews in Solid State Materials Sci. 7 (1978) 233
74. T.L. Einstein, Surface Sci. 83 (1979) 141
75. V.K. Medvedev, A.G. Naumovets and A.G. Fedorus,
    Sov. Phys. Solid State 12 (1970 301

76. M. Kaburagi and J. Kanamori, J. Phys. Soc. Japan
     43 (1977) 1686
77. V.K. Medvedev and I.N. Yakovkin, Sov. Phys. Solid
     State 19 (1977) 1515
78. E. Bauer, J. de physique 38 (1977) C4-146
79. L.A. Bol'shov, A.P. Napartovich, A.G. Naumovets
     and A.G. Fedorus, Sov. Phys. Usp.20 (1977) 432
80. A.G. Naumovets and A.G. Fedorus, Sov. Phys. JETP
     46 (1977) 575
81. O.V. Kanash, A.G. Naumovets and A.G. Fedorus,
     Sov. Phys. JETP 40 (1975) 903
82. D.A. Gorodetsky and Yu.P. Melnik, Surface Sci.62
     (1977) 647
83. L.A. Bol'shov, Sov. Phys. Solid State 13 (1971)1404
84. L.A. Bol'shov and A.P.Napartovich, Sov. Phys. JETP
     37 (1973) 713
85. G. Ertl and J. Küppers, Surface Sci. 21 (1970) 61
86. P.H. Kleban, Surface Sci. 83 (1979) L 335
87. P. Ducros, Surface Sci. 10 (1968) 118
88. G. Theodorou, Surface Sci. 81 (1979) 379
89. J.C. Buchholz, G.C. Wang and M.G. Lagally, Surface
     Sci. 49 (1975) 508
90. M.A. Van Hove and S.Y. Tong, Phys. Rev. Letters
     35 (1975) 1092
91. M.R. Barnes and R.F. Willis, Phys. Rev. Letters 41
     (1978) 1729
92. P.J. Estrup, J. Vac. Sci. Technol. 16 (1979) 635
93. G. Doyen, G. Ertl and M. Plancher, J. Chem. Phys.62
     (1975) 2957
94. R.A. Barker and P.J. Estrup, Phys. Rev. Letters 41
     (1978) 1307
95. M. Kramer and E. Bauer, submitted the Surface Sci.
96. E. Bauer, J. Japan. Assoc. Crystal Growth 5 (1978)49
97. G. LeLay, M. Manneville and R. Kern, Surface Sci.72
     (1978)405
98. R. Kern and G.LeLay, J. de Physique 38 (1977)C4-155
     and private communication
99. E. Bauer, H. Poppa, G. Todd and P.R. Davis, J.Appl.
     Phys. 48 (1977) 3773
100. F.C. Frank and J.H. van der Merwe, Proc. Roy. Soc.
     (London)A 198 (1949) 205, 216; A 200 (1950) 125;
     Disc. Faraday Soc. 5 (1949) 201
101. M. Schick, this conference
102. K.H. Lau and W. Kohn, Surface Sci. 65 (1977) 607
103. W. Kappus, Z. Physik B 29 (1978) 239
104. D.W. Bassett, Surface Sci. 53 (1975) 74
105. D.L. Kellogg and T.T. Tsong, Surface Sci.62 (1977)
     343
106. T.L. Einstein, Surface Sci.75 (1978) L161

107. A.M. Gabovich and E.A. Pashitskii, Sov. Phys.
     Solid State 18 (1976) 220
108. P. Johansson and H. Hjelmberg, Surface Sci. 80
     (1979) 171
109. N.R. Burke, Surface Sci. 58 (1976) 349
110. H. Gollisch and L. Fritsche, Z. Physik B 33 (1979)
     13
111. L.C.A. Stoop, Thin Solid Films 42 (1977) 33 and
     references therein; preprint
112. T.T. Tsong, Phys. Rev. Letters 31 (1973) 1207
113. D.W. Bassett and D.R. Tice, in E. Drauglis and
     R.I. Jaffee, edit., The physical basis of hetero-
     geneous catalysis, Plenum press, New York, 1975,
     p.231
114. H.-W. Fink, M.S. thesis, Clausthal 1979
115. G.L. Kellogg, T.T. Tsong and P. Cowan, Surface Sci.
     70 (1978) 485
116. D.W. Bassett, Thin Solid Films 48 (1978) 237 and
     references therein
117. R. Butz and H. Wagner, Proc. 7th Intern. Vac.
     Congr. & 3rd Internat. Conf. on Solid Surfaces,
     Vienna 1977, p.1289
118. J.G. Dash, Films on Solid Surfaces, Academic Press,
     New York 1975
119. G. Ertl, J. Vac. Sci. Technol. 14 (1977) 435; in
     L. Fiermans, J. Vennik and W. Dekeyser, edit.
     Electron and Ion Spectroscopy of Solids, Plenum
     Press, New York 1978, p. 144
120. D.P. Landau, in K. Binder, edit. Monte Carlo
     Methods in Statistical Physics, Springer, Berlin
     1979, p.337
121. H. Wagner, in Springer Tracts in Modern Physics,
     Springer, Berlin 1979, Vol. 85, p.151
122. A.M. Bradshaw and M. Scheffler, J. Vac. Sci. Technol.
     16 (1979) 447
123. L.D. Roelofs, R.L. Park and T.L. Einstein, J.Vac.
     Sci. Technol. 16 (1979) 478
124. A.R. Kortan, P.I. Cohen and R.L. Park, J. Vac.Sci.
     Technol. 16 (1979) 541

STUDIES OF LITERALLY TWO-DIMENSIONAL

MAGNETS OF MANGANESE STEARATE

Melvin Pomerantz

IBM T. J. Watson Research Center
Yorktown Heights
New York 10598, U.S.A.

I. Introduction

Phase transitions are fascinating physical phenomena. It is a remarkable fact that presumably short-range interactions between atoms can give rise to long-range order among them. The detailed explanation of such effects is a challenging problem in statistical mechanics. In order to simplify the problem and to provide test cases for theoretical methods, theoreticians have concentrated on model systems with low dimensionality; summations over one dimensional (1-d) or two-dimensional (2-d) spaces are usually easier than for 3-d. It has turned out that theories predict that low-dimensional systems will have special properties, rather unlike 3-d systems in many cases. The hope is that these theories can also be applied to the 3-d world. But, considering the simplifications of theories and the complications of Nature, one wonders if the theories are adequate. It seems that theorists have chosen to consider 1-d and 2-d models in order to make their work easier, but this has placed experimenters in the awkward position of trying to test the theories in a world that seems stubbornly three dimensional. Other lectures at this school have shown that it has proved possible to produce and study systems that closely resemble two-dimensional objects, in the form of films of atoms or molecules adsorbed on surfaces.

This lecture will describe the use of the Langmuir-Blodgett method for the preparation of materials which are *literally* two-dimensional magnets. The first question of interest is whether such a system makes a transition to long-range magnetic order. I will present some experimental results in Section III on the low temperature magnetic properties which address this question. In the remainder of this Introduction, I will review briefly the present theoretical results on the possibilities of magnetic ordering and some experimental results to provide background. In Section II, I will describe an experimental method, based on the Langmuir-Blodgett technique, by which 2-d magnets have been constructed. A variety of tests have been applied to the samples to verify that they are 2-d magnets; these will be surveyed briefly. Section IV is devoted to a discussion of the present experimental evidence for magnetic ordering in the 2-d Heisenberg magnet.

A convenient framework for the presentation of the results of the theory[1] of magnetic ordering follows from the hypothesis of universality[2]. This postulates that very few properties determine the critical points, or transitional behavior, of what seem like widely diverse physical systems. The properties that are believed to be important are (1) the range of the interactions between the particles (2) the dimensionality, d, of the spatial distribution of particles, (3) the dimensionality of the order parameter, n.

The range of interaction falls into two classes: long-range and short-range. Short-ranged interactions, it has been shown[3], are those for which the decrease in the potential is at least as rapid as $-\phi(r) \sim 1/r^{d+\sigma}$ with $\sigma > 2$. This would include exponentially decreasing interactions, such as exchange in magnets. The interaction is long-ranged in 1-d for $0 < \sigma < 1$, and in higher d, for $0 < \sigma < 2$. The important consequence of the range of interactions is that for long-ranged interaction or in the mean-field approximation, there is ordering in all spatial dimensionalities and it is independent of n. This result is illustrated in the bottom row of Fig. 1, as exemplified by the Weiss mean-field theory.

For short-ranged interactions, the possibility of ordering depends on the dimensionalities both of the structure and of the order parameter. This is illustrated in the upper three rows of Fig. 1. Spatial dimensionalities d = 1,2,3 are listed across the diagram and the numbers of spin components, n = 1,2,3, are listed downward. The diagram was drawn to help avoid confusion between the "dimensionality of space" and the "dimensionality of the order parameter" or worse, "spin space"; the important feature of the order parameter is better described as its symmetry. For spins with a single component order parameter the symmetry is axial; in magnetic terms it has an easy axis. A two-component order parameter has cylindrical symmetry; it has continuous symmetry and an easy-*plane* magnetically. A three component order parameter is spherically symmetric. Other familiar words describing the order parameter symmetry are "Ising" (n=1), x-y or planar (n=2) and "Heisenberg" (n=3). The first line in Figure 1 illustrates Ising models obtained by placing Ising spins on the sites of spatially 1-d, 2-d and 3-d lattices. Similarly we could draw 1-d x-y models, or 2-d x-y, etc. For each model, I have indicated the theoretical prediction on the possibility of a transition to long-range magnetic ordering. On the far right is an illustration of the expected behavior of the magnetic specific heats, C_m, for the Ising and Heisenberg models. When ordering occurs there is a discontinuity in either the slope or magnitude of C_m. It is predicted that none of the 1-d models can have magnetic order at T > 0 K. By contrast, all of the 3-d models can have magnetic ordering. In two dimensions, the possibilities depend on the kind of interaction. At one extreme is the Ising model in 2-d, for which Onsager[4] has produced a celebrated analytic solution which shows that there is a second order transition at a critical temperature $T_c = 2.27$ J/k, where J is the coupling constant in the interaction Hamiltonian $H = - J S_z^i S_z^j$, and k = Boltzmann's constant. The other extreme is the Heisenberg model where a proof has been given[5] that at T > 0 K long-range order is impossible. This theorem also forbids long-range magnetization for the x-y model in 2-d, but a state of bound magnetic vortices has been predicted[6] for this case. It has often been pointed out[6,7,8], but often is ignored[9], that

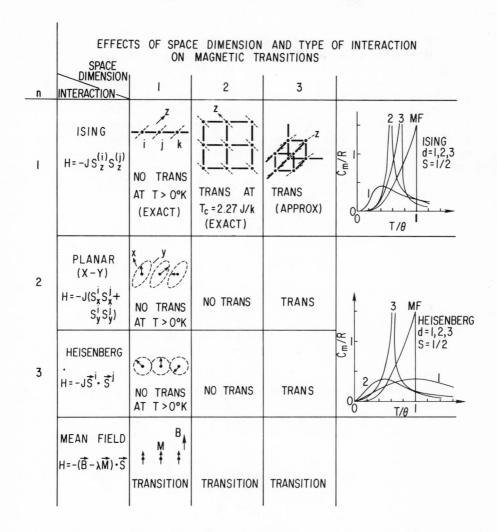

Fig. 1. Theoretical results on the possibility of transitions to long-range magnetic order at T > 0 K in model magnetic systems. Not all the structures for 2-d and 3-d are shown, but the 2-d x-y model is constructed by placing x-y interacting spins on the sites of the 2-d lattice, etc. The drawings of specific heat, C_m, *vs.* T illustrate that sharp anomalies are expected at transitions. The rows for n=1,2,3 assume short-range forces. Calculations of transition behavior in 3-d are approximate. The lowest row is for long-range forces or mean-field theories.

anisotropy can profoundly affect the possibility of ordering in the 2-d Heisenberg magnet. Thus the 2-d case is predicted to be the most varied and problematical.

It may be helpful to illuminate some of the physics behind these results by considering the Heisenberg model according to the spin-wave theory of F. Bloch. It was this work that originally raised the question about the effects of dimensionality on magnetic ordering. Bloch showed[10] that in a ferromagnet in which the coupling was by Heisenberg exchange of the form $H = -J\ S_i \cdot S_j$ that the excitations were spin-waves whose lower energies were $\varepsilon = D\sigma^2$, where D is proportional to JSa^2, σ is the wavevector and a is a lattice spacing. Each spin wave represented an excitation of one spin with reversed magnetization. With a ferromagnetic magnetization at T=0 K, M(0), all the spins were aligned. Bloch calculated the magnetization at T > 0 K by calculating the number of spin waves excited at T and subtracting that from M(0). The number of spin waves, N, is the integral over phase-space volume elements, dv_σ, of the probability that a spin wave of wavevector σ is excited. This is

$$N \ \propto \ \int_0^\infty \frac{dv_\sigma}{\exp{(D\sigma^2\beta)} \ - 1} \qquad\qquad 1.1$$

where $\beta = 1/kT$. Upon change of variables to $u^2 = D\sigma^2\beta$, this becomes

$$N \ \propto \ \left(\frac{kT}{D}\right)^{3/2} \int_0^\infty \frac{u^2 du}{\exp{u^2} - 1} \qquad\qquad 1.2$$

for 3-d space for which $dv_\sigma = 4\pi\sigma^2 d\sigma$. The integral converges, giving a constant. The result that the number of spin waves increases as $T^{3/2}$ (and M thereby decreases) is experimentally observed at low temperatures[11].

Bloch commented on the application of Eq. 1.1 to 2-d or 1-d lattices. In these cases, the volume elements are $2\pi\sigma d\sigma$ and $d\sigma$, respectively. These give integrals that do not converge at the lower limit. Bloch interpreted this as an instability of the magnetic structure: an infinite number of spin waves would be excited at finite temperature.

The physical reasons for this are two-fold: the energy of the spin wave varies continuously to zero, as σ^2. By contrast, if the energy did not go to zero, e.g., if it were of the form $\varepsilon = (D\sigma^2 + \text{constant})$, then even as $\sigma \to 0$ the integrals would converge and magnetic order could exist. Such a form for the spin wave energy arises if there is anisotropy or an external magnetic field. These create a gap in the spin wave spectrum, and there are no spin waves of infinitesimal energy. This points to the importance of anistropy in the Heisenberg model. It also suggests why there is order in the Ising model. The Ising interaction, $J\ S_z^i S_z^j$, has a built-in energy gap. The spin can only have components + 1/2 or -1/2. There is an energy change of J when a spin flips. There are no infintesimal excitations which can dissolve the order away at very low T. When the temperature is comparable to J/k, spins can flip and disorder occurs. Thus, the symmetry of the interaction is important because it permits, or prevents, excitations of infinitesimal energy.

The dimensionality is the other factor that influences stability. It enters via the element of phase space volume. At small σ, where spin waves might be excited at low

T, the phase space volume, dv_σ, is the larger, the smaller is the dimensionality. (For small numbers, $\sigma^2 < \sigma < 1$, which are the factors multiplying $d\sigma$.) Because the phase space is larger in 2-d than in 3-d, there are more spin waves excited in 2-d than in 3-d. This exemplifies the interesting feature of low-d spaces: they are breeding grounds for fluctuations, such as spin waves.

Experimental studies of the ordering in 2-d have been performed on "quasi" 2-d crystals in which magnetic ions are in planes that are relatively widely spaced. From these quasi 2-d crystals one can learn about interesting effects (e.g. fluctuations) in the approach to ordering. But the ordering, if it occurs, is almost aways to a 3-d magnetic structure. The one exception was in a crystal of $Rb_2Co_{0.7}Mg_{0.3}F_4$ where it was found[12] that the ordering was practically 2-d because there was correlation between no more than two planes. The anisotropy of Co places it in the Ising class.

To avoid the possibility of 3-d magnetic ordering, one can study *monolayers* of magnetic entities. A convenient system, O_2 adsorbed on grafoil has been measured by neutron diffraction[13], magnetic susceptibility[14] and specific heat[15]. These gave evidence for an antiferromagnetic structure in a monolayer. The range of order was limited, however, by the sizes of the uninterrupted surfaces of grafoil, to less than 100 nm. There was also a report[16] of a thermomagnetic effect due to a monolayer of O_2 adsorbed on gold, which was interpreted as caused by ferromagnetism of the oxygen. This needs further study because the temperature of the film was > 77 K, where O_2 is not a solid and where other studies indicated that it is paramagnetic, even as a monolayer. Thus, there has been at most one direct observation of long-range 2-d magnetic order: in the other cases either the ordering was three dimensional or it was of short range due to experimental conditions.

In the next section I will describe the Langmuir-Blodgett method for depositing molecular monolayers which was adapted to making literally 2-d magnets. If these have magnetic order it must be 2-d, by construction. It has proved possible to do experiments on macroscopic samples. The method has the advantage that a variety of different magnetic species can be utilized, in principle. Moreover, the structure is stable at room temperature, which is convenient. In particular, I shall describe how 2-d arrays, single atomic layers, of Mn^{+2} ions were prepared. Mn^{+2} was chosen because it has a closed $3d^5$ half-shell, and so should have no orbital angular momentum to first order. Spin-orbit coupling is thus very weak, and hence effects of anisotropy should be minimal. Compared to all the other transition-metal ions this should be the closest possible approximation to a 2-d Heisenberg magnet. Mn^{+2} has the advantage that its electron spin resonance, ESR, is observable at room temperature, which makes it easier to use this sensitive technique. It may seem fore-doomed, however, since the theory says that a 2-d Heisenberg magnet will not order, but that of course is what we wish to check.

It should be noted that a magnetic system for the study of 2-d effects has the advantage that in studying the *magnetic* properties the non-magnetic structure of the substrate or other surroundings are of no consequence for the dimensionality. The magnetic monolayer is suspended in a non-magnetic environment. In structural studies, by contrast, the 3-d substrate can strongly influence, or even determine, the monolayer crystal structure. Thus with magnetic monolayers, one can hope to achieve literally 2-d effects. Of course, by "literally 2-d" I do not mean the mathematical ideal of zero thickness and infinite extent in two orthogonal directions. A real material must have finite thickness and bounded area. The objective is to go to the physical limit: produce and study a material that is of macroscopic area and is exactly one magnetic atom thick.

II. The Langmuir-Blodgett technique for preparation of 2-d magnets.

The Langmuir-Blodgett (L-B) method is a well-known[17] technique for the deposition of organic monolayers. It is based on the simple principles that organic substances tend to dissolve in organic substances, and ionic substances in ionic, but that ionics do not tend to dissolve in organics. An interesting example of this is found in a family of molecules that have both organic and ionic parts, the so-called carboxylic acids, (HCH_2)-$(CH_2)_m$-COOH. Fig. 2a shows an example, with m = 16. They consist of a hydrocarbon chain (organic and hydrophobic) which is terminated at one end by a carboxylic acid, COOH. This end is ionic, indeed it is weakly acidic because the H tends to ionize in water, leaving behind a negatively ionized molecule. The acid end of the molecule is termed hydrophilic because it is attracted to water. Because of the duality of their attraction to water such molecules are called "amphiphilic". Which of their natures will dominate depends on the length of the hydrocarbon tail. We are all familiar with one of the shorter members of the series. The one for m=0, HCH_2-COOH, is acetic acid or vinegar. This we know is soluble in water; the ionic quality is dominant. In fact, for chains of length up to 12 carbons (m=10) the molecules are quite soluble in water. But at m=12 one finds a qualitative difference. Molecules spread on the surface of water remain on the surface. Thus, the hydrophobic quality is dominant for m > 12. The molecules used in our work have 18 carbons (m=16), (Fig. 2a), and are commonly known as stearic acid or more formally as octadecanoic acid, $C_{18}H_{36}O_2$. Stearic acid is readily soluble in organic solvents such as benzene, chloroform, and hexane. We use hexane (as the least noxious) and high purity stearic acid to make solutions of about 1 mg/ml.

The film-making procedes as illustrated in Fig. 2, b-g. Drops of this solution are placed on the surface of water and, if the surface area is large enough, the molecules disperse over the surface forming a dilute monolayer. The hexane evaporates leaving only the stearic acid (Fig. 2, b). A movable piston (a floating barrier) that traps the molecules can be used to compress them to a final area which Langmuir found[18] to be about 20 Å^2 per molecule (c). This corresponds to the cross sectional area of the chain when it floats perpendicular to the surface. If the molecules are further compressed, the film collapses. There is a region of surface pressure in which a film of low compressibility (solid) is stable. A typical surface pressure at this stage is about 20 dyne/cm, approximately the surface pressure applied by a barrier 20 cm. wide that is pulled by the weight of a paper clip. The contribution[19] of Katherine Blodgett was to show that the compressed film could be removed from the water by dipping a clean solid substrate through the film surface and pulling it out. If the substrates were hydrophilic, e.g. glass, usually no film was deposited on insertion (d), but if all went well, a film was deposited on the first retraction (e). Thus; nature, via Langmuir and Blodgett, has provided a simple means for depositing monolayers. The dipping can be repeated, and in fortunate cases multilayers can be built-up (f and g). The next step is to make them magnetic.

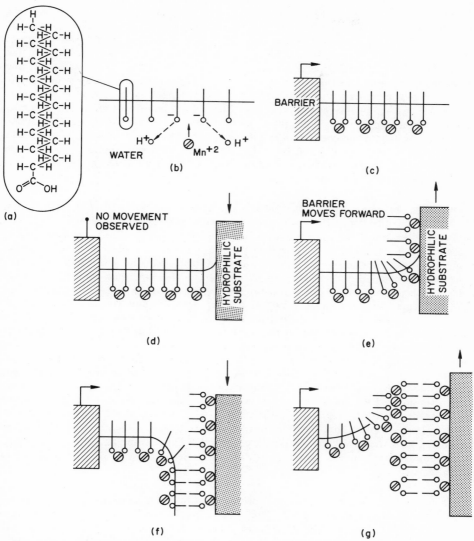

Fig. 2   (a) Structure of stearic acid, (b) - (g), the steps in producing 2-d magnets by
the Langmuir-Blodgett technique. In (b), stearic acid is spread on the water
containing Mn^{+2} ions. The ionized acid reacts with Mn^{+2} to produce $MnSt_2$.
(c) The monolayer is compacted. (d) A hydrophilic substrate (e.g., glass) is
inserted. No film is removed from the water. The barrier is stationary. If
the substrate were hydrophobic, (e.g. graphite) a layer would deposit, tail
toward the substrate. (e) Lifting the substrate, film is deposited. The barrier
moves forward. This produces a literally 2-d magnet. (f) Second layer is
deposited by reinserting the substrate. Layers attach tail-to-tail. The meniscus turns downward and barrier moves forward. (g) Third layer is deposited
on lifting the substrate. Many layers can be built up by repeating.

The L-B technique potentially has a solution for that problem, too. The carboxylic acid can be made to ionize in the water, and then positively charged ions can bind to the ends of the chains, as indicated in Fig. 2b. Such substances, metal ions bound to carboxylate chains, are also well known: they are soaps. To make a 2-d magnet it appears that one need merely attach magnetic ions (we used Mn^{+2}) to the chains and then deposit a monolayer of this film, as in Fig. 3a. Indeed there is yet a second way to form a 2-d magnetic structure: first deposit a non-magnetic soap film (e.g., cadmium stearate), and then two layers of magnetic film. Because of the principle of ionics attracting ionics, the two magnetic (ionic) ends become attached in a zipper-like fashion and a single surface of magnetic atoms is formed (Fig. 3b). The surface may not (or it may) be planar but it should still be 2-d. Multilayer films containing Cu have been prepared[20] in this way, with some complications. The ESR showed the anisotropy usually associated with Cu ions.

We must, of course, verify that such a procedure does work for Mn. 2-d or not 2-d, that is the question. First we must determine how to introduce Mn^{+2} into the films; secondly, find if the films will deposit; then show that the films remain as planar structures on the substrates, verifying that the Mn behaves as expected for a 2-d array. We must show that there is no Mn in 3-d inclusions. Finally, we will look for the occurrence of magnetic order in these monolayers. The fact that we wish to study single monolayers clearly restricts the types of experiments that can be utilized. In addition, because the monolayer is usually a much smaller quantity than the substrate, one must find substrates which do not interfere with the measurements. Thus, we are forced to use different substrates for different experiments. For example, the substrates used for magnetic experiments were suprasil quartz, because it has a sufficiently low concentration of background spins. For electron diffraction the substrates were standard grids supporting a film of 200 Å of graphite. For infrared transmission spectroscopy Si was used because it is transparent, whereas quartz is opaque in regions of interests in the I.R. In those cases in which the same measurements could be made with different substrates, e.g. x-ray diffraction, it was found that the films were identical. It is an assumption that the films are independent of the substrate, but this seems reasonable since the molecular entities are formed on the water and then are weakly bound (physisorbed) to the substrate. Moreover, the substrates were all amorphous materials.

Chemical composition: The first problem is to chemically bind Mn^{+2} to the carboxyl group of the stearic acid. This means that we wish to completely convert stearic acid to Mn stearate, Mn $(C_{18}H_{35}O_2)_2$, (abbr. $MnSt_2$). The preparation of L-B films with a variety of divalent ions has been described[17,21]. Generally, it was found that when the water bath was strongly acidic, no binding of metal ions occurred. Effectively, there is so much H^+ in the water that the stearic acid does not ionize. As the water is made less acidic, progressively more ions bind to the stearic acid. The limit on pushing this too far, and going to strongly basic baths, is that hydroxides of the divalent metals begin to form and precipitate. For our purposes, this would be deadly because we must avoid contamination by Mn that is in 3-d inclusions.

It was known[22] that myristic acid, the carboxylic acid with m = 12, spread on a water bath containing Mn^{+2}, reacted with the Mn^{+2} when the pH of the bath was greater than 5. Myristic acid films, however, collapsed at low surface pressure. We

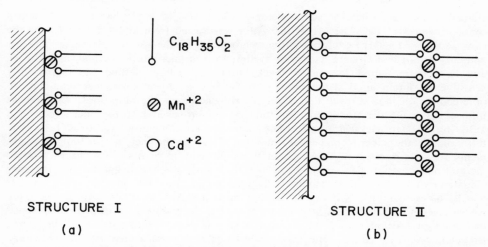

STRUCTURE I

(a)

STRUCTURE II

(b)

Fig. 3.   Literally 2-d magnets of (a) a single layer of $MnSt_2$ ("Structure I") (b) a layer of diamagnetic $CdSt_2$ and two layers of $MnSt_2$ ("Structure II").

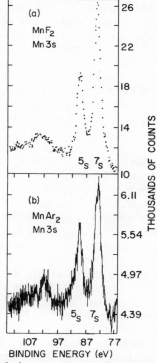

Fig. 4.   Mn 3s X-ray photoemission spectra of (a) MnF_2 from Ref. 25.   (b) Mn arachidate from Ref. 26.   The absolute values of the binding energies have been shifted into coincidence.   The important point is the near identity of the splitting between the ^5S and ^7S states in these compounds.

studied stearic acid films spread on water in which $10^{-3}M$ Mn^{+2} had been introduced by dissolving $MnCl_2$. The pH of the water was varied from about 5.3 to 7.5 by the slow addition of KOH. Multilayer films were pulled at the various pH values and the amounts of Mn and C were measured by the electron microprobe method, and Mn, C, and O were measured by nuclear backscattering. These methods are sufficiently sensitive to obtain good signal/noise with 50 layers. From these measurements we obtained the qualitative result that as the pH was increased, the ratio of Mn:C increased until at pH $\gtrsim$ 6.5 the atomic concentration ratio approached 1:36 (by microprobe) and 1:30 (by nuclear backscattering). The value for stoichiometric $MnSt_2$ is 1:36. The difference in the experimental values may be related to the fact that in both measurements there is damage to the films in ways difficult to assess. The backscattering method also gave at pH > 6.5 the ratio [O] : [Mn] = 4 which agrees with the stoichiometry of $MnSt_2$. Both methods detected no measureable amounts of K or Cl, which shows that little KSt or $MnCl_2$ are present. An important corollary of these measurements is that there is little *excess* Mn on the substrates, for example as trapped inclusions or precipitates. If this were not so, one would anticipate rather high, and irregular, values for Mn:C. But this was not observed in measurements on more than thirty samples.

We wanted to determine at an early stage if the Mn in the films was in its normal valence and spin states. The number of electrons and their energy levels have a profound influence on magnetism. It is unlikely, but conceivable, that the valence state might change in the bath. Also, the energy levels are affected by the crystal field in the film. The spin state of Mn^{+2} in weak crystal fields is ^6S, i.e., the five d electrons have parallel spin so that the total spin = 5/2. In strong crystal fields the d states may be split sufficiently that the electrons are redistributed such that a low spin state, S = 1/2, results[23]. This would indicate an unwelcome large anisotropy. Roger Pollak suggested that the spin state could be determined using ESCA. This is a favorable method since it senses only the upper 100 Å of a sample and is sensitive enough to measure a few monolayers.

The idea is to measure the spin state of the 3d electrons by its influence on the binding energies of 3s electrons. Suppose that an incident X-ray photon causes the emission of a 3s electron from Mn. The remaining 3s electron spin can couple with the 3d electrons either with spins parallel or anti-parallel. Due to *intra*-atomic exchange, the parallel state has the lower energy. Thus the energy needed to excite it is less; the binding energy of the emitted electron that leaves behind a parallel spin state is lower than if the anti-parallel state were excited. According to van Vleck[24], this exchange splitting of the binding energy is proportional to (2S+1), where S is the spin of the partially filled d shell. This method of measuring S has been verified in a series of Mn compounds, using ESCA. It was found that there is a characteristic splitting of about 6.6 eV for divalent Mn in weak crystal fields, (S=5/2). An example of this is MnF_2, whose 3s photo-electron spectrum is shown[25] in Fig. 4a. In Fig. 4b is the spectrum measured[26] in 8 layers on graphite of Mn arachidate. (The arachidate has hydrocarbon chains of 20 C, instead of 18 as in the stearate. The Mn binding energies should not be sensitive to small variations in chain length.). The splitting is nearly identical in (a) and (b), which shows that the Mn is divalent and the spin state is S=5/2. This means that the Mn in L-B films is in weak crystal fields: its isotropic, Heisenberg-like, nature is not grossly changed. Using ESCA we also searched for chloride impurities, but found none.

It proved very useful to study the infrared absorption in the films as a means of measuring the Mn concentration. The simplest technique is to prepare a film on the surface of the water and then skim a rod across the surface to crumple the film. The film is then picked up on a small piece of Si. It is dried in a dessicator to remove water. A surface of the order 100 cm^2 provides enough film to observe large absorptions by the CH_2 and carboxylate vibrations in the molecules. Any inorganic Mn salts trapped in the film do not contribute to the spectra, which are due to the vibrations of the organic molecule.

A series of spectra is shown in Fig. 5. In (a) are shown the absorptions of a skimmed film of stearic acid that had been on a bath of pH 6.5, but which contained no dissolved Mn. The sharp absorptions at about 3.5μ are characteristic of CH_2 vibrations. The absorption at about 6μ is due to C=O. This spectrum is also obtained[27] with stearic acid powder that has not been spread on water. In (b) is shown the spectrum of a skim from a bath of lower pH (5.8), but which contained 10^{-3} molar Mn^{+2}. It is found that, relative to the CH_2 absorption, the C=O is weaker and a new absorption at about 6.5μ has developed. It is known[28] that the formation of divalent metal soaps converts the double bond C=O to resonating single bonds C-O, and decreases the vibration frequency to the range around 6.5μ , depending on the attached metal ion. This absorption is thus a measure of the conversion of stearic acid to Mn stearate. In (c), in which the pH was increased to 6.8 with Mn concentration held at 10^{-3}M, one observes only the C-O absorption; the film is completely $MnSt_2$. From such measurements of the relative intensity of the C-O absorption peak compared to the CH_2 (which is unaffected by the reaction) we[29] found the fraction of Mn as a function of pH. This data is shown in Fig. 6. The I.R. measurements agree with the microprobe and nuclear backscatter in that the stoichiometry of $MnSt_2$ is reached when the pH exceeds about 6.5. The chemical measurements collectively show that the Mn is attached to the stearate film at the carboxylate group, and that there is little detectable excess Mn in other forms, e.g., inclusions of 3-d clumps. We also minimized the presence of precipitates by keeping the pH no larger than 7.2. The bath begins to discolor at higher pH, presumably because of the formation of $Mn(OH)_2$ precipitates.

It should be mentioned that IR spectra can be measured even of a single monolayer. An established method is total internal reflection.[30]. Recently, it was demonstrated[31] that infra-red surface waves propagating on the surface of a metal can be used to measure the spectrum with sensitivity to a monolayer.

Structure: We found that $MnSt_2$ films did deposit readily on clean substrates, as monitored by the movement of the surface barrier. The area of film leaving the surface was similar to the area of the plates being dipped. (Our apparatus did not permit a convenient measurement of the film area on the water, so that accurate measurements were not made.) The microprobe and nuclear backscatter measurements gave the concentrations of C on the substrates from which the area/molecule on the surface was derived. These came out to be typically 19 $\mathring{A}^2$ (microprobe) and 29 $\mathring{A}^2$ (backscattering). (The larger area/molecule from backscattering could arise from a loss of C due to the impact of the incident Mev He atoms). 19 $\mathring{A}^2$ is the known[17,18] sectional area across a hydrocarbon chain. The chemical density measurements thus indicate that the chains are packed like rods standing on end on the substrates, rather than lying down.

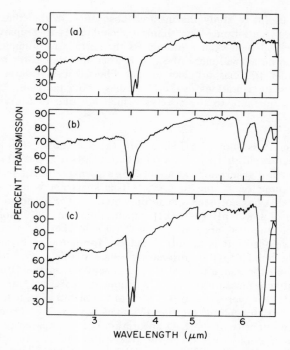

Fig. 5.   IR absorption by skims of stearic acid spread on water of (a) pH = 6.5, no
Mn (b) pH = 5.9, 10^{-3} molar Mn^{+2} (c) pH = 6.8, 10^{-3} molar Mn^{+2}.

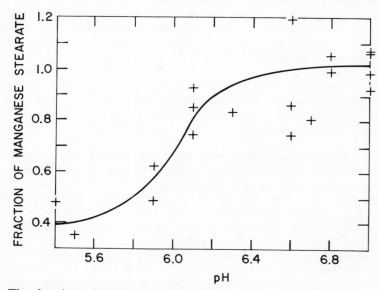

Fig. 6.   The fraction of $MnSt_2$ in skimmed films as a function of the bath pH,
determined from the amplitudes of IR absorptions of C-O ($6.5\mu m$) relative to
CH_2 ($3.5\mu m$).

The geometry of the in-plane packing was measured by electron diffraction. Electron microscope grids on which there was a 200 Å thick film of graphite bridging the grid openings were coated with L-B films of $MnSt_2$. On graphite, films deposit on the first insertion, (graphite is hydrophobic), so that a bilayer is formed on the first cycle. The films were examined in the electron microscope by passing the e-beam perpendicularly through the film. This gives information about the in-plane structure. Diffraction patterns were observed from a single bilayer and multiple bilayers. Even at the lowest practical exposures, the e-beam destroys the crystal order in about a minute. Nevertheless, adequate photographs could be made. An example is shown in Fig. 7. From such photographs, S. Herd[29] has deduced that the unit cell is either orthorhombic or monoclinic, with cell spacings $a_0 = 4.85$ Å, $b_0 = 7.73$ Å. This result is similar to that found for Pb stearate[32]. It was concluded that in $PbSt_2$ the Pb are located in planes, at the centers of the sides of the rectangular cell given by a_0 and b_0. Similar conclusions should apply to $MnSt_2$. This results in a nearest neighbor Mn-Mn distance of 4.56 Å. The space group is either $P112_1/b$ (monoclinic) or $P2_12_12$ (orthorhombic).

The observation of spots, albeit that they are broad, shows that there is average crystallinity across the area of the e-beam, $(10\mu m)^2$. Another observation made with the electron microscope was that the diffraction pattern in a given grid opening did not vary much in orientation as the grid was moved across the electron beam. This means that the crystallinity was about the same over the grid opening, a distance of about $80\mu m$. The significance is that eventually we will be seeking long-range magnetic order and thus long-range crytallographic order, or at least continuity, is required. A much larger limit on the area of continuity was established using electron tunneling. We have prepared films of area $1cm^2$ through which electron tunneling was observed[33]. This means that the relative area of metallic shorts was miniscule; there is film continuity over macroscopic distances.

Another feature of $MnSt_2$ films revealed by the electron microscope is that a second bilayer usually does not register with the first. We usually observe two sets of diffraction spots rotated from each other. With multilayers of many layers the pattern is a ring with some highlights, indicating a small degree of registry.

For our purposes it is important that the films remain attached to the substrates as planar layers. There have been reports[34] of L-B films forming into multilayers after deposition. In collaboration with A. Segmüller and F. Dacol, we have developed[35] X-ray methods for studying the layer structure of L-B films down to a single monolayer. The experiments were performed on an X-ray diffractometer in which the beam is monochromatized by diffraction from dislocation free Ge, and collimated by a series of slits. This confers sufficient resolution and parallelism that observations of the diffracted X-ray beam can be made to within 0.1° of the incident beam. Examples of this are shown in Fig. 8. In (a) the dots are experimental counts of the diffraction from one monolayer of $MnSt_2$ on a glass microscope slide. There is a single peak; this is not present on bare glass. The diffraction from a 9 layer film is shown in Fig. 8 (b). The maxima at grazing angles of $\phi = 0.85°$, $1.82°$, and $2.70°$ are the Bragg peaks corresponding to a lattice spacing $d = 49$ Å. This agrees with the expected unit cell size resulting from $MnSt_2$ molecules attaching ionic-to-ionic and organic-to-organic ends, and standing nearly perpendicular to the substrate. (The length of single stearate chain is about 25 Å and the unit cell is two such chains end to end.)

Fig. 7.   Electron diffraction from a 2 layer film of $MnSt_2$ produces the pattern of spots. The ring pattern is from amorphous Al, used for calibration. Courtesy of S. Herd.

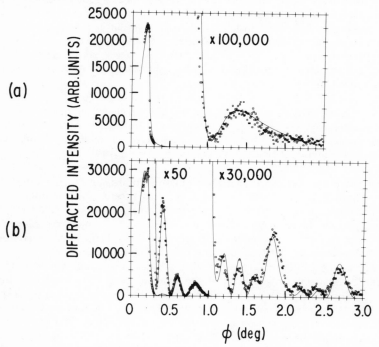

Fig. 8.   X-ray diffraction from $MnSt_2$ layers on glass: (a) one layer (b) nine layers. ϕ is the grazing angle in a ϕ, 2ϕ scan. The solid lines were calculated using a model described in the text.

The additional peaks in (b) are interference effects from the finite number of layers[36]. This effect is familiar from optical diffraction gratings with small numbers of slits, which have weak subsidiary maxima between the principal diffraction orders. In the case of L-B films, the "subsidiaries" can be larger than the principal peaks at low angles. The physical reason is that the X-rays for these low angle Bragg peaks are close to the critical angle for total internal reflection from the solid back into air. (The refractive index is smaller for solids than for air at X-ray frequencies.) Thus, the reflected beams are intense and their interference produces large variations in the reflected intensity.

A remarkably acccurate quantitative fit to the observed diffraction patterns can be obtained by computing the interference of X-rays reflected from a model structure consisting of uniform laminae corresponding to the various chemical constituents of the unit cell. The successive strata represent Mn, COO, $(CH_2)_{17}$, H, repeated for each deposited layer. The thicknesses are known from other experiments, and refractive indices were determined from our measured density of molecules on the surface. The interference patterns from such multi-laminar coatings, including the substrate, are shown as the solid lines in Fig. 8a, b. With minor adjustments of the parameters, a fit to the diffraction pattern that is within a factor of 2 over five orders of magnitude can be obtained[37]. The usefulness of these measurements is that different numbers of layers produce qualitatively different interference patterns. Thus one can verify the actual average number of layers on the substrate by this method. It has also proved possible[38] to observe the layer structure by neutron diffraction, which can give additional structural information.

Paramagnetic Properties: The chemical and structural studies show that L-B films can be prepared in which Mn is bonded to the COO^- ends of the molecules, and that these molecules can be deposited as monolayers or multilayers. Various tests were made, some of which were described above, by which Mn that was *not* in the 2-d arrays might be detected if it were sufficiently abundant. Mn that is in 3-d polycrystals picked up from the bath would be expected to be isotropic and not oriented with respect to the substrates. The Mn that is in a 2-d array should have characteristic anisotropic magnetic properties; it is these that I will describe next. Observation of these properties confirms that the Mn ions are predominantly in a 2-d array.

The experimental technique used was electron spin resonance[39] (ESR) because it has high sensitivity and it can reveal a variety of magnetic properties. Most of the experiments I shall describe were performed on a Varian 4500 X-band spectrometer with a cylindrical cavity operating in the TE_{011} mode. The experiments at 35 GHz (Q-band) were done on a Varian E-9 spectrometer also using a cylindrical cavity. In order to achieve sufficient signal/noise in the literally 2-d case, films were deposited on several 0.01 cm thick suprasil quartz plates. As many plates as possible were stacked with bits of 0.003 cm thick mylar as spacers. In the case of X-band measurements as many as 70 plates, 0.8 cm x 2.5 cm, were used. At Q-band ten plates, 0.3 cm x 0.8 cm, were possible. The temperature was measured by a calibrated carbon resistor attached to the cavity.

It had been observed by Boesch, et al[40], and Richards and Salamon[41] that in quasi 2-d crystals the line width, ΔH, of the ESR depends on the angle, θ, of the magnetic field with respect to the normal to the planes of the magnetic ions. The observed dependence,

$$\Delta H = A(3 \cos{}^2\theta - 1)^2 + B \qquad\qquad 2.1$$

was explained by Richards and Salamon by a mechanism which depends crucially on the dimensionality of the spin system. The origin of the line width is the dipole-dipole interaction among the spins. But the dipolar interaction in concentrated magnetic systems is affected by exchange interaction in two important ways. Exchange flips spins and thus modulates the dipolar interaction, leading to "exchange narrowing". It also leads to a diffusive flow of spin correlations. This spin diffusion has the effect of emphasizing particular components of the dipolar interaction. The extent to which this happens, and which component is emphasized, depends on the dimensionality, because diffusion itself depends on dimensionality. Richards and Salamon showed that the (3 $\cos^2\theta$ -1) dipole component was emphasized by spin diffusion in 2-d. In Fig. 9 is plotted the angular dependence of the line width of a sample of Structure II films (cf. Fig. 3), deposited on fifty separated plates. The fact that these data can be fitted by Eq. 2.1, as shown by the solid curve in Fig. 9, is strong evidence for a 2-d magnetic array.

Another ESR observation that was explained by the dimensionality was the angular dependence of the magnetic field at the peak of the absorption, H_0. A non-cubic distribution of dipoles can contribute an average field which can shift the resonance field. In particular, dipoles in a 2-d array contribute[41] an average field at a site i given by

$$H_{di} = \mu_z \sum_j (3 \cos{}^2\theta_{ij} - 1)r_{ij}^{-3} \qquad\qquad 2.2$$

where

$$\mu_z = \gamma^2 \hbar^2 S (S + 1) H_o/3k (T - \Theta) \qquad\qquad 2.3$$

and θ_{ij} = angle between H_0 and the position vector r_{ij}, γ = gyromagnetic ratio, k = Boltzmann's constant and Θ = Weiss temperature. These small demagnetization shifts are difficult to measure because the lines are wide and weak but data obtained on a sample of 35 multilayers (hence quasi 2-d) on a single plate, measured at 34.8 GHz and 80 K are shown in Fig. 10. Using Eq. 2.2, we calculated approximately the expected values of H_0, which are shown as the solid curve in Fig. 10. The value of Θ = -5 K used here will be justified below. The agreement, both as to form and magnitude, confirm that the paramagnetic behavior is two dimensional.

A qualitative explanation of this angular dependence of H_0 will be helpful to those not familiar with spin resonance, and will be important in understanding the results on the ordered magnetic state. The basis of ESR is that when a magnetic moment is in a magnetic field of magnitude H_m, it can absorb electromagnetic energy of frequency f, where f = $(\gamma/2\pi)H_m$. γ is a constant (a tensor, generally) that depends on the kind of atom and its environment. I have written H_m to emphasize that the magnetic moment responds to the local field it feels, which is a combination of external fields that may be

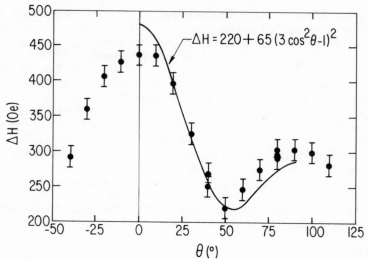

Fig. 9.   The measured ESR linewidth, (peak to peak of the derivative) *vs.* θ, the angle between H_0 and the film normal, of 2-d $MnSt_2$ of Structure II. The solid curve is the form $A + B (3 \cos^2\theta - 1)^2$ fitted to the data. T=80 K. f=9.3 GHz.

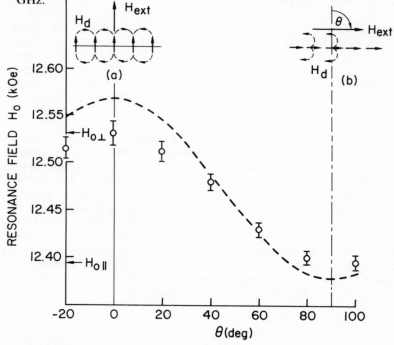

Fig. 10.   The ESR field, H_0, *vs.* θ, of 35 multilayers of $MnSt_2$. The dashed curve is Eq. 2.2, with $\Theta = -5$ K, and an approximation to $\Sigma\, r^{-3}$. T = 80 K. f = 34.8 GHz.

applied and internal fields. This is illustrated for a 2-d paramagnet of Mn^{+2} ions in the insets of Fig. 10. We applied a frequency of 34.8 GHz and the magnetic moments will absorb this frequency when the field on them is 12.43 kOe; they do not care where the field comes from. When the external field is applied perpendicular to the plane of spins, there is a small net alignment of dipoles in this direction. In Fig. 10, inset a, the moments are shown all aligned, for simplicity. The fields, H_d, of the dipoles on each other are opposite in direction to that of the external field, H_{ext}. The total field acting on the magnetic moments, H_m, $= H_{ext} - H_d$. Thus, in order for H_m to be 12.43 kOe, H_{ext} must be larger than this by an amount H_d, (given by Eq. 2.2). When H_{ext} is parallel to the plane, the net dipolar orientation is such that the dipolar fields add to the external field so that H_{ext} is less than H_m needed for resonance (cf., inset b of Fig. 10).

Shifts in the resonance field are important for this work because they result from internal fields. In a magnetically ordered state, the average internal field may be large and there may be large shifts of the resonance fields. This makes ESR a useful method for the detection of magnetic ordering.

III. Low Temperature ESR

To search for magnetic order[42] one reduces the temperature of the sample and looks for changes in the ESR signal. The results I shall describe in greatest detail were obtained on the sample used to obtain Figs. 9 and 10. It comprised 50 plates, each with 1 layer of $CdSt_2$ and 2 layers of $MnSt_2$ on each face (Structure II films of Fig. 3). I refer to it as "2-d $MnSt_2$". The plates were stacked in a quartz holder and placed in the ESR microwave cavity (9.3 GHz). A closed stainless steel can was placed around the cavity so liquid helium did not enter the cavity. This prevents the bubbling noise that He causes when it is inside a cavity and is not superfluid. (It turned out that this region of temperature, 2 K to 4 K, was of interest.) We also found that better signal/noise was obtained when the direct ESR absorption was recorded, rather than the usual derivative of the absorption.

In Fig. 11 is plotted the full width at half maximum of the absorption line, as a function of temperature. The data show characteristic behavior of the resonance line of a 2-d magnet approaching an antiferromagnetic (AF) transition. These characteristics are the decrease in anisotropy (i.e., the difference between the line widths with H parallel and perpendicular to the plate), and the increase of the widths themselves below a certain temperature. Similar variations were observed by Richards and Salamon[41] above the AF transition of K_2MnF_4, a quasi 2-d crystal. The line broadening was explained by them on the basis of increasing fluctuations at the antiferromagnetic wavevector. At T_N, the Neel temperature, this wavevector freezes into the AF state. These data do not agree with all the predictions of the Richards-Salamon theory, but not all experiments on quasi 2-d materials support the theory in all details. Of course, their theory assumes a specific Hamiltonian which may not be appropriate for all materials.

Another indication of a transition from the paramagnetic state is the change in the relative magnitudes of the resonance fields, when the field is perpendicular or parallel to the films, $H_{0\perp}$ and $H_{0||}$ respectively. I explained in Sec. II how the dipolar fields in the paramagnet caused $H_{0\perp}$ to be greater than $H_{0||}$ (cf. Fig. 10). As the temperature

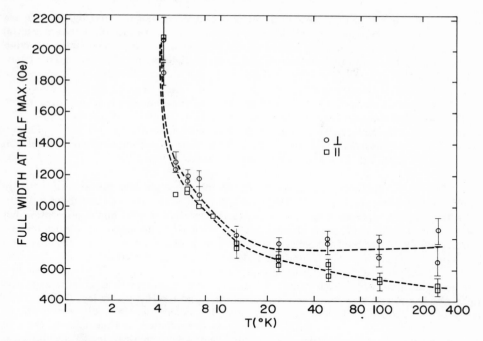

Fig. 11. ESR linewidth (full width at half max. of absorption), of 2-d MnSt$_2$ *vs.* temperature. $\perp$ and $||$ refer to the direction of the external field with respect to the film plane. f = 9.3 GHz.

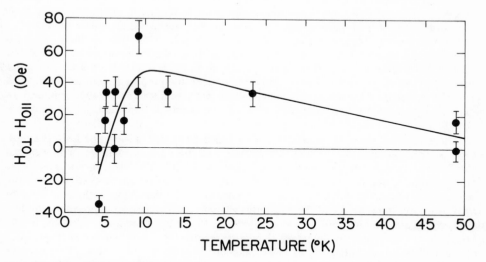

Fig. 12. The difference between the resonance fields $\perp$ and $||$ to the films *vs.* temperature for 2-d MnSt$_2$. f = 9.3 GHz.

is decreased it is found that, instead of steadily increasing as predicted by Eq. 2.3, the difference $H_{0\perp} - H_{0||}$ begins to decrease near 10 K and eventually changes sign at about 5 K, as shown in Fig. 12. This is a clear departure from paramagnetic behavior. The quality of the data precludes making precise calculations, but the qualitative change is unmistakable.

To give an appreciation of the primary data in the paramagnetic phase, I reproduce in Fig. 13a experimental absorption curves with H $\perp$ and H $||$ to the films. The absorption peaks are very similar and centered near $g = 2$. On the scale of fields shown here the difference $H_{0\perp} - H_{0||}$, of order of tens of oersteds, is barely visible. It is of the sign expected for a paramagnet; the temperature was 17 K.

When the temperature is reduced to 1.5 K, the measured absorption curves change drastically, as shown in Fig. 13b. There remain small peaks, centered at about 3.3 kOe, as in the higher temperature spectra, but with $H_\perp$ the largest part of the absorption has shifted downfield and peaks at about 1.5 kOe. With $H_{||}$ the maximum absorption has shifted downfield to about 2.5 kOe. The shifted peak of absorption were measured as a function of angle at T = 1.4 K, and the field at the peak is plotted in Fig. 14. One sees that the resonance field is not a monotonic function of the angle. It passes through a maximum at about 60°. At no angle, however, is the resonance field as large as it is in the paramagnetic state, 3.3 kOe = ω/γ. Thus the trend for $H_{0\perp}$ to be less than $H_{0||}$ as temperature decreases, first observed in Fig. 12, culminates in a difference $H_{0\perp} - H_{0||} \approx$ - 1 kOe at low temperature. The angular dependence, and particularly the sign of the shifts in H_0, cannot arise from paramagnetic demagnetization. They are clear signs of a magnetically ordered state. The small, unshifted peak is presumably due to poorly formed film or perhaps some Mn impurity.

Another characteristic of magnetic ordering is a relatively rapid onset as the temperature is decreased. The measured temperature dependence of the shift of $H_{0\perp}$ is shown in Fig. 15. I have plotted the difference of $H_{0\perp}$ in the paramagnetic region, $H_{0\perp}$ at 77 K was chosen, from its values at low temperatures. This, then, is the down-field shift of $H_{0\perp}$. One sees that the shift begins gradually as the temperature decreases, but at about 2.0 K there is a shift of about 1 kOe within a few tenths of a K. In other samples the shift is not so abrupt, but in all carefully prepared samples there is a relatively rapid shift in the resonance position with temperature. The temperature dependence of the shift is another indication of a cooperative transition.

There are two possible mechanisms that can cause large and abrupt ESR changes: one is a structural transition which affects the ESR because of changes in the crystal field[23]; this would not require ordering of the magnetic system. The second possibility is a transition to a state of magnetic order. The most direct method of determining magnetic structures is neutron scattering. Attempts to observe magnetic scattering of neutrons in bulk powders of $MnSt_2$ were unsuccessful.[43] The reason is probably the low density of magnetic atoms in the sample. Even the improvement afforded by totally deuterating $MnSt_2$, thereby reducing the large incoherent scattering by protons, does not seem likely to be sufficient to make magnetic scattering observable. The neutron diffraction did, however, give information on the possibility of a structural transformation. Bragg scattering from the nuclei was observed both above 77 K and at 1.9 K. No changes in the lattice spacings were discerned between these temperatures. Thus, there

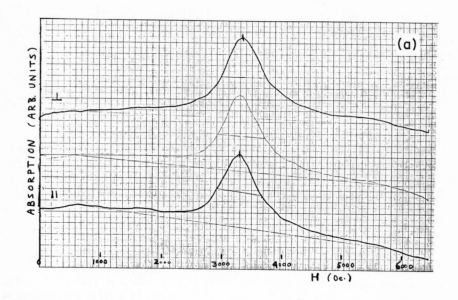

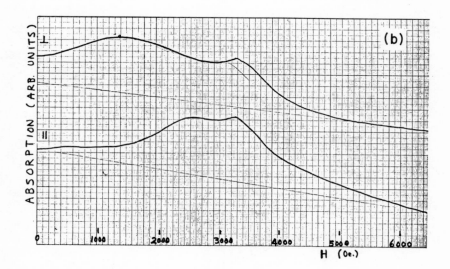

Fig. 13. Experimental graphs of ESR absorption of 2-d $MnSt_2$ $vs.$ external fields, $\perp$ and | | to the film, at (a) 17 K (b) 1.5 K. f = 9.3 GHz.

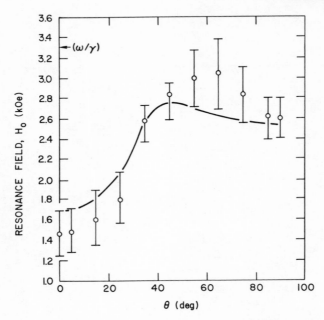

Fig. 14. Angular dependence of resonance fields of 2-d $MnSt_2$, at $T = 1.4$ K . The solid curve is a best fit to the theory of resonance in a weak ferromagnet in which **D** is parallel to H_K, and both are normal to the film.

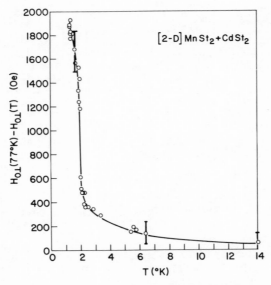

Fig. 15. Temperature dependence of the down-field shift of $H_{0\perp}$,the peak of ESR absorption in perpendicular orientation, for 2-d $MnSt_2$. $f = 9.3$ GHz.

are no large changes in the structure of powders of $MnSt_2$. It is unlikely that small distortions of the structure could cause the observed magneto-crystalline anisotropy, considering that Mn^{+2} is an S state ion. The structure of films of $MnSt_2$ has not been measured below 2 K, but it is likely to be similar. Thus, the evidence does not support the hypothesis of a structural transition. We are lead to seeking an explanation based on a transition to magnetic order.

Other experiments on bulk powder gave insight into the kind of order that might occur in films. These were magnetic susceptibility, χ, by the Faraday balance method[44] and a. c. methods[45,] and ESR[44]. The Faraday balance measurements showed linear dependence of M $vs.$ H, and straight line Curie-Weiss plots of $1/\chi$ vs T for 25 K $<$ T $<$ 295 K. The Weiss temperatures varied among various samples tested, but were of the order of -50 K, i.e., antiferromagnetic. At low temperatures, M vs H showed a steepened rise at low H and linearity at high H. The measured χ was greater than extrapolated from the Curie-Weiss plot. The ESR of powders showed line broadening and downfield shifts of H_0 at temperatures below 10 K. A. C. methods did not reveal any anomalies in χ down to about 2 K. Thus, at least some measurements on powders indicate antiferromagnetic interaction but some deviation from this at low temperature. The measurements of Haseda, et.al.,[45] indicated that ordering occurred at 0.5 K, rather lower than the ESR shifts. This difference is unexplained at present.

The temperature dependence of the magnetic susceptibility can also be measured from the ESR of films. The strength of the ESR signal depends on the magnitude of the magnetic moment that is absorbing energy. More precisely, the area under the ESR absorption curve is proportional to the magnetization and thus to the susceptibility in a fixed field, H_o. To get an idea of the dependence of χ on T in films, we measured roughly the area of the ESR absorption as a function of T on a multilayer sample of 120 layers. The results, shown in Fig. 16 (a) are an approximately Curie-Weiss behavior at high T, but at low temperatures, as shown in (b), there is an enhanced χ ($1/\chi$ is reduced) compared to the extrapolated Curie-Weiss behavior. Note also that the intercept is at (small) negative temperature, corresponding to antiferromagnetic interactions. The value of $\Theta \approx$ -5 K indicated by Fig. 16 b was used to compare Eq. 2.3 to the data of Fig. 10. A slightly larger negative value of Θ would improve the fit.

Thus there are numerous measurements which indicate that the interaction is dominantly antiferromagnetic: the good fit of Eq. 2.2 to the data of Fig. 10, using $\Theta <$ 0; the increase of the ESR line width at low temperatures; the negative Weiss temperatures found from the temperature dependences of the susceptibilities of powders and multilayers. Note also that the usual interaction between Mn ions is antiferromagnetic. We can also readily show that the contrary assumption, ferromagnetic interaction, leads to contradiction with the experimental results shown in Fig. 14. Kittel showed[46] that demagnetizing effects in films could produce large shifts in H_0 if the films were ferromagnetic. For example, with H perpendicular to the plane of a film, the resonance field is increased by $4\pi M$, where M is the magnetic moment per unit volume. The concept of magnetization, M, is not appropriate here because this 2-d magnet has no volume; there is no "inside". We can easily predict qualitatively the effect of ferromagnetic ordering by the same argument that explained the angular dependence of H_o, Fig. 10. If the magnetic dipoles were ordered ferromagnetically, as shown in Fig. 10, inset a, the dipoles would produce fields upon each other which oppose the exernal field in

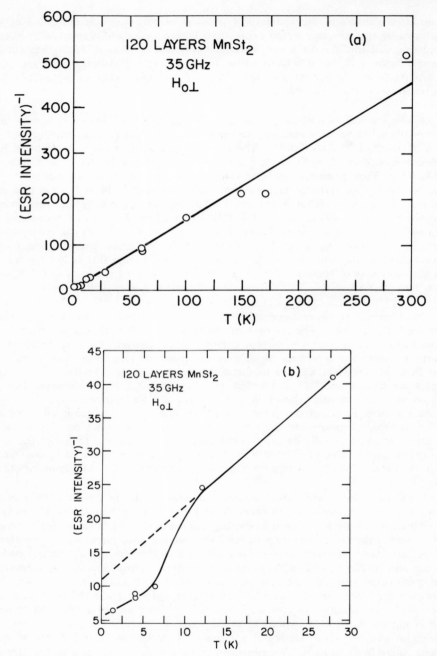

Fig. 16. "Curie-Weiss" plot of the inverse ESR intensity (proportional to χ^{-1}) *vs.* temperature for a sample of 120 multilayers of $MnSt_2$ on one plate. f = 35 GHz. (a) 2 < T < 300 K (b) Expanded view of region 2 < T < 25 K shows deviations from linear dependence.

the perpendicular orientation. Thus, a field *larger* than $H_o = \omega/\gamma$ is needed to produce sufficient field at the spins to give resonance. This is opposite to the observed field shift, which is to much lower fields in the perpendicular orientation. (Fig. 14). Thus, the direction of the shift of the ESR is incompatible with ferromagnetism. We return to antiferromagnetism in seeking an explanation of Fig. 14.

The difficulty with an explanation based on pure antiferromagnetism is that, below the Neel temperature of an AF, χ should decrease whereas the data of Fig. 16b showed that χ increased at low temperature. This suggests ferromagnetic ordering. An explanation that we advanced[42,44] is that MnSt$_2$ is a "weak-ferromagnet" (WF). A WF is dominantly an antiferromagnet in which the spins in opposing sublattices are canted with respect to each other to produce a small net magnetic moment. The interaction that produces canting is of the form of anti-symmetric exchange, $D \cdot M_1 \times M_2$. It was proposed as an explanation of WF by Dzyaloshinsky[47], and calculated quantum mechanically by Moriya.[48] The presence of such an interaction in an AF causes the moments of opposing sublattices to turn toward each other producing a small ferromagnetic, FM, moment. This is able to account for the fact that the susceptibility below T_N is larger than in the paramagnetic or the pure AF states.

Dzyaloshinsky showed that only in certain low symmetry crystals could an interaction of the form $D \cdot M_1 \times M_2$ not vanish by symmetry. I checked that the two possible crystal structures found for the films in Sec. II are compatible with WF. Following the procedure used by Borovik-Romanov[49], one considers the effect of performing all the symmetry operations of the structure, including time reversal, on a proposed magnetic structure. WF is possible if there is a set of symmetry operations that transforms *both* an AF and a FM structure into themselves. I find that in the $P2_12_12$ structure that the AF arrangement with moments in the c plane, pointing alternately in the + b and -b directions has the same transformation properties as a FM arrangement with spins in the c plane, but all pointing in the a direction. Thus, there is a set of operations under which both the FM vector along the a axis $F = M_1 + M_2 = F_a$ and the AF vector along the b axis $A = M_1 - M_2 = A_b$ are unchanged. The Dzyaloshinsky interaction can be written $D \cdot F \times A/2 = D \cdot F_a \times A_b/2$. Thus D must have a component in the c direction, which is normal to the films in order to produce WF. It is not clear if the other suggested space group, $P112_1/b$, could be formed in a bilayer of MnSt$_2$ because that structure has ions in planes separated by c/2, but we do not deposit two planes of Mn ions. However, the $P112_1/b$ structure also is found to be compatible with several WF arrangements. Thus, I find that the symmetries of both structures are compatible with D normal to the film.

Spin resonance has been observed in a number of WF compounds; the study of KMnF$_3$ by Saiki[50], and Yoshioka and Saiki[51] was of particular interest. They measured the anisotropy of the resonance field and found a non-monotonic angular dependence, reminiscent of MnSt$_2$ (cf. Fig. 14.). They explained their data by a model in which the anisotropy field, H_K, and D were parallel. Eq. 3.1 is the result of their calculation for the resonance frequency, ω, as a function of external field H.

$$\left(\frac{\omega}{\gamma}\right)^2 = (3\cos^2\phi - 1)H_E H_K - H_D{}^2 + \frac{1+\alpha^2}{2}H^2 - \left(\frac{1+\alpha^2}{2} + \frac{-H_E H_K + H_D{}^2}{H_{\|c}^2}\right)H^2 \sin^2(\phi+\Theta)$$

$$-\frac{1-\alpha}{\alpha}H_D{}^2 \sin^2\phi \pm \left[(1+\alpha)^2\left(H^2\cos^2(\phi+\Theta)\right.\right.$$

$$-\frac{H_D{}^2}{\alpha^2}\sin^2\phi\right)\left\{2H_E H_K - H_D{}^2 + \frac{H_E H_K - H_D{}^2}{H_{\|c}^2}H^2\sin^2(\phi+\Theta)\right.$$

$$+\left(-3H_E H_K + \left(\frac{(1-\alpha)^2}{2\alpha^2}+1\right)H_D{}^2\right)\sin^2\phi\right\}$$

$$+\frac{1}{4}(1-\alpha^2)^2\left(H^2\cos^2(\phi+\Theta) - \frac{H_D{}^2}{\alpha^2}\sin^2\phi\right)^2$$ \hfill (3.1)

$$+\left\{\frac{H_E H_K}{H_{\|c}^2}H^2\sin^2(\phi+\Theta) + \left(\frac{(1-\alpha)^2}{2\alpha^2}H_D{}^2 - H_E H_K\right)\sin^2\phi\right\}^2$$

$$+4H_D{}^2 \sin^2\phi\left\{\frac{(1-\alpha)^2}{4\alpha^2}H_D{}^2 + \frac{1-\alpha}{2\alpha}H_E H_K\right\}\left\{-\frac{H^2}{H_{\|c}^2}\sin^2(\phi+\Theta)\right.$$

$$\left.\left. +\frac{1}{\alpha}\sin^2\phi\right\}\right]^{1/2}.$$

The parameters in Eq. 3.1 are the exchange field $H_E = JM$, Dzyaloshinsky field $H_D = DM$ (where M is the sublattice magnetization), $H_{\|c}{}^2 = 2H_E H_K - H_D{}^2$, $\alpha = 1 - H_E \chi_{\|}/M \approx (1-T/T_N)$ (where $\chi_{\|}$ is the susceptibility parallel to M), θ = angle between H and H_K, and ϕ is the solution of $H_{\|c}{}^2 \sin 2\phi = \alpha H^2 \sin 2(\phi + \theta)$ when M is not normal to the easy axis. To see if such a model could explain our observations, the data points of Fig. 14 were fitted to Eq. 3.1 using a multi-parameter fitting program[52]. The best fit is shown by the solid curve in Fig. 14. The parameters which gave the fit are $H_E H_K = 3.6$ k0e^2, $H_D = 1.6$ k0e and $\alpha = 0.34$. It is difficult to get independent values of H_E and H_K. The value of H_D for MnSt$_2$ falls between the values of 1.1 k0e in KMnF$_3$[51] and 4.4 k0e in MnCO$_3$[53]. The value found for α, 0.34, is to be compared to $1-T/T_N$. If $T_N = 2.0$ K is taken from Fig. 15, then the computer says the measurement was made at $T = 1.3$ K, whereas it was actually made at $T = 1.4$ K, which is good agreement. The conclusion is that the model of weak ferromagnetism with D and H_K normal to the films gives a fairly good quantitative fit to the anisotropy of H_0, with a value of H_D that is comparable to that measured in other Mn compounds. In addition, the temperature dependence of H_0, Fig. 15, is similar to that of known weak ferromagnets such as KMnF$_3$[50] crystals and MnCO$_3$ powder[54]. It is observed that the resonance field decreases with temperature; it does not disappear. Thus the model of weak ferromagnetism can explain all the observed properties of the ESR at low temperature: intensity (i.e., χ) vs. T, H_0 vs. T, and H_0 vs. θ, quantitatively with reasonable parameters or qualitatively.

A difficulty with the explanation based on a WF is the large ratio of H_D/H_E. A value of $H_E = 25$ k0e was obtained by Ferrieu[55] from analysis of the ESR anistropy in the paramagnetic state. This gives $H_D/H_E = 0.06$. The theory of Moriya[48] predicts that $H_D/H_E \approx \Delta g/g_0$, where Δg is the difference between the actual g value and the free spin g_0. For Mn compounds this is typically $\Delta g/g_0 \approx 10^{-3}$, and I find $\Delta g/g < 0.02$ for MnSt$_2$. There is a numerical discrepancy here. Also, in analyzing the anisotropy of H_0 I have not included the influence of the resonance at $-\omega$ which could produce some distortion of the line observed at $+\omega$, especially near $\theta = 0$. Here the linewidth is about equal to H_0, so some effect of the $-\omega$ resonance could be present. An analysis including the $-\omega$ line did not seem warranted considering the difficulty in obtaining data at low H, where the baseline of the ESR is not well determined. (cf. Fig. 13b).

For our objective, to see whether 2-d magnetic order occurs, it is not of primary importance what the actual kind of magnetic ordering is. For example, besides weak ferromagnetism, I can suggest another possibility that deserves further investigation. The model is an antiferromagnet with missing Mn ions, i.e., a slightly dilute antiferromagnet. Such a structure is plausible considering the method of fabrication. When such an AF orders some of the spins on one sublattice are not cancelled because spins are missing from the other sublattice. If the number of empty sites is N_e one expects a statistical preponderance on one sublattice of about $N_e^{1/2}$. This model would have at least some of the observed properties: an antiferromagnetic interaction, and a net moment below T_N. Such a system has been discussed by Neel[56], but I do not know if the theory of its ESR has been produced. Experiments with controlled dilution are needed to test this hypothesis.

The ESR of 2-d magnets of Structure I (Fig. 3) was also observed down to 1.4 K. A total of 70 plates was used at 9 GHz. The variation of the line width with temperature was very similar to that observed with Structure II films (Fig. 11). However, the peak of the absorption had no shift as the temperature was reduced. I interpret this to mean that there is a tendency toward AF ordering in the Structure I films, but that the ordered state does not set in above 1.4 K. This can be understood because the density of Mn^{+2} in Structure I films is less than in Structure II, so that the exchange interactions will be weaker. The influence of the unsymmetrical environment of the Mn due to the substrate here is difficult to evaluate.

IV. Discussion and Conclusion

The objective of this study was to produce a literally two-dimensional magnetic structure of Heisenberg-like spins, and to check whether or not it would order magnetically. We have achieved the 2-d structure on a macroscopic scale. The magnetic interaction did not turn out to be purely Heisenberg in $MnSt_2$, according to the presently favored interpretation of the data, based on weak ferromagnetism. Although Mn is a simple ion there are additional interactions: the Dzyaloshinsky interaction, anisotropy and dipolar fields. Our success is also limited by the fact that so far only spin resonance has been used to study the magnetic properties of the 2-d magnets. This requires the application of an external field and is subject to questions of interpretation. One would prefer static, thermodynamic, zero field measurements in order to be able to compare with the properties usually predicted by theories. Such measurements are currently being prepared.

However, despite producing the best yet physical approximation to the theoretical model one wonders if it will ever be possible to actually test the theory of the stability of the 2-d Heisenberg magnet. Firstly, there is the question of how large a 2-d object one must have before one is free of finite-size effects[57]. I have been told[58] of estimates that the sample might be required to be the order of the size of the planet! This would strain our budget.

Then there is the question of producing an interaction that is exactly Heisenberg, i.e., spherically symmetric. This seems to me to be extremely unlikely in a 2-d magnetic system. Clearly a 2-d structure is highly anisotopic. The dipolar interactions, which are inevitable in magnets, impose an easy plane anisotropy in the plane of a ferromagnet. There are almost surely interactions with the crystalline environment which are anisotropic because the magnetic ions must see a different environment in the plane of the magnets, compared to the environment perpendicular to this surface. This is by definition of a two-dimensional magnet. Thus it is physically almost a contradiction in terms to assume spherically symmetric objects coupled in a 2-d array.

One might hope that the anisotropy might be made small by some clever means, or by lucky cancellations. But how small would the anisotropy have to be? Calculations by Lines[8], of the Neel temperatures of 2-d Heisenberg antiferromagnets with various amounts of uniaxial anisotropy, showed a rather weak dependence of T_N on the anisotropy. Pelcovits and Nelson[59] have given an approximate relation, $T_N \propto |\log(H_K/H_E)|^{-1}$. This agrees roughly with Lines' results. Quantitatively, Lines found that the transition temperature at high anisotropy was $T_N \approx JS(S+1)/k$ in 2-d, which is within a factor of two less than T_N in a 3-d lattice. As the anisotropy was decreased, at constant J, he found that T_N decreased proportional to $|\log(H_K/H_E)|^{-1}$ in 2-d, and was independent of H_K in 3-d, roughly. But the log dependence is so weak that T_N in 2-d remains quite high. For example, if $H_e = 10^6$ Oe and H_K is decreased from, say, 10^5 Oe which is enormous, to 10^2 Oe which is modest, T_N in 2-d decreases by about a factor 3; it is only a factor 4 less than the 3-d ordering temperature. This suggests that it will be very difficult, perhaps impossible[6], to realize an anisotropy that is sufficiently small that it does not have a profound influence on the ordering.

These considerations should affect the way in which theorists explain experiments. It would be misleading, when discussing an actual experiment, to say that the sample was equivalent to a Heisenberg magnet and, therefore, will not order. It should be borne in mind that the 2-d Heisenberg model is only a model and that, for magnets at least, one may never find a physical embodiment that is sufficiently either 2-d or Heisenberg.

This does not mean that the study of 2-d magnetism is futile. Indeed there are predictions[6] of interesting properties of 2-d systems, such as topological ordering, cross-over effects in both the dimensionality and the dominant interaction. Magnetic films prepared by the Langmuir-Blodgett technique may provide useful samples, with control over the third dimension.

V. Acknowledgments

I wish to express my thanks to numerous collaborators whose expertise has contributed to this work: A. Aviram, J. Axe, E. Burstein, Ch. Chen, W. P. Chen, F. Dacol, R. Evans, F. Ferrieu, W. Hammer, S. Herd, A. Hjortsberg, J. Kirtley, H. Lillienthal, R. Linn, R. Nicklow, R. Pollak, A. Segmüller, E. Simonyi and A. Taranko. Others whose helpful discussions have been important include P. Brosious, T. McGuire, F. Mehran, R. Pelcovits, T. D. Schultz, B. Scott, B. D. Silverman, J. Slonczewski and K.W.H. Stevens. The early support by N. Shiren is appreciated. The rapid production of this paper is due to the cooperation and enthusiasm of D. Kahn, L. Kristiansen and our ever loving E. Marino.

VI. References

1. Theory is extensively reviewed in the series Phase Transitions and Critical Phenomena, Edited by C. Domb and M. S. Green. See also Ref. 6, and Proceedings of Fermi Schools, Courses LI (1971), LIX (1976), North-Holland.
2. M. E. Fisher in Essays in Physics Vol. 4:43, (1972) Academic Press, R. B. Griffiths, Phys. Rev. Lett. 24: 1479 (1970).
3. M. E. Fisher, S. K. Ma, B. G. Nickel, Phys. Rev. Lett. 29:9l7 (1972).
4. L. Onsager, Phys. Rev. 65:117 (1944).
5. N. D. Mermin, and H. Wagner, Phys. Rev. Lett. 17:1133 (1966).
6. J. M. Kosterlitz and D. J. Thouless, Prog. in Low Temp. Phys. VIIB, p. 371 Ed. D. F. Brewer, North-Holland, (1978).
7. C. Herring and C. Kittel, Phys. Rev. 81:869 (1951).
8. M. E. Lines, J. Phys. Chem. Solids 31:101 (1970).
9. Names are withheld to protect the guilty.
10. F. Bloch, Z. Phys. 61:206 (1930).
11. B. E. Argyle, S. H. Charap, and E. W. Pugh, Phys. Rev. 132:2051 (1963).
12. H. Ikeda, M. T. Hutchings and M. Suzuki, J. Phys. C, 11:L 359 (1978). For a review of quasi 2-d experiments, see L. J. de Jongh and A. R. Miedema, Adv. Phys. 23:1 (1974).
13. J. P. McTague and M. Neilsen, Phys. Rev. Lett. 37:596 (1976).
14. S. Gregory, Phys. Rev. Lett. 40:723 (1978).
15. O. Vilches, unpublished results presented at this school.
16. V. D. Borman, B. I. Buttsev, V. A. Konakov, B. I. Nikolaev, and V. I. Troyan, JETP Lett. 27:527 (1978).
17. Reviewed by G. L. Gaines, Insoluble Monolayers at Liquid-Gas Interfaces Interscience, New York (1966).
18. I. Langmuir, J. Am. Chem. Soc. 39:1848 (1917).
19. K. B. Blodgett, J. Am. Chem. Soc. 57:1007 (1935).
20. J. Messier and G. Marc, J. de Physique 32:799 (1971). cf. also P. A. Chollet, J. Phys. C 7:4127 (1974).
21. D. W. Deamer, D. W. Meek and D. G. Cornwell, J. of Lipid Res. 8:255 (1967).
22. G. A. Wolstenholme and J. H. Schulman, Proc. Farad. Soc. 46:475 (1950).
23. C. J. Ballhausen, Introduction to Ligand Field Theory, McGraw-Hill, New York, 1962.
24. J. H. van Vleck, Phys. Rev. 45:405 (1934).
25. S. P. Kowalczyk, L. Ley, R. A. Pollak, F. R. McFeely and D. A. Shirley, Phys. Rev. B7:4009 (1973).
26. M. Pomerantz and R. A. Pollak, Chem. Phys. Lett., 31:602 (1975).
27. M. Avram and G. D. Mateescu, Infra Red Spectroscopy, Wiley Interscience, (1972).
28. B. Ellis and H. Pyszora, Nature 181:181 (1958).
29. M. Pomerantz, S. Herd and E. E. Simonyi, Bull. Am. Phys. Soc. 23:431 (1978), and to be published.

30.   N. J. Harrick, J. Phys. Chem. 64:1110 (1960), Internal Reflection Spectroscopy,Wiley and Sons, N. Y., (1967).

31.   A. Hjortsberg, W. P. Chen, E. Burstein and M. Pomerantz, Optics Comm. 25:65 (1978).

32.   J. F. Stephens and C. Tuck-Lee, Appl. Crystallogr. 2:1 (1969). Note that Figs. 4 and 6 are switched.

33.   J. Kirtley and M. Pomerantz (unpublished).

34.   F. Kopp, U. P. Fringeli, K. Mühlethaler and H. H. Günthard, Biophys. Struct. Mechanism 1:75 (1975).

35.   M. Pomerantz, F. Dacol and A. Segmüller, Phys. Rev. Lett. 40:246 (1978),

36.   D. C. Bisset and J. Iball, Proc. Phys. Soc. London, Sect. A 67:315 (1954).

37.   M. Pomerantz and A. Segmüller, Thin Solid Films (to be published).

38.   R. M. Nicklow, M. Pomerantz, and A. Segmüller, Bull. Am. Phys. Soc. 24:488 (1979).

39.   ESR in quasi 2-d is reviewed by P. M. Richards, in Proceedings of the International School of Physics "Enrico Fermi", Course LIX, (1976).

40.   H. P. Boesch, U. Schmocker, F. Waldner, K. Emerson, and J. E. Drumheller, Phys. Lett. 36A:461 (1971).

41.   P. M. Richards and M. Salamon, Phys. Rev. B 9:32 (1974).

42.   M. Pomerantz, Solid State Comm., 27:1413 (1978).

43.   J. Axe, private communications.

44.   M. Pomerantz and A. Aviram, Solid State Comm., 20:9 (1976).

45.   T. Haseda, H. Yamakawa, M. Ishizuka, Y. Okuda, T. Kubota, M. Hata, and K. Amaya, Solid State Comm. 24:599 (1977). Also N. Giordano and D. Prober, private communications.

46.   C. Kittel, J. Phys. Rad., 12:149 (1951).

47.   I. Dzyaloshinsky, J. Phys. Chem. Solids 4:241 (1958).

48.   T. Moriya, Magnetism: 1, 85 Ed. Rado and Suhl, Academic Press. (1963).

49.   A. S. Borovik-Romanov in Elements of Theoretical Magnetism, Ed. S. Krupic-ka, and J. Sternberk, p. 193, Iliffe Books Ltd. (London) (1968).

50.   K. Saiki, J. Phys. Soc. Japan 33:1284 (1972).

51.   H. Yoshioka and K. Saiki, J. Phys. Soc. Japan 33:1566 (1972).

52.   The APL/360 program was written by R. Evans, assisted by R. Linn.

53.   A. S. Borovik-Romanov, N. M. Kreines, and L. A. Prozorova, Sov. Phys. JETP 18:46 (1964).

54.   M. Pomerantz and A. Taranko, unpublished.

55.   F. Ferrieu, private communication.

56.   L. Neel, Rev. Mod. Phys. 25:58 (1953).

57.   Y. Imry, Ann. of Phys. 51:1 (1969) and references therein.

58.   D. Goodstein and S. K. Ma, semi-private communications.

59.   R. A. Pelcovits and D. R. Nelson, Phys. Lett. 57A:23 (1976).

ON A STANDARDIZED MEASURE OF SUBSTRATE UNIFORMITY

J. G. Dash

Department of Physics
University of Washington
Seattle, WA 98195

Thin film studies are now being carried out with a variety of relatively uniform substrates; exfoliated graphite powders and compacts, graphitized carbon black, lamellar halides, alkali halides, oxides, and others. These adsorbents, although quite uniform, are not ideal, for they have varying densities and types of imperfections. Their heterogeneity can be ignored for certain kinds of investigations, but are important in others, especially in the neighborhood of phase transitions, where the films may have divergent compressibilities. While it is obviously desirable to use the most uniform substrate in every experiment, other practical requirements usually demand a compromise, involving a less ideal adsorbent. The variety of adsorbents, coupled with the sensitivity of films to heterogeneity, creates a need for some common method for measuring and specifying the heterogeneity of every experimental adsorbent and installation. The measurement should be relatively simple to carry out and interpret, inexpensive and adaptable to all types of experimental system. It is proposed here that the vapor pressure isotherms of Kr at 77K, specifically the riser of the second atomic layer, can satisfy these requirements.

The Kr/graphite system is one of the most widely and thoroughly studied.[1-12] Kr has also been used with a variety of other uniform substrates.[13-17] In each of these systems there is distinct layering at liquid nitrogen temperatures, the vapor pressure isotherm showing a succession of nearly vertical steps. The second layer step is especially easy to examine, since the vapor pressure is on the order of 1 Torr at 77K. These steps indicate first order transitions, which in ideal systems would be exactly vertical, i.e. with chemical potential $\mu(P,T)$ = constant. In real systems the chemical potential varies due to substrate imperfections and size effects. If the variations are attributed completely to size effects the substrate quality can be expressed by a single quantitative parameter, the "average domain size" of uniform regions.

In an ideal system the chemical potential remains constant during phase condensation in the film, but if the condensed phase is formed in small islands of typical dimension r, the chemical potential is shifted by an amount[17,18]

$$\delta\mu = A\sigma/r \tag{1}$$

where A is the molecular area and σ is the boundary line tension between the condensed and dilute 2D phases. If the 3D vapor is sufficiently dilute to be approximated by an ideal gas,

$$\delta\mu = kT\ell nP(r)/P(\infty). \tag{2}$$

The line tension can be estimated from the bulk heat of sublimation q_s, as[17]

$$\sigma = 0.16 \ q_s/A^{1/2}. \tag{3}$$

The molecular area of the condensed phase must depend to some extent on the substrate. On graphite the 2D solid condenses into the registered $\sqrt{3} \times \sqrt{3}$ structure. In this event the equations (1) – (3) are combined to give the average radius of 2D islands

$$r = \frac{1.1 \times 10^{-21}}{kT\left(\delta P/P\right)}, \tag{4}$$

where we have expanded the logarithm in (2) for small $\delta P/P$. We identify r with the average size of uniform domains of the substrate.

The accuracy of the method has been checked in principle with N_2 monolayers on Grafoil. The average size of 2D solid islands in equilibrium with low density 2D gas was estimated from the shift of the melting point temperature with coverage,[18] and found to be in agreement with the coherence length of the islands, as directly determined by neutron diffraction.[19,20] Kr has a first layer phase diagram similar to N_2 but shifted to higher temperatures. The first layer of Kr at 77K undergoes 2D condensation at a pressure of a few millitorr, which can be difficult to measure with precision. The pressure of the second layer condensation at 77K ranges from about 0.1 to 1 Torr on various substrates, and this pressure can be measured accurately with simple apparatus.

In conclusion we emphasize that the proposed standard offers a simple convention for specifying substrate imperfection but cannot take the place of more complete characterization. In many current studies the heterogeneity of substrates and films can be determined in much greater detail. It is to be hoped that additional conventions will eventually arise from these newer techniques.

REFERENCES

1. J. H. Singleton and J. D. Halsey, Jr., J. Phys. Chem. 58, 330,1011 (1954).
2. G. Ehrlich and F. G. Hudda, J. Chem. Phys. 30, 493 (1959).
3. A. Thomy and X. Duval, J. Chim. Phys. Physicochim. Biol. 66, 1966 (1969); 67, 286, 1101 (1970).
4. T. Engel and R. Gomer, J. Chem. Phys. 52, 5572 (1970).
5. F. A. Putnam and T. Fort, Jr., J. Phys. Chem. 79, 459 (1975).
6. J. A. Venables, H. M. Kramer, and G. L. Price, Surf. Sci. 55, 41 (1976); 57, 782 (1976).
7. J. Suzanne and M. Bienfait, J. Phys. (Paris) 38 Suppl. C4,93 (1977).
8. M. D. Chinn and S. C. Fain, Jr., Phys. Rev. Lett. 39, 146 (1977).
9. C. Marti, B. Croset, P. Thorel, and J. P. Coulomb, Surf. Sci. 65, 532 (1977).
10. P. M. Horn, et al., Phys. Rev. Lett. 41, 961 (1978); P. W. Stephens, et al., Phys. Rev. Lett. 43, 47 (1979).
11. D. M. Butler, J. A. Litzinger, G. A. Stewart, and R. B. Griffiths, Phys. Rev. Lett. 42, 1289 (1979).
12. M. Bienfait, J. G. Dash, and J. Stoltenberg, Phys. Rev. (to be published).
13. Y. Larher, J. Chim. Phys. 68, 796 (1971).
14. T. Takaishi and M. Mohri, J. Chem. Soc. Faraday Trans. I68, 1921 (1972).
15. Y. Larher and D. Haranger, Surf. Sci. 39, 100 (1973).
16. J. G. Dash, R. Ecke, J. Stoltenberg, O. E. Vilches, and O. J. Whittemore, Jr., J. Phys. Chem. 82, 450 (1978).
17. Y. Larher, Mol. Phys. 38, 789 (1979).
18. T. T. Chung and J. G. Dash, Surf. Sci. 66, 559 (1977).
19. J. K. Kjems, L. Passell, H. Taub, J. G. Dash, and A. D. Novaco, Phys. Rev. B13, 1446 (1976).
20. W. F. Brooks, Brookhaven Informal Report, BNL 22617 (1967) (unpublished).

SCIENTIFIC PROGRAMME

Monday, June 11

R. B. Griffiths; Thermodynamics of Surface Systems I

June 12

M. Bienfait; Classical van der Waals Systems I
A. Boato; Molecular Scattering from Surfaces
A. Thomy; Uniform and Nonuniform Adsorption
R. B. Griffiths; Thermodynamics of Surface Systems II
D. L. Goodstein; Thermodynamic Technique in Adsorption

June 13

M. Nielsen; Neutron Diffraction and Inelastic Scattering I
M. Bienfait; Classical van der Waals Systems II
M. Schick; Theory of Helium Monolayers I
R. B. Griffiths; Thermodynamics of Surface Systems III
M. Rasetti; Theorems on Phase Transitions

June 14

M. Nielsen; Neutron Diffraction and Inelastic Scattering II
M. Bienfait; Classical van der Waals Systems III
A. Luther; Commensurate and Incommensurate Phases I
M. Schick; Theory of Helium Monolayers II

Short Communications

C. E. Campbell; Liquid-gas phase equilibrium in 2D ^{4}He.
C. Deutsch; Free energy of low density 2D matter.
G. Meissner; Collective excitations and liquid-solid transition of
surface electrons.

June 15

M. Nielsen; Neutron Diffraction and Inelastic Scattering III

A. Luther; Commensurate and Incommensurate Phases II
S. C. Fain; LEED Studies of Commensurate and Incommensurate Phases
J. Villain; Theory of Orientational Disorder
M. G. Richards; Nuclear Magnetic Resonance and Mössbauer
                Spectroscopy I

Short Communication

Y. Larher; Critical points in physisorbed layers.

June 16

M. Schick; Theory of Helium Monolayers III
L. Nosanow; 2D and 3D Theory of Corresponding States
J. M. Kosterlitz; Long Range Order in Two Dimensions I

June 18

M. G. Richards; Nuclear Magnetic Resonance and Mössbauer
                Spectroscopy II
J. M. Kosterlitz; Long Range Order in Two Dimensions II
B. A. Huberman; Topological Defects in 2D and Quasi 2D Systems
L. Passell; Recent Neutron Scattering Results from Brookhaven
P. Thorel; Recent Neutron Scattering Results from Institut von Laue-
           Langevin
P. W. Stephens; X-ray Scattering from Monolayer Films

June 19

J. M. Kosterlitz; Long Range Order in Two Dimensions III
J. D. Reppy; Superfluidity in Thin ^{4}He Films I
E. K. Riedel; Anisotropic N-vector Models in 2D
M. T. Béal-Monod; ^{3}He Magnetic Ordering at Surfaces
M. G. Richards; Nuclear Magnetic Resonance and Mössbauer
                Spectroscopy III

June 20

J. D. Reppy; Superfluidity in Thin ^{4}He Films II
D. O. Edwards; ^{3}He-^{4}He Liquid Films
B. I. Halperin; Melting in 2D
A. Holz; Melting of Registered and Incommensurate Films
H. Shechter; Mössbauer Spectroscopy of Diffusion and Melting

June 21

J. D. Reppy; Superfluidity in Thin ^{4}He Films III
R. A. Hallock; Persistent Currents and Dissipation
M. Bretz; Helium on Graphite
J. G. Dash; Evolution of Bulk Properties
S. B. Doniach; Topological Excitation in Thin Film Superconductors

June 22

E. Bauer; Phases of Chemisorbed Films I
M. Pomerantz; Studies of Two Dimensional Magnets
Y. Imry; Lattice Gas Model of Solid-Liquid-Gas Transitions

Short Communications

O. E. Vilches; Realization of 2-state and 3-state Potts models.
G. A. Stewart; Heat capacity evidence for a triple point in Xe/gr.
E. Lerner; Calorimetry of first and second layers of Ne/gr.
K. A. Penson; T=0 phase transitions in 2D quantum systems.
H. Taub; Neutron spectroscopy of monolayers.
M. Kerszberg; Landau-Ginzburg-Wilson model for roughening transitions

June 23

E. Bauer; Phases of Chemisorbed Films, II and III

Short Communications

P. H. Kleban; Lattice gas model for O on Ni (110).
H. J. Lauter; Nucleation of solid ^{4}He on graphite.
D. L. Goodstein; Phonon reflection spectroscopy.
J. G. Dash; On a standardized measure of substrate uniformity.

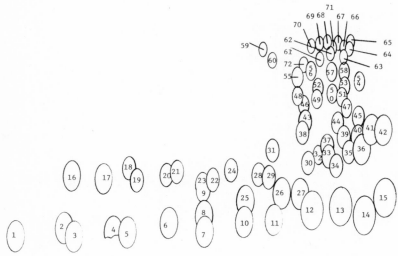

PARTICIPANTS

1. B. Neudecker
2. E. Bauer
3. H. Taub
4. J. Quateman
5. M. Manninen
6. M. Rasetti
7. A. Walker
8. R. Beaume
9. M. Kerszberg
10. L. Mistura
11. H. Schechter
12. P. Kleban
13. S. Doniach
14. P. Thorel
15. G. Carneiro
16. J. Tuul
17. J. Laheurte
18. M. Schick
19. K. Penson
20. D. Bittner
21. E. Lerner
22. M. Nielson
23. Z. Galasiewicz
24. H. Lauter

25. T. Engel
26. A. Stella
27. M. Montambaux
28. J. Lorenc
29. H. Weichert
30. G. Torzo
31. S. Fain
32. D. Butler
33. M. Godfrin
34. M. Béal-Monod
35. M. Bretz
36. J. Ruvalds
37. L. Passell
38. D. Goodstein
39. D. Baeriswyl
40. O. Vilches
41. T. Regge
42. J. Dash
43. M. Richards
44. Y. Imry
45. M. Feigenblatt
46. J. Mochel
47. M. Héritier
48. A. Holz

49. E. Riedel
50. A. Thomy
51. G. Meissner
52. R. Hallock
53. P. Stephens
54. J. Owers-Bradley
55. M. Bienfait
56. J. Connolly
57. K. Uzelac
58. L. Nosanow
59. R. Desai
60. P. Nightingale
61. L. Bruch
62. G. Stewart
63. R. Griffiths
64. T. Burkhardt
65. B. Halperin
66. S. Hikami
67. S. Hurlbut
68. J. Timonen
69. D. Edwards
70. G. Agnolet
71. J. Reppy
72. A. Levi

PARTICIPANTS

G. AGNOLET, Physics Department, Cornell University, Ithaca, NY
14853, USA

D. BAERISWYL, RCA Laboratories Ltd., Badenerstrasse 569, 8048 Zurich,
Switzerland

E. BAUER, Physikalisches Institut, Teknische Universitat Clausthal,
Leibnizstrasse 4, 3392 Clausthal-Zellerfeld, West Germany

M.T. BÉAL-MONOD, Institut Max von Laue-Paul Langevin, 156 X Centre
de Tri, 38042 Grenoble Cédex, France

R. BEAUME, Department de Physique, Univ. d'Aix-Marseille II, 13288
Marseille, France

M. BIENFAIT, Department de Physique, Univ. d'Aix-Marseille II,
13288 Marseille, France

D.N. BITTNER, Physics Department, University of Michigan, Ann Arbor,
MI 48109, USA

G. BOATO, Istituto di Scienze Fisiche, Viale Benedetto XV, 16132
Genova, Italy

M. BRETZ, Physics Department, University of Michigan, Ann Arbor,
MI 48109, USA

L.W. BRUCH, Department of Physics, University of Wisconsin, 1150
University Ave., Madison, Wisconsin 53706, USA

T.W. BURKHARDT, Institut fur Festkorperforschung der
Kernforschunsanlage, Postfach 1913, 5170 Julich 1, West Germany

D.M. BUTLER, Department of Physics, Carnegie-Mellon University,
Schenley Park, Pittsburgh, PA 15213, USA

C.E. CAMPBELL, Department of Physics, University of Minnesota,
Minneapolis, MN 55455, USA

G.M. CARNEIRO, Departamento de Fisica, PUC/RJ, Rua Marques de Sao
Vicente 225, Gavea - CEP 22451 Rio de Janeiro RJ, Brasil

J.W.D. CONNOLLY, Condensed Matter Sciences, Division of Materials
Research, National Science Foundation, Washington, DC 20550, USA

J.G. DASH, Physics Department, University of Washington, Seattle,
Washington 98195, USA

R.C. DESAI, Department of Physics, University of Toronto, Toronto, Ontario M5S 1A7, Canada

C. DEUTSCH, Université de Paris XI, Centre d'Orsay, Laboratoire de Physique des Plasmas, 91405 Orsay Cédex, France

S.B. DONIACH, Applied Physics, Stanford University, Stanford, CA 94305, USA

D.O. EDWARDS, Department of Physics, Ohio State University, 174 W. 19 Ave., Columbus, OH 43210, USA

T. ENGEL, IBM Research Laboratory, Saumerstrasse 4, 8803 Ruschikon, Zurich, Switzerland

S.C. FAIN, Physics Department, University of Washington, Seattle, WA 98195, USA

M.Y. FEIGENBLATT, Department of Physics, Tel Aviv University, Ramat Aviv, Tel Aviv, Israel

Z. GALASIEWICZ, Institut of Theoretical Physics, University of Wroclaw, W. Cybuskiego 36, 50-205 Wroclaw, Poland

M.H. GODFRIN, CNRS 166x, Centre de Tri, 38042 Grenoble Cedex, France

D.L. GOODSTEIN, Physics Department, California Institute of Technology, Pasadena, CA 91109, USA

R.B. GRIFFITHS, Department of Physics, Carnegie-Mellon University, Pittsburgh, PA 15123, USA

R.B. HALLOCK, Department of Physics, University of Massachusetts, Amherst, MA 01003, USA

B.I. HALPERIN, Lyman Laboratory of Physics, Harvard University, Cambridge, MA 02138, USA

M. HÉRITIER, Université de Paris-Sud, Centre d'Orsay, Lab. de Physique des Solides, Batiment 510, 91405 Orsay, France

S. HIKAMI, Research Institute of Fundamental Physics, Kyoto University, Kyoto 606, Japan

A. HOLZ, Institut fur Theoretische Physik, Universitat der Saarlandes, 6600 Saarbruken, West Germany

B.A. HUBERMAN, Xerox Palo Alto Research Center, 3333 Coyote Hill Rd., Palo Alto, CA 94304, USA

S. HURLBUT, Physics Department, University of Washington, Seattle, WA 98195, USA

Y. IMRY, Physics Department, Tel Aviv University, Tel Aviv, Israel

S.J. KNAK JENSEN, Department of Physical Chemistry, Chemical Institut, Aarhus University, Aarhus 8000 C., Denmark

M. KERSZBERG, Department of Electronics, Weizmann Institute of Science, Rehovot, Israel

P.H. KLEBAN, Department of Physics and Astronomy, University of Maine at Orono, Bennett Hall, Orono, Maine 04469, USA

J.M. KOSTERLITZ, Department of Physics, University of Birmingham, Birmingham B15 2TT, UK

M. KRUSIUS, Department of Physics, University of Turku, 20500 Turku 50, Finland

J.P. LAHEURTE, Lab. de Physique de la Matière Condensée, Université de Nice, 06034 Nice Cédex, France

Y. LARHER, CEN Saclay, 91190 Gif-sur-Yvette, France

H.J. LAUTER, Institut Max von Laue-Paul Langevin, Ad. Poste 156 x, 38042 Grenoble Cédex, France

E. LERNER, Instituto de Fisica, Caixa Postal 68020, Universidade Federal de Rio de Janeiro, Rio de Janeiro, Brasil

A.C. LEVI, Libera Università di Trento, Facoltà di Scienze, Trento, Italy

J. LORENC, Institute of Theoretical Physics, University of Wroclaw, Cybulskiego 36, 50-205 Wroclaw, Poland

A. LUTHER, NORDITA, Blegdamsvej 17, 2100 Copenhagen Ø, Denmark

M. MANNINEN, Low Temperature Laboratory, Helsinki University of Technology, 02150 Espoo 15, Finland

N.D. MERMIN, Physics Department, Cornell University, Ithaca, NY 14853, USA

G. MEISSNER, Fachrichtung 11.1, Theoretische Physik, Universitaat der Saarlandes, 66 Saarbruken, West Germany

L. MISTURA, Facoltà di Ingegneria, Istituto di Fisica, Città Universitaria, Piazzale delle Scienze, 5, 00185 Roma, Italy

J.M. MOCHEL, Department of Physics, University of Illinois, Urbana, IL 61801, USA

M.G. MONTAMBAUX, Université de Paris-Sud, Centre d'Orsay, 91405 Orsay, France

B. NEUDECKER, Ubierstrasse 2, 5450 Neuwied 11, West Germany

M. NIELSEN, Risø National Laboratory, 4000 Roskilde, Denmark

P. NIGHTINGALE, Physics Department, Technische Hogeschule, Delft, Holland

L. NOSANOW, Condensed Matter Sciences, Division of Materials Research, National Science Foundation, Washington, DC 20550, USA

J.R. OWERS-BRADLEY, Department of Physics and Astronomy, Northwestern University, Evanston, IL 60201, USA

L. PASSELL, Physics Department, Brookhaven National Laboratory, Upton, NY 11973, USA

K.A. PENSON, Université de Paris-Sud, Batiment 510, 91405 Orsay, France

M. POMERANTZ, IBM Research Laboratories, Yorktown Heights, NY 10598, USA

J.H. QUATEMAN, Physics Department, University of Michigan, Ann Arbor, MI 48109, USA

M. RASETTI, Istituto di Fisica, Università di Torino, Torino, Italy

T. REGGE, Istituto di Fisica, Università di Torino, Torino, Italy

J.D. REPPY, Physics Department, Cornell University, Ithaca, NY 14853, USA

M.G. RICHARDS, University of Sussex, School of Mathematical and Physical Sciences, Falmer, Brighton BN1 9QH, UK

E.K. RIEDEL, Physics Department, University of Washington, Seattle, WA 98195, USA

J. RUVALDS, Department of Physics, University of Virginia, Charlottesville, Virginia 22901, USA

M. SCHICK, Department of Physics, University of Washington, Seattle, WA 98195, USA

H. SHECHTER, Physics Department, Technion, Haifa, Israel

A.F. SILVA-MOREIRA, Low Temperature Physics, California Institute
of Technology, Pasadena, CA 91125, USA

A. STELLA, Katholieke Universiteit Leuven, Instituut voor
Theoretische Physica, Celestijnenlaan 200 D, 3030 Heverlee, Belgium

P.W. STEPHENS, M.I.T., Room 13-2138, Cambridge, MA 02139, USA

G.A. STEWART, Department of Physics, University of Pittsburgh,
Pittsburgh, PA 15260, USA

Z. SZPRYNGER, Institut of Low Temperature Physics, Polish Academy
of Science, Pl. Kate, Wroclaw, Poland

H. TAUB, Department of Physics, University of Missouri-Columbia,
Columbia, MO 65211, USA

A. THOMY, CNRS - Lab. Maurice Letort-Larings, Route de Vandoeuvre,
B.P. no. 104, 54600 Villers-Nancy, France

P. THOREL, Max von Laue - Paul Langevin, 156 X Centre de Tri,
38042 Grenoble Cedex, France

J. TIMONEN, NORDITA, Blegdamsvej 17, 2100 Kobenhavn Ø, Denmark

G. TORZO, Unità Basse Temperature, Istituto di Fisica, Università
di Padova, Padova, Italy

J. TUUL, Laboratoire de Physique Moleculaire des Hautes Energies,
B.P. no. 2, 06530 Peymeinade, France

K. UZELAC, Université de Paris-Sud, Centre d'Orsay, 91405 Orsay,
France

O.E. VILCHES, Department of Physics, University of Washington,
Seattle, WA 98195, USA

J. VILLAIN, Laboratoire de Diffraction Neutronique, Centre d'Etudes
Nucléaires, 38041 Grenoble Cedex, France

A. WALKER, Department of Theoretical Physics, University of Oxford,
1 Keble Road, Oxford OX1 3NP, UK

H. WIECHERT, J. Gutenberg-Universitat, Fachbereich Physik (18),
Institut fur Physik, Postfach 3980, 6500 Mainz, Germany

SUBJECT INDEX

Acetic acid 322
Adsorbates 126
   Ar 144, 150, 152-154
   CD_4 144
   D_2 139, 150, 156
   ^3He 142
   ^4He 142
   H_2 139, 142, 150
   Kr 144, 154
   N_2 143, 152
   O_2 145, 154
   Xe 144
Adsorption-desorption
   equilibrium 275-286,
   292, 298, 299
Adsorption
   heterogeneity 347
   isotherms 30, 31, 39,
   277, 347
   kinetics 56, 277, 278
Alkalis 294, 306, 309
Anisotropy, magnetic 318, 320,
   341, 343, 344
Antiferromagnetism of $MnSt_2$
   334 ets.
Ar 43, 50, 55
Auger electron spectroscopy 30,
   31, 39, 277, 279, 286, 303, 304

Band structure 72
Barium 275, 294-296
Bound vortex pairs 215
Butane 49

Cadmium stearate 324

Cahn, J. W., 2, 15, 19, 28
Carbon monoxide 270, 271, 298
Carboxylic acid 322
CH_4 51, 52, 53
Chemisorption 267-311
Clean surfaces 273-287
Coherence length 133, 152, 262
Commensurate layer 41, 277,
   295, 303, 305
Commensurate phase
   of helium 99
   of krypton 108
   of nitrogen 109
Compression 296-299
Continuous transition 296-299
Conventions in surface
   thermodynamics, 4ff, 8, 10f,
   14f, 18ff, 25f
Copper 279, 304, 305, 308, 309
Copper stearate 324
Correlation time 169, 172
Critical exponents 82
Critical point 23
Critical temperature (2D) 32,
   33, 38, 56
Crystal fields in magnets 326,
   336
Crystals in two dimensions 210
Curved interfaces 20f

Debye Waller factor 47, 137
Degrees of freedom 194
Densities 6, 14
Desorption kinetics 58, 278-280,
   303, 304